Pallavi Saxena • Anil Kumar Gupta • Rogert Sorí
Editors

# Fog: Formation, Chemistry, Dynamics, Impact and Dissipation

Springer

*Editors*
Pallavi Saxena
Assistant Professor, Department of Environmental Sciences
Hindu College, University of Delhi
Delhi, Delhi, India

Rogert Sorí
Senior Researcher, Environmental Physics Laboratory (EPhysLab) - CIM
University of Vigo
Ourense, Spain

Anil Kumar Gupta
PI & CEO, Integrated Centre for Adaptation, Disaster Risk-resilience and Sustainability, ICARS
Indian Institute of Technology – IIT Roorkee
Greater Noida, Uttar Pradesh, India

ISBN 978-981-96-4421-6 ISBN 978-981-96-4422-3 (eBook)
https://doi.org/10.1007/978-981-96-4422-3



This Springer imprint is published by the registered company Springer Nature Singapore Pte Ltd.
The registered company address is: 152 Beach Road, #21-01/04 Gateway East, Singapore 189721, Singapore

# Preface

Fog forms when air reaches its saturation point, either by gaining moisture or losing heat. Several prevailing factors including the climatology of area, time of year, thermal characteristics of air, boundary layer stability, vertical and horizontal wind speeds, inorganic and organic constituents, dew point depression, topography and the characteristics of condensation nuclei are important for fog prediction and assessment. The presence of fog affects a number of environmental processes, including global and local climate, air quality, water quality, air–surface interactions and the thermal and radiative budget of the atmosphere. Additionally, fog affects social life, leading to an increase in road and rail accidents, flight delays and both direct and indirect adverse impacts on human beings and crop yield. Therefore, with the changing climate and environment, the nature of fog has also changed. Given its impact on humans and various systems, studying fog has become essential.

*Fog: Formation, Chemistry, Dynamics, Impact and Dissipation* describes about the fundamental concepts of fog introduction, chemistry, formation, dynamics, impacts and dissipation. The chapters provide a holistic overview of the basic concepts of fog formation, the role of meteorology in fog formation, the chemistry of fog, how different chemical constituents present in fog affect its quality and the surrounding atmosphere, impacts on plant species, human health, visibility and the economy as well as the dissipation of fog by cost-effective techniques. This book serves as a valuable tool for students in environmental science, biological science, medical science and agriculture. It will be unique to environmental consultants, meteorologists, atmospheric scientists, researchers and other professionals involved in air quality, meteorology and research related to plants and humans.

Delhi, Delhi, India — Pallavi Saxena
Greater Noida, Uttar Pradesh, India — Anil Kumar Gupta
Ourense, Spain — Rogert Sorí

# Acknowledgement

The editors gratefully acknowledge the contributions of all authors who wrote chapters on key aspects of the book, helping to make it a comprehensive and holistic reference publication. We would also like to extend our sincere gratitude to Ms. Nupur Verma, Associate Editor at Springer, for her cooperation and help in completing this assignment.

Dr. Pallavi Saxena extends her sincere thanks to Prof. Anju Srivastava, Principal, Hindu College, University of Delhi, for the encouragement and support in the research field.

Dr. Saxena is highly grateful to the iLEAPS community and SERB (SR/FTP/ES-74/2013) for providing international exposure in the field of atmospheric sciences, which motivated the idea and framework of this manuscript. She also acknowledges her parents, Dr. Akhilesh Saxena and Dr. Neelima Priyadarshini, for their constant support and encouragement, who have been an inspiration to her in working for science and society. Prof. Anil Kumar Gupta expresses his gratitude for the encouragement received from Dr. Akhilesh Gupta, Senior Advisor and Head of the Climate Change Programme at DST-GOI, as well as to his team at the IIT Roorkee, Greater Noida Campus, UP. Dr. Rogert Sorí would also like to extend his sincere regards to his team at the Universidade de Vigo, Spain, and to his family members.

# Contents

# Editors and Contributors

## About the Editors

**Pallavi Saxena** is an Assistant Professor in the Department of Environmental Science, Hindu College, University of Delhi, Delhi, India, and serves as a member of the SSC for iLEAPS and GEIA. Her areas of interest include air pollution, nature-based solutions and sustainability indicators, and she has published about 60 publications. Dr. Saxena has been selected as a Lead Author for the GEO-7 UNEP Report and has authored an Independent Expert Report on "Climate Change and Biodiversity" for the European Commission in 2024. Dr. Saxena has also been selected as an Expert Member in various international meetings and forums, including Stockholm +50, COP28, COP29, IPBES 9, 10 and 11 and KCI 2022.

**Anil Kumar Gupta** is currently Full Professor of Policy-strategy and Capacities and Chief Executive of ADRES Network at ICARS IIT Roorkee - Greater Noida campus (NCR) India. He was Head of Divisions - Environment, Climate and DRM, Policy-Planning and Training at NIDM, India for over 13 years. He was Programme Director of CECR and several projects, viz. High-CAP, CAP-RES, NADMP, NEFC-DMP, HER-CAP, and was Head of International Cooperation. He was Expert Team Member of WMO Climate Statement and South Asia Core-Group Member of IUCN-CEM. He is principal editor of the Springer-Nature Book series on Disaster Resilience and Green Growth. He was recipient of the National Science Day Young Scientist Award, Excellence Award by the SEOH, IDRC Canada's Thank Tank Initiative Senior Fellowship, Outstanding Contribution Award of UN-PEDRR, and the recent one - WCDM DRR Award for Disaster Preparedness. Prof. Gupta is popularly known as NBS man of India, is founder of SAMYUKTI-ICARSS and 'Accelerating Transformation' group.

**Rogert Sorí** is a Senior Researcher at the Environmental Research Group (EPhyslab), University of Vigo, Spain. He holds a BSc in Meteorology (2010) from the Higher Institute of Technology and Applied Sciences (InSTEC), Havana, Cuba, an MSc in Climate Sciences (2012) and a PhD in Marine Sciences (2018), both from the University of Vigo. His research focuses on climatology, particularly the occurrence and attribution of extreme hydrometeorological events, such as droughts and intense rainfall, exploring the origins of precipitation and the influence of the atmospheric branch of the hydrological cycle on these phenomena.

## Contributors

**Albenis Pérez-Alarcón** Environmental Physics Laboratory (EPhysLab), Centro de Investigación Mariña, Campus de Ourense, Universidade de Vigo, Ourense, Spain

Facultade de Ciências, Instituto Dom Luiz (IDL), Universidade de Lisboa, Lisboa, Portugal

**Rimsha Arshad** Department of Space Science, University of the Punjab, Lahore, Pakistan

**Oscar A. Benedico-Rodríguez** Provincial Meteorological Centre of Ciego de Ávila, Institute of Meteorology, Avenida de los Deportes S/N, Ciego de Ávila, Cuba

**Jakob Ernst** Environmental Physics Laboratory (EPhysLab), Centro de Investigación Mariña, Campus de Ourense, Universidade de Vigo, Ourense, Spain

**Eliani Ezani** Department of Environmental and Occupational Health, Universiti Putra Malaysia, Serdang, Selangor, Malaysia

**Luis Gimeno** Environmental Physics Laboratory (EPhysLab), Centro de Investigación Mariña, Campus de Ourense, Universidade de Vigo, Ourense, Spain

**Anil Kumar Gupta** Integrated Centre for Adaptation, DRR and Sustainability (ICARS, GOI-DST Joint COE), IIT, Roorkee, India

**Mojtaba Heydarizad** State Key Laboratory of Marine Geology, Tongji University, Shanghai, China

**Madhavi Jain** Urban Systems Lab, The New School, New York, NY, USA

**Reena Jain** Department of Chemistry, Hindu College, University of Delhi, Delhi, India

**Saumya G. Kutty** School of Environmental Sciences, Jawaharlal Nehru University, New Delhi, India

**Merlyn Mathew** Environmental Studies and Resource Management, TERI School of Advanced Studies, Delhi, India

**Raquel Nieto** Environmental Physics Laboratory (EPhysLab), Centro de Investigación Mariña, Campus de Ourense, Universidade de Vigo, Ourense, Spain

**Bhanu Pandey** CSIR-Central Institute of Mining and Fuel Research, Barwa Road Campus, Dhanbad, Jharkhand, India

Academy of Scientific and Innovative Research (AcSIR), Ghaziabad, Uttar Pradesh, India

**Gautam Pathare** HighCAP Project, National Institute of Disaster Management, Ministry of Home Affairs, GoI, Delhi, India

**Atar Singh Pipal** Indian Institute of Tropical Meteorology, Ministry of Earth Sciences, Pune, India

**Pradhi Rajeev** Department of Civil and Environmental Engineering, Indian Institute of Technology Patna, Patna, India

**Kalpana Rajouriya** Department of Chemistry, Dr. Bhimrao Ambedkar University, Agra, India

**Pallavi Saxena** Department of Environmental Science, Hindu College, University of Delhi, Delhi, India

**Diksha Singh** CSIR-Central Institute of Mining and Fuel Research, Barwa Road Campus, Dhanbad, Jharkhand, India

**Srishti Sinha** DFPD-DMPD Project National Institute of Disaster Management, Ministry of Home Affairs, GoI, Delhi, India

**Saurabh Sonwani** Department of Environmental Studies, Zakir Husain Delhi College, University of Delhi, New Delhi, India

**Atisha Sood** Centre for Disaster & Health, National Institute of Disaster Management, Ministry of Home Affairs, GoI, Delhi, India

**Rogert Sorí** Environmental Physics Laboratory (EPhysLab), Centro de Investigación Mariña, Campus de Ourense, Universidade de Vigo, Ourense, Spain

**Anju Srivastava** Department of Chemistry, Hindu College, University of Delhi, Delhi, India

**Milica Stojanovic** Environmental Physics Laboratory (EPhysLab), Centro de Investigación Mariña, Campus de Ourense, Universidade de Vigo, Ourense, Spain

**Ajay Taneja** Department of Chemistry, Dr. Bhimrao Ambedkar University, Agra, India

**Salman Tariq** Department of Space Science, University of the Punjab, Lahore, Pakistan

**Alexey Valero-Jorge** Department of Agricultural Systems, Forestry and Environment (Unit Associated to EEAD-CSIC Soils and Irrigation), Center for Agri-Food Research and Technology of Aragon (CITA), Zaragoza, Spain

**Jurgen van der Heijden** TNO (Dutch Organisation for Applied Scientific Research), The Hague, The Netherlands

**Marta Vázquez** Environmental Physics Laboratory (EPhysLab), Centro de Investigación Mariña, Campus de Ourense, Universidade de Vigo, Ourense, Spain

# Abbreviations

| | |
|---|---|
| AE | Angstrom exponent |
| AERMOD | Atmospheric dispersion module |
| AI | Artificial intelligence |
| AIRS | Atmospheric infrared sounder |
| AOD | Aerosol optical depth |
| AMSU | Advanced microwave sounding unit |
| AQPs | Aquaporins |
| AROME | Application of research to operations at mesoscale |
| AVHRR | Advanced very high-resolution radiometer |
| BGI | Blue-green infrastructure |
| BTD | Brightness temperature difference |
| CALIOP | Cloud-aerosol Lidar with orthogonal polarization |
| CASCC | Caltech Active Strand Cloud Water Collector |
| CCN | Cloud condensation nuclei |
| CO | Carbon monoxide |
| CrIS | Cross-track infrared sounder |
| DPR | Dual-frequency precipitation radar |
| EC | Elemental carbon |
| EnKF | Ensemble Kalman filter |
| EUMETSAT | European Organisation for the Exploitation of Meteorological Satellites |
| EV | Electric vehicle |
| FGD | Flue gas desulfurization |
| GMI | GPM Microwave Imager |
| GPM | Global precipitation measurement |
| GSOD | Global Surface Summary of the Day |
| HIRLAM | High-resolution limited area model |
| $HNO_3$ | Nitric acid |
| HRRR | High-resolution rapid refresh |
| IASI | Infrared atmospheric sounding interferometer |
| IC | Ion chromatography |

| | |
|---|---|
| IGP | Indo-Gangetic Plain |
| IIR | Imaging infrared radiometer |
| LIDAR | Light detection and ranging |
| LWC | Liquid water content |
| ML | Microwave links |
| MODIS | Moderate resolution imaging spectrometer |
| NIDM | National Institute of Disaster Management |
| NOx | Nitrogen oxides |
| NTSB | National Transportation Safety Board |
| NWP | Northwest Pacific |
| O3 | Ozone |
| OC | Organic carbon |
| OM | Organic matter |
| PAN | Peroxyacetyl nitrate |
| PBL | Planetary boundary layer |
| PDO | Pacific Decadal Oscillation |
| PM | Particulate matter |
| RAC | Rotating arm collector |
| RH | Relative humidity |
| ROS | Reactive oxygen species |
| RSL | Received signal level |
| $SO_2$ | Sulfur dioxide |
| SYNOP | Surface synoptic observations |
| TEM | Transmission electron microscope |
| TKE | Turbulent kinetic energy |
| UCIs | Urban cool islands |
| UCL | Urban canopy layer |
| UDI | Urban dry island |
| UHI | Urban heat island |
| VIIRS | Visible Infrared Imaging Radiometer Suite |
| VOCs | Volatile organic compounds |
| WHO | World Health Organization |
| WMO | World Meteorological Organization |
| WRF | Weather Research Forecasting |
| WSOC | Water-soluble organic carbon |

# Chapter 1
# An Introduction to Fog: Formation, Chemistry, Dynamics, Impact and Dissipation

**Pallavi Saxena, Anil Kumar Gupta, and Rogert Sorí**

## 1.1 General Introduction

Fog is defined as low visibility due to the suspension of tiny water droplets in the air (Beiderwieden et al. 2005). According to Indian Meteorological Department (IMD) it is defined as visibility reduction of less than 1 km (Climate Research and Services, n.d.). Fog formation is defined as achieving the saturation point by air either in the absence of heat or in the presence of moisture. Low temperatures, high relative humidity, calm wind speed and a number of air pollutants can act as cloud condensation nuclei which are responsible for controlling fog formation and its life cycle (Smith et al. 2024; Lakra and Avishek 2022; Sonwani et al. 2022). The saturation value required for the fog formation can be driven by aerosol number concentration (Mazoyer et al. 2019). Elevated aerosol concentration is directly proportional to an increase in small-sized cloud droplets that causes droplet radius to decrease significantly and results in rising fog's optical thickness and albedo (Fritz et al. 2021). Latent heat emitted during the condensation increases aerosol concentration which can accelerate the Brownian collision, further enhancing the fog formation processes (Guo et al. 2015). Aerosol is an important constituent for fog droplet formation; in its absence, more than 100% relative humidity is required to form fog droplets (Ahrens and Henson 2021). The age of an aerosol particle and the

P. Saxena (✉)
Department of Environmental Sciences, Hindu College, University of Delhi, Delhi, India
e-mail: pallavisaxena@hinducollege.ac.in

A. K. Gupta
Integrated Centre for Adaptation, DRR and Sustainability (ICARS, GOI-DST Joint COE), IIT, Roorkee, Gr. Noida Campus (NCR), India

R. Sorí
Environmental Physics Laboratory (EPhysLab), Centro de Investigación Mariña, Campus de Ourense, Universidade de Vigo, Ourense, Spain

 
P. Saxena et al. (eds.), *Fog: Formation, Chemistry, Dynamics, Impact and Dissipation*,
https://doi.org/10.1007/978-981-96-4422-3_1

transformation of an aerosol to a fog droplet are governed by nucleation, coagulation and condensation.

Fog chemistry is also correlated with deliquescence property of aerosols. Hydroscopic aerosols will act as CCN via nucleation, while hydrophobic aerosols are responsible for droplet formation through scavenging via collision and coalescence. Increased organic carbon is found in polluted urban sites, while decreased carbon content is found in remote sites (Herckes et al. 2013). Increased air pollution originating from anthropogenic sources is one of the main culprits of smog formation and dense fog (Saxena et al. 2020). Aerosol particles and secondary aerosol particles which are formed due to combination of two or more primary air pollutants play the role of cloud condensation nuclei and provide a substratum for the water molecules to condense on its surface. The CCN acts as a catalyst for visibility reduction, either by contributing through fog formation, haze, or smog. In addition, there are other air pollutants like particulate matter, NO, $NO_2$, $NO_x$, $NH_3$, $SO_2$, CO, $O_3$, $C_6H_6$ and $CH_3C_6H_5$, which have an important role in visibility reduction (Wagh et al. 2023). Sulphur dioxide and particulate matter are the main pollutants in industrial smog which are mainly released from coal burning and oil, while $NO_x$, $O_3$ and VOCs are contributing to photochemical smog formation. Ammonia ($NH_3$) and nitric acid ($HNO_3$) are responsible for the formation of ammonium nitrate ($NH_4NO_3$) which is formed via secondary formation from its precursors and is highly hygroscopic in nature and formed during day and night from NOx through oxidation of hydroxyl radical (OH) and $O_3$ (Xie et al. 2023). Sulphur dioxide helps in the formation of sulfate species like ammonium bisulphate ($NH_4HSO_4$) and ammonium sulfate ($(NH_4)_2SO_4$) which are hygroscopic compounds perfect for the formation of CCN (Chen et al. 2019) and severe fog episodes (Hameed et al. 2000).

At worldwide level, fog affects a number of daily routine life activities. These also lead to a number of economic losses. Fog adversely affects human health both directly and indirectly. Aerosol particles, like nitric acid, sulphur dioxide, sulphur oxide, increase the morbidity and mortality due to respiratory diseases related to cardiopulmonary, throat irritation, cardiovascular, muscular and lung cancer (Tanaka et al. 1998; Pope 1989). Exposure to $SO_2$ causes respiratory tract infections, further causing asthma and bronchitis, whereas exposure to $NO_2$ has little impact on respiratory system. Exposure to acidic fog often leads to deterioration of respiratory system which can further lead to cancer (Avol et al. 1988; Lavorini et al. 2001). Several studies have reported that hospital visits were increased during foggy days as compared to non-foggy days (Hu et al. 2015; Wielgosinski and Czerwinska 2020; Gilardoni and Fuzzi 2017). Tanaka et al. (1998) observed that bronchial epithelium problems might have occurred due to various possible mechanisms related with $H^+$, $O_3$, $H_2SO_4$ and $HNO_3$, and that this process happens since airway resistance caused by acidic pollutants increases because of less absorption capacity of hydrogen ions in the airway mucus (Hackney et al. 1985). Fog also affects transportation activities, leading to economic losses, inconvenience and even death (Croft 2003). The total financial and human life loss from them is approximately the same as that of losses that occurred during natural hazards like floods, drought, cyclones, etc. (Gultepe et al. 2009). Bad visibility is one of the main reasons for causing approximately

35% of weather-related accidents in civil aviation, multiple-vehicle crashes and about 168 casualties per year. During dense foggy days, a large number of flights got cancelled due to security reasons, according to reports from Canada, the United States and India. This may lead to heavy economic loss to airlines between USD 5000 and USD 25,000 for any delayed or cancelled flight. Moreover, due to dense fog, there is a huge loss of human lives: 50 people die annually in Canada due to road accidents. In the United States, 13,720 deaths between 1994 and 2004 were happened due to road accidents and 1122 from fatal air accidents (Aviation Safety Network 2021; Governments of India, Ministry of Road Transport & Highways, Transport Research Wing, New Delhi; *Road Accidents in India*, 2019). Fog also affects military rescue operations activities which cause property damage, supply delays and event cancellation. Fog also affects the functioning of schools and colleges (Gultepe et al. 2007; Forthun et al. 2006). Fog also causes smog formation, especially in urban areas. There are mainly two types of smog: the classical smog (London type) (Allaby 2008) and the photochemical smog (Los Angeles type) (Osibanjo et al. 2021). Trace gases like $NO_x$ and HCs were trapped near the earth's surface and gave rise to the secondary formation of tropospheric ozone ($O_3$). Consequently, these trapped air pollutants in the boundary layer remain suspended and cause damage to ecosystem and climate change (Forthun et al. 2006).

A number of studies were conducted to develop methods to remove fog artificially. Fog dissipation can be effectively done by increasing its droplet size through coalescence processes. Adding specific amounts of hygroscopic particles like sodium chloride (NaCl) improves the visibility and provides an effective solution for reducing fog (Houghton and Radford 1938; Jiusto et al. 1968). Plank et al. (1971) suggested another way for fog dissipation where dry air (situated above the inversion layer) mixes thoroughly with the saturated air in the fog by a helicopter's downwash. In this case, the air is completely saturated, and the droplet size increases (droplet coagulation) under the influence of the acoustic field leading to fog precipitation (Czyz and Markowski 2007).

## 1.2 Summary of Chapters

A number of points discussed above have clearly depicted that with the changing environmental conditions, the nature of fog has also changed. This chapter describes about the fundamental concepts of fog introduction, chemistry, formation, dynamics, impacts and dissipation. This chapter provides a holistic overview of basic concepts of fog formation, role of meteorology in fog formation, chemistry of fog, how different chemical constituents present in fog affects its quality and surrounding atmosphere, impacts on plant species, human health, visibility and economy and dissipation of fog by cost-effective techniques. Chapter 2 offers a comprehensive overview of various types of fog, delving into the physical and dynamic processes that influence its formation, development and dissipation. The analysis focuses on key factors, including microphysics, condensation nuclei, synoptic conditions and

surface-atmosphere interactions. Besides, the chapter explores state-of-the-art forecasting techniques and detection methods, emphasizing recent advancements in numerical modelling, satellite remote sensing and artificial intelligence applications. Chapter 3 summarizes about the chemistry behind the fog formation and factors influencing its chemistry. Further, the chemical characteristics of fog and related recent observations studies about their ionic composition and concentrations are highlighted. The fog chemistry in urban and rural settings and the role of fog in the scavenging of gas and particle phase organics have also been reviewed. Chapter 4 synthesizes existing knowledge about the chemical elements of acidic fog, shedding light on techniques, conclusions and implications for environmental science and public health. This chapter findings have implications for environmental policy and future research paths, highlighting the importance of ongoing monitoring and research into the characterization of chemical constituents in acidic fog. Chapter 5 highlights the different facets of urbanization, the feedback mechanisms through which it modifies the local weather and climate of cities, and how it modulates the urban fog dynamics. This chapter also explores different aspects of the urban environment and their role in modifying the urban boundary layer and creating urban microclimates, the processes governing fog lifecycle and its impact on visibility degradation while emphasizing the role of urban boundary layer, microclimates, anthropogenic emissions and their combined effects on atmospheric conditions conducive or detrimental for urban fog and highlighting the importance of urban planning in preventing and mitigating the losses due to urban fog. Chapter 6 summarizes a quick discussion of all the topics associated with fog dynamics in affecting air quality and climate. Fog formation, fog forecasting and detection, the impact of fog on human health, transportation and economy, ecology and the environment are also among the topics covered in this chapter. Chapter 7 discusses the variety of fogs—radiation, advection and upslope—which characterizes different cooling mechanisms and discusses the impact of atmospheric aerosols on fog. It provides excellent contributions for water supply, plant growth and nutrient uptake, relieving water stress within the ecosystems, whereas acidic fog formed due to pollution threatens plant health and soil quality. This chapter also provides an in-depth analysis of the metabolic responses of plants to fog and illustrates the impact of genetic modulations on their growth patterns. Chapter 8 aims to trace the historical significance of fog, from the medieval miasma theory to modern epidemiological studies, by examining its multifaceted effects on human health, its role in trapping hidden air pollutants, mechanisms of acute and chronic exposure, and its broader implications for transportation. Chapter 9 examines the economic, financial and legal challenges of governance in addressing air pollution and fog, particularly the division of responsibilities among different government departments. It presents case studies that illustrate how multifunctional solutions can address environmental issues through the interplay of water, soil, nature and more. Chapter 10 describes about satellites and sensors pivotal role in the prediction and monitoring of fog, ability to understand and forecast these complex weather phenomena and details about the accuracy of fog prediction model. Chapter 11 explores cost-effective approaches to remediate and control fog, with a focus on minimizing health impacts and

enhancing public safety. This chapter also offers a framework for addressing fog-related challenges in a manner that is both effective and environmentally responsible. Chapter 12 highlights the necessity of reducing aerosol concentrations by implementing stricter emissions regulations, promoting cleaner energy sources, improving public transportation and expanding urban greenery. It also calls for heightened public awareness, advanced fog forecasting models and cross-border cooperation to address transboundary pollution issues.

## 1.3 Conclusions

Fog is normally considered a local phenomenon that expands up to 100 kms in many places. Meteorological parameters like ambient temperature, wind speed, characteristics of surface and air mass, physico-chemical properties of CCN, behaviour of boundary layer, cloud cover, relative humidity and terrain affect the formation, evolution and dissipation of fog. To understand and assess different segments of fog, research has been done on a broader basis; however, the research is still lacking in fog forecasting area. New techniques are increasingly developed, particularly artificial intelligence. Proper understanding of fog properties, mechanisms and processes would help in improving the prediction capabilities and developing new algorithms in forecasting.

This chapter aims to provide a brief overview of different aspects of fog, including its controlling factors like formation, type, techniques for detection and prediction, and its effect on human and transport.

## References

Ahrens CD, Henson R. Meteorology today: an introduction to weather, climate, and the environment. Cengage learning; 2021.

Allaby MF. Smog, and poisoned rain. New York: Facts on File; 2008.

Aviation safety network. Aviation Safety Database 2021. Available online: https://aviation-safety.net/database/

Avol EL, Linn WS, Wightman LH, Whynot JD, Anderson KR, Hackney JD. Short-term respiratory effects of sulfuric acid in fog: a laboratory study of healthy and asthmatic volunteers. JAPCA. 1988;38:258–63.

Beiderwieden E, Wrzesinsky T, Klemm O. Chemical characterization of fog and rain water collected at the eastern Andes cordillera. Hydrol Earth Syst Sci. 2005;9(3):185–91.

Chen J, Li Z, Lv M, Wang Y, Wang W, Zhang Y, et al. Aerosol hygroscopic growth, contributing factors, and impact on haze events in a severely polluted region in northern China. Atmos Chem Phys. 2019;19(2):1327–42. https://doi.org/10.5194/acp-19-1327-2019.

Climate Research & Services. MET Glossary. Pune: IMD Indian Meteorological Department; n.d.. Retrieved June 28, 2023 from https://www.imdpune.gov.in/Reports/glossary.Pdf

Croft PJ. Fog. In: Holton JR, Curry JA, Pyle JA, editors. Encyclopedia of Atmos. Sciences. San Diego: Academic; 2003. p. 777–92.

Czyz H, Markowski T. Acoustic method of airport fog precipitation. Aviation. 2007;11:26–30.

Forthun GM, Johnson MB, Schmitz WG, Blume J, Caldwell RJ. Trends in fog frequency and duration in the Southeast United States. Phys Geogr. 2006;27:206–22.

Fritz J, Meyer H, Tseng W, Lin NH, Klemm O. Covariation of droplet size distribution and air humidity in fog: a methodological approach. J Hydrol. 2021;594:125934. https://doi.org/10.1016/j.jhydrol.2020.125934.

Gilardoni S, Fuzzi S. Chemical composition of aerosols of different origin. In: Tomasi C, Fuzzi S, Kokhanovsky A, editors. Atmos, aerosols. Hoboken: Wiley; 2017. p. 183–221.

Gultepe I, Tardif R, Michaelides SC, Cermak J, Bott A, Bendix J, Muller MD, Pagowski M, Hansen B, Ellrod G, et al. Fog research: a review of past achievements and future perspectives. Pure Appl Geophys. 2007;164:1121–59.

Gultepe I, Pearson G, Milbrandt JA, Hansen B, Platnick S, Taylor P, Gordon M, Oakley JP, Cober SG. The fog remote sensing and modeling field project. Bull Am Meteorol Soc. 2009;90:341–60.

Guo L, Guo X, Fang C, Zhu S. Observation analysis on characteristics of formation, evolution and transition of a long-lasting severe fog and haze episode in North China. Sci China Earth Sci. 2015;58(3):329–44. https://doi.org/10.1007/s11430-014-4924-2.

Hackney JD, Linn WS, Avol EL. Potential risks to human respiratory health from "Acid Fog": evidence from experimental studies of volunteers. Environ Health Perspect. 1985;63:57–61.

Hameed S, Mirza MI, Ghauri BM, Siddiqui ZR, Javed R, Khan AR, Rattigan OV, Qureshi S, Husain L. On the widespread winter fog in northeastern Pakistan and India. Geophys Res Lett. 2000;27(13):1891–4. https://doi.org/10.1029/1999gl011020.

Herckes P, Valsaraj KT, Collett JL Jr. A review of observations of organic matter in fogs and clouds: origin, processing and fate. Atmos Res. 2013;132:434–49. https://doi.org/10.1016/j.atmosres.2013.06.005.

Houghton HG, Radford WH. On the local dissipation of warm fog. Papers Phys Ocean Meteor. 1938;6:63.

Hu Y, Lin J, Zhang S, Kong L, Fu H, Chen J. Identification of the typical metal particles among haze, fog, and clear episodes in the Beijing atmosphere. Sci Total Environ. 2015;511:369–80.

Jiusto JE, Pilié RJ, Kocmond WC. Fog modification with giant hygroscopic nuclei. J Appl Meteorol. 1968;7:860–9.

Lakra K, Avishek K. A review on factors influencing fog formation, classification, forecasting, detection and impacts. Rendiconti Lincei Scienze Fisiche e Naturali. 2022;1–35:319. https://doi.org/10.1007/s12210-022-01060-1.

Lavorini F, Fontana GA, Pantaleo T, Camiciottoli G, Castellani W, Maluccio NM, Pistolesi M. Fog-induced respiratory responses are attenuated by Nedocromil sodium in humans. Am J Respir Crit Care Med. 2001;163:1117–20.

Mazoyer M, Burnet F, Denjean C, Roberts GC, Haeffelin M, Dupont JC, Elias T. Experimental study of the aerosol impact on fog microphysics. Atmos Chem Phys. 2019;19(7):4323–44. https://doi.org/10.5194/acp-19-4323-2019.

Osibanjo OO, Rappengluck B, Retama A. Anatomy of the March 2016 severe ozone smog episode in Mexico-City. Atmos Environ. 2021;244:117945.

Plank VG, Spatola AA, Hicks JR. Summary results of the Lewisburg Fog Clearing Program. J Appl Meteorol. 1971;10:763–79.

Pope CA. Respiratory disease associated with community air pollution and a steel mill, Utah Valley. Am J Public Health. 1989;79:623–8.

Governments of India, Ministry of Road Transport & Highways, Transport Research Wing, New Delhi. Road Accidents in India. 2019. Available online: www.morth.nic.in

Smith DK, Reka S, Dorling SR, Ross AN, Renfrew IA, Jayakumar A, Anurose TJ, Parde AN, Ghude SD, Rumbold H. Forecasts of fog events in northern India dramatically improve when weather prediction models include irrigation effects. Communications Earth & Environment. 2024;5(1):141. https://doi.org/10.1038/s43247-024-01314-w.

Sonwani S, Saxena P and Srivastava A. Carbonaceous aerosol variability and SOA formation during foggy days in Delhi, India. Front Sustain Cities 2022;4:951340. https://doi.org/10.3389/frsc.2022.951340

Saxena P, Sonwani S, Sharma SK, Kumar P, Chandra N. Carbonaceous aerosol variations in foggy days: A critical analysis during the fireworks festival. Fresenius Environ. Bull. 2020;29:6639–56.

Tanaka H, Honma S, Nishi M, Igarashi T, Teramoto S, Nishio F, Abe S. Acid fog and hospital visits for asthma: an epidemiological study. Eur Respir J. 1998;11:1301–6.

Wagh S, Kulkarni R, Lonkar P, Parde AN, Dhangar NG, Govardhan G, et al. Development of visibility equation based on fog microphysical observations and its verification using the WRF model. Modeling Earth Systems and Environment. 2023;9(1):195–211. https://doi.org/10.1007/s40808-022-01492-6.

Wielgosinski G, Czerwinska J. Smog episodes in Poland. Atmosphere. 2020;11:277.

Xie F, Su Y, Tian Y, Shi Y, Zhou X, Wang P, ... Lü C. The shifting of secondary inorganic aerosol formation mechanisms during haze aggravation: The decisive role of aerosol liquid water. Atmos. Chem. Phys. (2023);23(4):2365–78.

# Chapter 2
# Atmospheric Processes and Mechanisms Involved in Fog Formation

**Rogert Sorí, Milica Stojanovic, Albenis Pérez-Alarcón, Jakob Ernst, Marta Vázquez, Mojtaba Heydarizad, Alexey Valero-Jorge, Oscar A. Benedico-Rodríguez, Raquel Nieto, and Luis Gimeno**

## 2.1 The Phenomenon of Fog

Fog is a complex meteorological phenomenon defined as a suspension of minute, typically microscopic, water droplets in the atmosphere near the Earth's surface (WMO 2017), which reduces visibility to less than 1 km (NOAA 2024). It is a phenomenon that has been widely investigated; however, the complexity of its formation, development, dissipation, and accurate prediction makes its study a challenge for the scientific community (Gultepe et al. 2007; Westerhuis et al. 2020; Parde et al. 2024). Some of the factors involved in these steps are atmospheric stability, radiation balance, moisture availability, aerosols, turbulence, advection, topography, and land use change (Bari et al. 2023). The physical mechanisms driving fog

R. Sorí (✉) · M. Stojanovic · J. Ernst · M. Vázquez · R. Nieto · L. Gimeno
Environmental Physics Laboratory (EPhysLab), Centro de Investigación Mariña, Campus de Ourense, Universidade de Vigo, Ourense, Spain
e-mail: rogert.sori@uvigo.es

A. Pérez-Alarcón
Environmental Physics Laboratory (EPhysLab), Centro de Investigación Mariña, Campus de Ourense, Universidade de Vigo, Ourense, Spain

Facultade de Ciências, Instituto Dom Luiz (IDL), Universidade de Lisboa, Lisboa, Portugal

M. Heydarizad
State Key Laboratory of Marine Geology, Tongji University, Shanghai, China

A. Valero-Jorge
Department of Agricultural Systems, Forestry and Environment (Unit Associated to EEAD-CSIC Soils and Irrigation), Center for Agri-Food Research and Technology of Aragon (CITA), Zaragoza, Spain

O. A. Benedico-Rodríguez
Provincial Meteorological Centre of Ciego de Ávila, Institute of Meteorology, Avenida de los Deportes S/N, Ciego de Ávila, Cuba



P. Saxena et al. (eds.), *Fog: Formation, Chemistry, Dynamics, Impact and Dissipation*,
https://doi.org/10.1007/978-981-96-4422-3_2

formation vary depending on meteorological conditions, often leading to its classification into different types, such as radiation fog, advection fog, and upslope fog.

Meteorological studies in ancient Greece, by the outstanding philosopher and scientist Aristotle, described the phenomenon of fog (Neumann 1989). In the most recent history, early investigations about fog primarily focused on the characterization of its frequency and geographical distribution. Notably, research in the United States during the early twentieth century (Stone 1936) sought to map regions most prone to fog. Other studies investigated the geographical distribution of fogs at local and regional scales in different continents (Hardwick 1973). However, these early efforts often lacked detailed analyses of the underlying atmospheric processes governing fog formation. Subsequent studies have progressively advanced, focusing on the physical modelling of fog dynamics (Matveev 1984), providing important insights into the thermodynamics of fog formation.

The study of fog has gained significant attention due to its broad range of impacts. Fogs not only represent a substantial hazard for transportation systems, maritime navigation, and road traffic but also have profound implications for human health and agriculture (Pérez-Díaz et al. 2017; Manisalidis et al. 2020; Du et al. 2021). Despite its negative effects, positive impacts are also known; for example, fog is considered an important source of water in many drylands (Adhikari and Wang 2020). In addition, fog plays a significant role in biogeochemical cycles. Studies have shown that fog can contribute to atmospheric deposition processes, particularly in forested mountain regions, where fog water droplets capture pollutants and nutrients from the atmosphere, which are then deposited onto vegetation and soils, especially in ecosystems where precipitation is limited, and fog water constitutes a major source of moisture input (Hůnová 2024). Thus, the study and prediction of fog results in a critical area of meteorological research (Bari et al. 2023). Contemporary efforts, driven by advances in observational technology and numerical modelling, have enhanced our understanding of fog processes. Modern fog research employs sophisticated tools such as remote sensing, satellite imagery, and high-resolution numerical weather prediction (NWP) models (Guo et al. 2021), permitting one to understand the role of atmospheric processes and mechanisms involved in fog formation, development, and decay.

## 2.2 Types of Fog: Atmospheric and Dynamic Conditions for the Formation, Development, and Dissipation

### *2.2.1 Types of Fog*

Fog has been classified into different types depending on the environmental conditions that lead to its formation, and the atmospheric processes involved. Over time, various classifications of fog have been proposed, but today, six main types are widely recognized: radiation fog, advection fog, upslope fog, evaporation fog, ice fog, and precipitation fog. The formation and forecasting of fog encompass a wide array of physical processes, ranging from the microscopic scale of aerosols to the synoptic scale (Lakra and Avishek 2022), but typically form in aerosol-laden surface air under conditions of high relative humidity, ranging from undersaturation to slight supersaturation (Pruppacher and Klett 1997). The life cycle of fog is highly sensitive to factors like surface temperature gradients, wind speeds, and the presence of aerosols that can either enhance or inhibit droplet formation by providing condensation nuclei (Gultepe et al. 2007). Even though most studies reveal that the importance of moisture in the atmospheric layer closest to the surface is crucial, there are results also showing that water vapour from the soil layers is transported through the vadose zone to the surface, contributing to the formation of fog in fog-dominated arid areas (Adhikari and Wang 2020). Specific information for each type of fog is detailed below.

#### 2.2.1.1 Radiation Fog

Radiation fog forms typically during the end of the night or early morning hours, when the air near the ground cools through radiative heat loss. Figure 2.1 shows the radiation fogs observed in Bogojevce, Leskovac, Serbia, at different timings in a day. During the night, the ground loses heat by emitting long-wave radiation, and this cooling is transferred to the layer of air just above the surface (Hoch et al. 2011). The absence of clouds facilitates radiative heat loss, while light winds help mix the cooled air just enough for saturation to occur, without dispersing the moisture (Andersen et al. 2023; Weedon et al. 2024). As the surface air cools, its temperature can reach the dew point, and the saturation vapour pressure decreases, altering the amount of water vapour the air can hold, allowing water vapour to condense into tiny droplets (Price 2019; Lakra and Avishek 2022). In addition to the

**Fig. 2.1** Radiation fogs were observed in Bogojevce, Leskovac, Serbia, on 10 October 2024 at 06:30 (left), credit by Snezana Stojanovic; 13 October 2024 at 06:00 (middle), and 22 October 2024 at 06:00 (right), taken by Dusan Stojanovic

**Fig. 2.2** Valley fog observed in Roddino, Italy, on 6 September 2024 at 07:47. (Credit: Elisabeth Ernst)

cooling rate, the aerosol concentration also impacts the characteristics of the fog and its formation (Gandhi et al. 2024). However, a key condition for radiation fog formation and persistence is the high amount of water vapour and temperature decrease that permit moist air to reach saturation concerning water, meaning the relative humidity must be close to 100%, but also low wind speed (Smith et al. 2018; Price 2019).

Radiation fogs are more frequent during winter months as the nights are longer, cooling is stronger, and air tends to be moist following recent rain fog (Stull 1988). Another factor that increases the likelihood of radiation fog formation is the presence of an inversion layer. Surface heating and vigorous mixing lead to the dissipation of fog. This type of fog is also usually called valley fog, because it is frequently observed in valleys and low-lying areas where cooler air tends to pool (Zhong et al. 2001), which is prevalent in valleys, where the air carrying the fog is heavier and denser, causing it to settle on the bottom of the valley (Lakra and Avishek 2022). The following image (Fig. 2.2) shows an example of this type of fog, clearly identified by the fog bank observed in the valley.

A useful formula for estimating the radiative cooling process that leads to radiation fog formation is the Stefan-Boltzmann Law for blackbody radiation. According to Li et al. (2023), the temperature ($T_0$) at which water vapour condensates is given by:

$$To = \sqrt[4]{\frac{LWout - (1-\varepsilon) * LWin}{\varepsilon * \sigma}} - 273.15 \tag{2.1}$$

where *LWout* and *LWin* represent the long-wave and short-wave outgoing and incoming radiation; σ is Stefan-Boltzmann constant at $5.67 \cdot 10^{-8}$ W m$^{-2}$ K$^{-1}$; and the emissivity of the wet vegetation surface (ε) of 0.98 following, according to López et al. (2012).

Observational and modelling studies have shown that various atmospheric factors influence radiation fog. Processes such as drainage currents and turbulence can affect the development of fog (Li and Pu 2022), but without the presence of radiative cooling, the fog does not form. Other situations like urban heat island effects can inhibit the formation of radiation fog by maintaining higher surface temperatures, while rural areas with extensive vegetation may promote the retention of moisture necessary for fog development (Gultepe et al. 2007). Various studies agree (depending on the region) that wind speed should be lower than 3 m/s for the formation of fog (Pérez-Díaz et al. 2017; Gandhi et al. 2024). Other methods have also proved to be valuable to investigate the radiation fog evolution. For example, the water vapour isotope dynamics associated with radiation fog evolution were found to induce variability of similar amplitude as the variability induced by large-scale cold frontal clouds (Li et al. 2023). Finally, radiation fog typically dissipates quickly after sunrise as solar radiation warms the ground and the surrounding air. As the ground and air warm, the temperature rises above the dew point, causing the fog droplets to evaporate.

#### 2.2.1.2 Advection Fog

Advection fog forms when warm, moist air moves horizontally over a cooler surface, causing the air to cool to its dew point, leading to condensation and fog formation (Gultepe et al. 2007). Several atmospheric and surface conditions, such as temperature differences, relative humidity, and wind speed, are involved in the formation process. (Liu et al. 2016). Advection fogs are typically found in coastal areas, where warm ocean air masses contact cooler land or water surfaces (Fernando et al. 2021), but it can also occur inland, especially during winter when warm air flows over cold, snow-covered ground (Gultepe et al. 2007; Mohan et al. 2020). In such situations, advection fog forms with a slight or moderate pressure gradient directed from water towards land. With such a flow, the surface wind decreases slowly at night under the combined influences of increasing stability caused by surface cooling over the land and the tendency towards production of a land breeze which counteracts the general pressure gradient to some extent (George 1940). Therefore, the persistence of this type of fog is usually controlled by the availability of moist air and the cooling power of the underlying surface.

Advection fog typically forms under specific synoptic conditions, such as a high-pressure system that allows horizontal advection of warm and moist air, controlling the timing of advection fog occurrence (Dorman et al. 2021; Lakra and Avishek 2022). The subsidence related to positive high-pressure anomalies enhances the thermal and moisture stratification between the marine atmospheric boundary layer and the free atmosphere through adiabatic warming (Zhou et al. 2015; Yang et al. 2018). The dry air above helps cool the fog layer by enhancing the long-wave radiative cooling at the fog top and the vertical mixing beneath, causing air temperature to drop below sea surface temperature (Yang et al. 2018). In coastal areas like the California coast, advection fog typically forms due to sea breezes. These breezes transport warm, moist air masses from the central Pacific Ocean over the cooler upwelling waters of the Pacific. This process continues until the air reaches the North American coast (Lewis et al. 2003).

#### 2.2.1.3 Upslope Fog

According to the WMO (2017), upslope fog forms when moist air is forced to ascend along the slope of a mountain or elevated terrain. This process resembles cloud formation but occurs near the ground, adhering to the slopes of the terrain. As the air rises, it is adiabatically cooled, governed by the dry adiabatic lapse rate (9.8 °C per km) until the air reaches its dew point, where condensation starts, forming fog. This process resembles cloud formation but occurs near the ground, staying on the slopes of the terrain. Upslope fog is common in mountainous regions and can create extensive fog layers that persist along the slopes and valleys (NWS, n.d.). Upslope fog is a type of hill fog that appears differently depending on perspective; what is seen as a stratus by someone in a valley may be considered fog by someone at a higher elevation (Venne et al. 1997). The formation of upslope fog is typically driven by synoptic-scale atmospheric conditions that induce a steady flow of moist air towards higher terrain. Synoptic systems such as high-pressure or cold fronts can generate winds that push air masses up a slope in complex local topographical regions like ridges (Wolfe 2022; Jentsch and Weidinger 2022). Upslope fog can cover large vertical areas, creating dense layers that significantly reduce visibility across a wide region, as appreciated in Fig. 2.3. The studies show that upslope fog's persistence and density can significantly affect visibility in mountainous regions, with consequences for transportation and the local climate (Ohashi and Suido 2021; Chen et al. 2021).

**Fig. 2.3** Upslope fog observed completely obscuring the visibility of a mountain on the horizon. Innsbruck, Austria. Taken on 25 September 2024 at 08:27. (Credit: Jakob Ernst)

#### 2.2.1.4 Evaporation (Steam) Fog

Evaporation fog occurs when there is a significant temperature contrast between the water surface and the overlying air, for example, from cold air advection over a warmer ocean surface (Heo et al. 2010). Figure 2.4 shows evaporation fog observed near the southeast coast of Kujalleq, Groenlandia, in different timings in a day. The warm water evaporates, increasing the humidity in the cooler air. As evaporation continues, the air near the surface becomes saturated, and fog forms. Indeed, the difference in vapour pressure between the air and the water surface leads to evaporation, and mixing water vapour into the cold air leads to supersaturation and fog (Gultepe et al. 2007). The wind plays a crucial role afterwards; light winds allow moisture to accumulate leading to condensation, and spread horizontally, while stronger winds can disperse the moist air before fog forms (Stull 1988; Walker 2003). Several thermodynamic processes, such as the latent heat flux, the saturation vapour pressure, the heat and energy transfers, and the condensation, are involved in the formation of evaporation fog (Cox 1973; Heo et al. 2010). The formation process is particularly efficient in cold regions like the Arctic (Didenko et al. 1994).

#### 2.2.1.5 Ice Fog

When air is in a supersaturated state at temperatures above freezing, moisture can condense into tiny liquid droplets suspended in the air, creating a familiar phenomenon known as fog. If the air temperature is below freezing, ice fog may form (Mago and Sherif 2005). Ice fog, often termed pogonip (derived from the Shoshone Native American word "payinappih"), generally forms at low temperatures and consists

**Fig. 2.4** Evaporation fog observed near the southeast coast of Kujalleq, Groenlandia, on 7 July 2023 at 12:41 pm and 21:11 respectively. (Credit: Jakob Ernst)

solely of ice crystals (Gultepe et al. 2015). Thus, it most often occurs in high latitudes or high elevations, being most common in Arctic and subarctic regions during winter, and can be seen in populated areas where moisture sources such as exhaust from vehicles or industry are present. Robinson et al. (1957) reviewed various studies and described the ice fog as a problem of air pollution in the Arctic, as pollutants act as a source of ice nucleating particles. The process of sublimation of atmospheric water vapour on sublimation nuclei begins at temperatures less than −15 to –20 °C (Petterssen 1940; Gultepe et al. 2014). Thus, unlike regular fog, which forms from water vapour condensing into liquid droplets, ice fog develops when water vapour sublimates directly into ice crystals. The amount of water vapour in the air, as well as its capacity to hold moisture, decreases significantly at low temperatures. The formula for the saturation vapour pressure over ice, given by Eq. 2.2, is directly related to the formation of ice fog, as this phenomenon depends on the saturation vapour pressure over ice under extremely cold conditions.

$$ei = 6.11 * e^{\left(\frac{22.46 * T}{T + 272.62}\right)} \tag{2.2}$$

$ei$ represents the saturation vapour pressure over ice (in hPa) and $T$ is the air temperature (in °C). This equation indicates that the saturation vapour pressure over ice is much lower than over water, meaning that fog can form in conditions with very low absolute humidity, as long as the temperature is low enough (Curry and Webster 1999). Thus, this equation is essential for determining if the air is near the saturation point over ice and for calculating the relative humidity concerning ice, which shows weather conditions are conducive to ice fog formation. Once formed, the crystals can be as small as 5–10 μm and usually have sizes less than 200 μm, depending on their shape and density (Koenig 1971; Gultepe et al. 2014, 2015; Gultepe et al. 2017a, b). It can remain suspended in the air for an extended time and may produce a very low rate of precipitation compared to light snow and other snowy conditions (Gultepe et al. 2014).

Ice fog occurs under stable atmospheric conditions, where radiative cooling is intense and there is minimal wind. Clear skies lead to rapid heat loss from the surface at night, causing the air near the ground to reach extremely low temperatures (Gultepe et al. 2020). In urban areas, fog often occurs alongside industrial zones or cities, where exhaust from vehicles and heating systems adds extra water vapour to the air, promoting fog formation (Lill et al. 2024). Additionally, topography plays a role in ice fog development. Valleys or low-lying areas are more prone to ice fog because cold, dense air tends to settle in these regions, keeping the temperatures low and allowing fog to persist for longer periods.

#### 2.2.1.6 Precipitation (Frontal) Fog

Precipitation fog, also known as frontal fog, occurs when precipitation falls into a layer of cooler air near the surface. Frontal systems, especially warm fronts, where warm, moist air overrides cooler air near the ground in extensive areas ahead of warm fronts (Houssos et al. 2012; NWS, n.d.). As the warm, moist air of the front moves over a cooler and unsaturated layer of air, precipitation can occur, and the evaporated precipitation adds moisture to the cooler air below, further increasing its humidity. The cooling effect from evaporation often brings the air temperature down to its dew point, promoting condensation (Lakra and Avishek 2022). Stable atmospheric conditions often prevent vertical mixing, allowing moisture to accumulate near the surface and sustain fog. Additionally, the presence of aerosols or condensation nuclei in the atmosphere can influence the size and density of fog droplets, impacting the fog's optical properties and persistence (Mazoyer et al. 2019).

## 2.3 Atmospheric Factors Influencing Fog Formation and Dissipation

### *2.3.1 Mechanism Influencing for Formation*

#### 2.3.1.1 Temperature, Radiative Cooling, Humidity, and Saturation

Several factors influence the complex process of fog formation and development, such as temperature, humidity, pressure, wind patterns, and atmospheric stability. However, many types of fog are primarily driven by radiative cooling, where the Earth's surface loses heat through the emission of infrared radiation. As already mentioned, clear skies increase the radiative loss, while cloudy conditions tend to inhibit it, limiting the condensation of water vapour near the surface and the formation of fog (Cotton and Anthes 1992; Wærsted et al. 2017). Similarly occurs with advection fog, which also depends on warm, moist air moving over a cooler surface, causing the air to cool to its dew point, leading to condensation (Liu et al. 2016; WMO 2017). Precipitation fog, for example, forms when rain falls into a cooler air layer, increasing the moisture content and leading to fog. Additionally, evaporation fog (also known as steam fog) forms when cold air passes over a warm body of water, increasing the air's moisture content and leading to saturation (Gultepe et al. 2007).

#### 2.3.1.2 Synoptic and Local Weather Patterns, Atmospheric Stability, and Wind Patterns

Synoptic-scale systems play a crucial role in determining the intensity and extent of a fog event, particularly marine fog (Lakra and Avishek 2022). Larger synoptic weather systems, such as high-pressure zones, favour fog formation by fostering clear skies, calm winds, and stable atmospheric conditions. These conditions suppress vertical air movement, trapping moisture near the ground, which allows the ground to cool efficiently (Bowling et al. 1968; Gultepe et al. 2007). Under stable atmospheric conditions, the limited vertical air movement helps retain moisture near the surface, favouring temperature inversions and the formation of fog (Dorman et al. 2020; Lakra and Avishek 2022). For example, valley fog often forms in these settings, as cooler air settles into low-lying areas, promoting condensation (Oke 1987). However, wind plays a dual role in fog formation. Light winds (generally <3 m/s) are optimal for fog development as they allow for the mixing of air without dispersing moisture too rapidly, favouring the surface air to reach its dew point (Croft 2003; van der Velde et al. 2010; Pérez-Díaz et al. 2017). Conversely, stronger winds tend to prevent fog formation by stirring up the air too much, keeping the atmosphere unsaturated (Stull 1988; Pérez-Díaz et al. 2017). Atmospheric stability driven by synoptic conditions (e.g. high geopotential heights and strong anticyclonic systems) enhances the descent of free-tropospheric air and controls the surface aerosol particle concentrations, affecting the occurrence of fog (Maalick et al. 2016; Guy et al. 2021).

On the other hand, the turbulence, convection, and precipitation associated with low-pressure systems generally prevent fog formation by stirring and drying the atmosphere. However, some findings have revealed that under certain conditions a low-pressure system can induce the formation of fog. During the coastal fog field project that took place onshore and offshore of the Avalon Peninsula in Newfoundland, Canada, from August 25 to October 8, 2018, it was observed that stratus clouds expanded due to cloud-top radiation, precipitation, and warm-air advection caused by a cyclonic system. This expansion led to an increase in surface temperature and relative humidity around the low-pressure area, resulting in the cloud base extending downward to the surface and forming fog (Dorman et al. 2021).

#### 2.3.1.3 Topography and Microclimates

Indeed, on a more local scale, topography and microclimates also influence the formation and development of drought (Scholl et al. 2002; Ball and Tzanopoulos 2020; Rita et al. 2021). Local studies have shown that altitude significantly affects the probability of fog occurrence, through changes in thermodynamic characteristics of the air mass, turbulent mixing, and exchange fluxes between the surface and the atmosphere, but the effect is non-linear (Cuxart et al. 2021; Hůnová et al. 2021). It is observed in coastal regions, valleys, and mountainous areas, which also tend to have microclimate characteristics. For example, upslope fog forms as moist air is forced to ascend elevated terrain, cooling adiabatically as it rises and condensing to form fog (WMO 2017).

### *2.3.2 Fog Dissipation Key Factors*

Atmospheric conditions also induce fog dissipation, which occurs when moisture is removed from the air or when the air temperature rises sufficiently to evaporate the small water droplets. Several studies and reports have documented the main factors that drive fog dissipation. The National Weather Service (NWS n.d.) has listed the most important, which are listed and explained below, also considering other studies and examples:

**Moisture Reduction Mechanisms**

1. Turbulent Transfer of Moisture to the Surface
   The downward moisture flux induces the condensation of water vapour on the surface as dew or frost (Li et al. 2021). This process removes moisture from the fog layer, making it harder for the fog to persist. Turbulence near the surface promotes this transfer by pushing moisture downward (Huang et al. 2015; NWS, n.d.).
2. Turbulent Mixing with Drier Air
   Strong turbulence also enhances the mixing of fog and dry air (Rodhe 1962). This process dilutes the moisture in the fog, causing evaporation and thinning of the fog layer (Khoury et al. 2023).
3. Advection of Drier Air
   The advection of drier air into a region with fog decreases the relative humidity, causing fog to evaporate. This technique has been used for dissipating fog (Christensen and Frost 1980).
4. Condensation into Clouds
   Moisture from fog can sometimes be transported into the atmosphere, where it condenses to form clouds. As moisture rises and forms clouds, it reduces the amount of water vapour near the surface, allowing fog at lower levels to dissipate (NWS, 202; Zhang et al. 2024).

**Heating of the Air and Winds**

1. Turbulent Transport of Heat from the Ground
   After sunrise, the ground warms up, and turbulent heat transfer helps transport this warmth to the cooler air above. As the fog layer warms, the air temperature rises above the dew point, leading to the evaporation of fog droplets (Mazoyer et al. 2022; Parde et al. 2024).
2. Advection of Warmer Air.
   Warm air moving horizontally into a foggy area can increase the temperature of the fog layer, thereby reducing the relative humidity and leading to the evaporation of fog droplets (Dione et al. 2023).
3. Transport Over a Warmer Surface
   Fog may dissipate when air moves over a warmer land or water surface. As the air absorbs heat from the surface, it warms up, leading to a drop in relative humidity and causing the fog to evaporate (NWS n.d.).
4. Adiabatic Warming via Subsidence
   Adiabatic warming occurs when air is compressed during descending movements, such as in subsidence or downslope winds (Nicholls and Pielke Sr 2018). As the air compresses and warms, the fog evaporates due to reduced relative humidity. This process is especially common in mountainous areas, where downslope winds can quickly clear fog (Stull 1988; NWS n.d.).

5. Turbulent Mixing with Warmer Air Aloft
   Another factor in fog dissipation is the mixing of cooler, foggy air with warmer air from above. As turbulence brings warmer air into the fog layer, the fog starts to evaporate (NWS n.d.). This typically occurs when an inversion layer breaks down, allowing warmer air to mix downward (Gultepe et al. 2007).
6. Latent Heat Release from Cloud Formation
   When fog is lifted and forms clouds, the release of latent heat during condensation can further warm the surrounding air. This additional heat accelerates the dissipation of fog at lower altitudes, helping to clear the air near the surface (NWS n.d.).

## 2.4 Modelling and Prediction of Fog

### *2.4.1 Overview of Models Used for Fog Prediction*

The negative impacts of fog and associated conditions on human activities such as aviation, road transportation, and marine and military operations (Gultepe et al. 2007, 2015) warrant the accurate prediction of fog events (Singh and Harun 2023). Given the complex processes involved in fog formation, evolution, and dissipation (Steeneveld et al. 2015; Román-Cascón et al. 2016; Martinet et al. 2020), continuous efforts have been made over time to predict fog and visibility using numerical weather prediction (NWP) models. These models simulate atmospheric conditions based on physical equations governing meteorological processes (Coiffier 2011; Krishnamurti and Bounoua 2018) and provide valuable information about atmospheric circulation patterns and associated weather phenomena, including fog. While many of the earlier studies (e.g., Bergot and Guedalia 1994; Clark and Hopwood 2001) used one-dimensional models (1D) for fog prediction, the limitations of these models in representing large-scale meteorological conditions affecting fogs (Singh et al. 2018) pointed out the advantages of three-dimensional ones such as the Fifth-Generation NCAR/Penn State Mesoscale Model (MM5), Weather Research Forecasting (WRF), Application of Research to Operations at Mesoscale (AROME), High-Resolution Limited Area Model (HIRLAM), and Met Office's Unified Model (Fu et al. 2006; Van der Velde et al. 2010; Hu et al. 2017a; Smith et al. 2018; Martinet et al. 2020; Taylor et al. 2021). In general terms, these models are tailored to accurately simulate microphysical processes, wind conditions, humidity, temperature gradients and inversions, surface cooling, and radiative heat losses, which are the key contributors to fog formation and evolution (Steeneveld and de Bode 2018; Duconge et al. 2020; Yang and Gao 2020; George et al. 2023; Li et al. 2023).

The ability of NWP models to simulate and predict fog depends on the type of fog formed because different processes drive different fog types (Bari et al. 2015; Lakra and Avishek 2022). For example, while fog originating from advection depends on dynamic processes, radiative fog mainly depends on radiative processes, interaction between the Earth's surface and the atmosphere and turbulence. In addition, the reliable simulation of fog by NWP models is highly dependent on the horizontal model resolution (Boutle et al. 2016, 2022; Jayakumar et al. 2018; Pithani et al. 2019; Smith et al. 2021), microphysics and planetary boundary layer schemes (Lin et al. 2017; Martinet et al. 2020; Fernando et al. 2021), the initial and boundary conditions (Lewis et al. 2004; Gao et al. 2007, 2010; Bari 2019; Lakra and Avishek 2022), vertical resolution (Philip et al. 2016), water deposition (Tav et al. 2018), and aerosol activation (Mazoyer et al. 2019). Various specialized techniques, such as ensemble forecasting and data assimilation, are used to improve the accuracy of fog forecasts within these models. Ensemble forecasting, for example, runs multiple model simulations with slightly different initial conditions to capture the range of possible future states, which is particularly useful for predicting low-visibility events like fog that are highly sensitive to small changes in the atmosphere (e.g., Price et al. 2015; McCabe et al. 2016; Koračin 2017; Smith et al. 2021; Kim et al. 2023; Wang et al. 2023). Meanwhile, data assimilation integrates real-time observations (e.g., from satellites, weather stations, and aircraft) to improve forecasts (e.g., Liu et al. 2011; Wu et al. 2017; Bell et al. 2021; Gao et al. 2024; Sim et al. 2024).

### 2.4.2 *Limitations and Challenges in Fog Forecasting*

Fog forecasting is challenging due to the complexity, nonlinearity, and small scale of the physical processes driving fog formation and evolution (Gultepe et al. 2007). While NWP models are skilful in providing a broad framework for fog prediction, they usually fail to predict the exact location and time evolution of the fog life cycle (Bergot 2013; Román-Cascón et al. 2016; Boutle et al. 2018). One of the primary challenges is the spatial, vertical, and temporal resolution of the models (e.g., Boutle et al. 2022; Jayakumar et al. 2018; Pithani et al. 2019; Smith et al. 2021). Fog formation is often a localized phenomenon influenced by microscale processes like surface cooling, radiative transfer, and moisture condensation (Zhong et al. 2001; Price 2019; Bergot and Koracin 2021). Generally, NWP models run on grid scales that are often too coarse to resolve these small-scale processes accurately. In this context, high-resolution models can partially resolve surface and topographic heterogeneities and represent the thermodynamic and small-scale processes throughout the fog life cycle, but their widespread implementation is limited due to the need for high computational resources (Yadav et al. 2022). However, some works have noted that even high-resolution models fail to simulate the fog onset and development (Steeneveld et al. 2015).

Another challenge is the representation of physical processes in the model equations, particularly those involving cloud microphysics, radiation, and boundary layer interactions (e.g., Lin et al. 2017; Mazoyer et al. 2019; Zhang et al. 2020; Wang et al. 2023). Fog typically forms in the atmospheric boundary layer, which is highly dynamic and difficult to model. Whilst the fog onset is more sensitive to the planetary boundary layer scheme, the dissipation depends on the microphysics parameterization (Steeneveld et al. 2015). Therefore, the choice of the parameterization schemes is a crucial step in model configuration to improve fog prediction. Many uncertainties in the fog microphysical aspects are associated with the characteristics, distribution, and evolution of the cloud condensation nuclei, which are not available and thus cannot be adequately included in the NWP models (Lakra and Avishek 2022). It should also be noted that finding one single parametrization combination that reliably simulates many cases is still a challenge. Additionally, successful numerical fog forecasting requires accurate initial and lateral boundary conditions (Rémy et al. 2012; Bari et al. 2015; Pithani et al. 2020), which can be sparse or of low quality in certain regions. The correct representation of land properties and land-atmosphere interaction is key for modelling fog (Steeneveld and de Bode 2018). In particular, reliable representation of soil properties (Maronga and Bosveld 2017; Steeneveld and de Bode 2018; Ganeshi et al. 2023), land use (Jayakumar et al. 2018), thermal roughness (Weston et al. 2019), and albedo and snow depth (Zhang and Pu 2019) in NWP is critical for improving numerical fog predictions (Smith et al. 2021).

The chaotic nature of the atmosphere presents a challenge for all forms of weather forecasting, including fog prediction. Some recent studies (Pu et al. 2016; Chachere and Pu 2016; Zhang and Pu 2019) have highlighted that the wrong forecasting of the near-surface atmospheric conditions, mainly due to a lack of observations in the atmospheric boundary layer (Pu et al. 2016), is associated with the highest errors in numerical fog prediction. Overall, among the above factors attempting to accurately predict numerical prediction of fog events, the differences in processes driving the types of fog contribute to a challenge in forecasting (Lin et al. 2017; Lakra and Avishek 2022).

### 2.4.3 Data Assimilation and Observation

Data assimilation plays a crucial role in improving the accuracy of modelling, by integrating real-time observational data into NWP models, especially for nowcasting and short-range forecasting of different weather events (e.g. Strajnar et al. 2015; Li et al. 2016; Bergot and Lestringant 2019; González-Jardines et al. 2022). As fog formation and evolution are highly sensitive to micro-scale atmospheric conditions such as temperature, humidity, wind patterns, and accurate and real-time observations are essential for reducing uncertainties in the initial state of the atmosphere and improving the model performance in forecasting near-surface conditions (e.g., Zhang et al. 2015; Pu 2017) and, consequently, fog events. Some studies (Wang et al. 2014, 2023; Hu et al. 2017a; Martinet et al. 2020) have demonstrated that improving the initial temperature and humidity profiles using data assimilation could potentially play a greater role in the accuracy of fog forecasts than assimilating surface wind observations. In this line, reducing bias in the initial state of soil conditions improves near-surface air temperature and humidity profiles by changing the boundary layer thermodynamic structure (Massey et al. 2014; Dy and Fung 2016). These improvements in the initial conditions are directly translated into reducing the large errors in fog simulations (e.g., Liu and Pu, 2019; Kim et al, 2020). Numerical maritime fog forecasting has also been beneficial by assimilating radar and satellite data, and by improving the temperature stratification and water vapour distribution in the maritime atmospheric boundary layer (Liu et al. 2011; Wang et al. 2014, 2017). In particular, data assimilation techniques, such as 3DVar (three-dimensional variational assimilation), 4DVar (four-dimensional variational assimilation), and EnKF (ensemble Kalman filter), have been shown to significantly improve fog forecasts (Wang et al. 2014; Hu et al. 2017a, b; Smith et al. 2024; Bell et al. 2022; González-Jardines et al. 2022). Overall, these methods involve incorporating data from various sources (e.g., satellites, radar, surface stations and aircraft observations) and optimizing the model's initial state based on these real-time observations.

In particular, satellite and radar data have become an important source of real-time observations in numerical fog modelling, providing extensive spatial coverage and offering insights into the atmospheric conditions that contribute to fog formation. Specifically, satellites have the advantage of providing data where there is an absence of in situ observations or poorly instrumented areas, such as oceans or mountainous regions. This is critical for fog prediction, as fog can form in areas where conventional ground-based observations are sparse (Koračin 2017; Lakra and Avishek 2022). Remote sensing is an effective tool for monitoring fog (Wu et al. 2015, 2017). Some studies have used the Advanced Very High-Resolution Radiometer (e.g., Baldocchi and Waller 2014), Moderate Resolution Imaging Spectroradiometer (e.g., Wu and Li 2014), Cloud-Aerosol Lidar and Infrared Pathfinder Satellite Observation (e.g., Xiao et al. 2019), and polar-orbiting satellite data (e.g., Chaurasia and Jenamani 2017; Egli et al. 2017; Weston and Temimi 2020) for fog detection. Overall, the combined data assimilation from cloud radar and satellite observations may improve the initial conditions of temperature, humidity, and liquid water content in NWP models, leading to a more accurate fog prediction (e.g., Martinet et al. 2020; Bell et al. 2022).

## 2.5 Observed Changes in Fog Formation and Development and Their Attribution

Climate change can significantly impact fog formation through alterations in parameters such as temperature, humidity, and atmospheric circulation patterns in different regions worldwide (Hůnová et al. 2020; Maurer et al. 2019; Schalkwyk and Dyson 2013). A review by Klemm and Lin (2016) indicated that various studies confirmed a reduction in fog occurrence up to that date. A more recent study of fog-day occurrences using data from over 8000 Global Surface Summary of the Day (GSOD) ground stations for the period 1975–2017 also confirmed a notable decline in fog-day frequency, particularly observed in the United States (USA) with a reduction of approximately 20–25% and much of Europe with a decrease of around 3–5% (Lavigne and Liu 2022). However, some stations have reported an increase in fog frequency or no significant trends (Niu et al. 2010; Quan et al. 2011; Syed et al. 2012). Rising air temperatures, with lower relative humidity, are often suggested as a primary factor for the decrease in fog frequency (Ding and Liu 2014). On the other hand, elevated aerosol concentrations have been linked to increased fog frequency (Vautard et al. 2009; Quan et al. 2011; Gray et al. 2019). Recent studies have explored the impacts of climate change on marine fog patterns and their broader climatic implications, showing that these patterns are likely to change with increasing temperatures, which may shift fog formation towards higher altitudes or cause a decline in the frequency of fog events affecting certain regions in many parts of the world (Gultepe et al. 2017; Kawai et al. 2016). Climate model simulations suggest that these changes are primarily driven by alterations in large-scale circulation patterns and horizontal temperature advection near the surface, particularly changes in the high-pressure system and associated wind patterns over the North Pacific are expected to control fog occurrence (Kawai et al. 2018). Globally, marine fog occurrence is generally low (0.2%), with higher concentrations in specific regions such as the Northwest Pacific, Northwest Atlantic, and coastal upwelling zones influenced by sea surface temperatures, atmospheric stability, and large-scale climate oscillations like the North Pacific Oscillation and North Atlantic Oscillation (Dorman et al. 2019).

Various studies have been conducted on the trends observed in fog formation across the continents, while for others (e.g. Africa, Australia), we found few studies about fog, but basically related to climatological studies without trend analysis. There are various studies about fog trends and variability in North America. According to Gaffen and Ross (1999) in the period 1961–1995, an increase in specific humidity and a decrease in general visibility have been observed in the United States. Findings of Forthun et al. (2006) for a longer study period (1940–2000), revealed that 12 of 26 weather stations in the southern United States exhibited a decreasing annual trend in fog days, while only the recorded series of 2 stations experienced an increasing trend. Westcott (2002) identified a decline in fog frequency around the Great Lakes region, but an increase in the midwestern United States in the 1948–1996 period. In the area of Alaska, it also has been found that there was a 60–70% decrease in fog hours over the study period 1948–2022. Their findings highlight the greatest reduction during the 1970s and 1980s, with ice fog hours closely following the decreasing trend in cold temperature hours, possibly associated with changes in air pollution or near-surface moisture.

An analysis using satellite data also indicated a strong decrease in fog frequency in California's Central Valley between 1981 and 2014, with climate change, improved air quality, and changes in agricultural practices mentioned as potential contributing factors (Baldocchi and Waller 2014). A study by Herckes et al. (2015) also shows a decrease of dense fog by approximately 50% in this region of California. Coastal fog has also experienced a negative trend on the southwestern coast of the

United States due to a decrease in aerosol concentration, an increase in the sea surface temperature, urbanization, and changes in large-scale circulation (Carswell 2015; LaDochy and Witiw 2010, 2012; Williams et al. 2015). In agreement with these findings, and focusing on a more local scale, some authors analysed fog trends in the city of Los Angeles for the period 1948–2004, showing a clear decrease in hours of dense fog, particularly at stations near the Pacific coast (Witiw and LaDochy 2008; LaDochy, 2005).

Regarding South America, most studies have been conducted on the São Paulo metropolitan region, the largest urban agglomeration in Brazil. Gonçalves et al. (2002) assessed climate change signals in São Paulo from 1950 to 2000, demonstrating that rising air temperatures, especially minimum temperatures, can directly influence fog formation. These authors also identified a connection between urban development, sea surface temperatures and temperature trends, which may contribute to a decrease in fog frequency. These align with the findings of Gonçalves et al. (2008), whose results indicated a decreasing trend in fog events from 1933 to 2005 for this region. These authors commented that the rise in air temperature and the corresponding drop in relative humidity are the most likely causes of the observed trend. A recent study conducted by Mühlig et al. (2020) for the period 1998–2018 also showed that the decreasing trends in annual fog occurrences were associated with increasing temperatures and decreasing relative humidity over the Metropolitan Area of São Paulo.

Decreasing trends in fog frequency have also been reported for the Europe continent (Van Oldenborgh et al. 2010; Vautard et al. 2009). A study conducted by Avotniece et al. 2015 analysed the frequency of fog, long-term changes in fog frequency, and the atmospheric conditions that favour fog events in Latvia over 52 years. The results showed a decreasing trend in fog frequency, possibly associated with the observed increase in air temperature and improvements in air quality since industrialization. Bokwa et al. (2018) assessed the long-term fog variability in Kraków, Poland, from 1966 to 2015. This study revealed a significant reduction in fog frequency, with the average number of foggy days decreasing by approximately 60%. This decline was linked to decreasing trends in days with low wind speeds and high relative humidity. A decrease in the occurrence of radiation fogs in urban areas of Southern Germany from 1949 to 1990 was also reported and associated with the urban heat island effect and a moisture deficit (Sachweh and Koepke 1995). Hůnová et al. (2020) found a decrease in fog formation from 1961 to 2018 across 12 different sites in the Czech Republic but did not evaluate the possible causes.

Numerous studies have examined fog trends in East Asia, especially in China. Chen et al. (2006) examined the connection between foggy days and temperature variations in China from 1961 to 2003, highlighting a significant overall decline in fog frequency. The most substantial decreases were observed in West China, North China, the Yunnan-Guizhou Plateau, and parts of the hilly areas in Southeast China. However, small regions with increasing fog were identified in Eastern Sichuan, Northern Hubei, and the North China Plain in the lower reaches of the Yellow River. Based on statistical analyses, the authors attributed these fog trends to increases in temperature. However, many previous studies on fog variations in

China have shown inconsistent results. Niu et al. (2010) found that the number of fog events in wintertime over eastern-central China during the 1976–2010 period doubled over the past, attributing this to significant increases in relative humidity and the frequency of wind events. The weakening of the East Asian winter monsoon system appears to be responsible for these changes. In a more recent study, Ding and Liu (2014) observed that the number of fog days in China from 1961 to 2011 exhibited a two-stage variation. There was an increasing trend before 1980, followed by a decrease after 1990. They identified the reduction in surface relative humidity as a key factor contributing to this trend but also indicated that fog trends are highly sensitive to the study region, period, and fog classification, reflecting the complex effects of various influencing factors. A study focused on the regional scale published in the same year found that the number of fog days increased before 1995, remained stable from 1995 to 2003, and then decreased significantly in the North China Plain region (Fu et al. 2014). They argue that the reduction in fog, particularly during the winter heating period, is primarily due to improvements in air quality since 2003. Similarly, Quan et al. (2011) observed an increase in fog during the 1970s and 1980s, followed by a decline after 1999 for the same region.

According to Maurer et al. (2019), urban areas of Taiwan experienced improved visibility and a decline in the number of fog days, while rural coastal regions exhibited either no significant trend or a reduction in visibility (Maurer et al. 2019). This study found that the combination of climate change and urbanization has contributed to rising temperatures (averaging +0.035 °C per year) and an overall decrease in relative humidity (averaging −0.125% per year), leading to improved long-term visibility in Taiwan's cities (Maurer et al. 2019). The findings of Belorid et al. (2015) identified a significant decline in radiation fog occurrences between 1989 and 2013 at 10 of the 23 stations studied across South Korea. The most notable reductions in radiation fog were observed at stations where trends showed a steep decline in maximal cooling rates, a decrease in mean specific humidity, and substantial population growth over the past two decades before 2013. Another study for Japan showed a decline in summer sea fog occurrences on Hokkaido Island, associated with anomalies in synoptic circulation patterns (Sugimoto et al. 2013). A rise of winter fog over the Indo-Gangetic Plains and South Asia during the 37 years from 1977/78 to 2013/14 (Kutty et al. 2020) attributed to cooler air temperatures, likely due to more favourable large-scale circulation patterns along with an increase in nucleating aerosols (Syed et al. 2012; Kutty et al. 2020). In India, a great country with a complex climate due to the effects of the monsoon, the recent positive trends of fog are more likely the result of a complex interaction between urban expansion and the associated rapid change in aerosol loading, resulting in impacts on radiation balance, microphysics, and heat-island processes (Smith et al. 2023). Another study for Iran, in southwestern Asia, showed declining fog trends in almost all the country's stations from 1985 to 2005. According to this study, rising temperatures could have been the responsible factor for these trends.

## 2.6 Current and Future of Fog Research and Prediction

Improvements in computational resources and the development of high-resolution numerical weather prediction (NWP) models have greatly enhanced the study and prediction of fog by accurately resolving small-scale atmospheric processes such as advection, turbulence, and surface heterogeneities (Boutle et al. 2016; Gultepe et al. 2016; Ducongé et al. 2020). Efforts to refine microphysics schemes in NWP models are ongoing, addressing gaps in representing the properties of fog (Nelli et al. 2024). Recent studies have improved the understanding of fog microphysics, including cloud droplet formation, concentration, and evaporation, along with advancements in parameterization techniques (Gultepe et al. 2017a, b; Mazoyer et al. 2022; Field et al. 2023). This enhanced understanding can contribute to developing new microphysics and planetary boundary layer models to address the complexities of fog formation and its variability (Smith et al. 2018). However, further progress is needed in understanding fog microphysics, particularly the interaction between large- and small-scale factors (Smith et al. 2018; Wagh et al. 2023).

Ensemble forecasting, which accounts for uncertainties in model physics and initial conditions, has shown promise for simulating fog variability and near-surface conditions (Smith et al. 2021; Alaoui et al. 2022; Parde et al. 2022; Kim et al. 2023; Wang et al. 2023). Furthermore, soil states, including soil moisture and temperature, influence fog prediction through land-atmosphere feedback, with coupled models enhancing boundary layer simulations (Trigo et al. 2015; Sgoff et al. 2020). Advanced data assimilation techniques, such as 4DVar and EnKF, further refine initial conditions by integrating real-time observations from diverse sources (Moore et al. 2020; Shao and Nerger 2024).

Machine learning (ML) methods have also emerged as valuable tools for fog understanding and prediction, leveraging large datasets and uncovering complex relationships among variables (Durán-Rosal et al. 2018; Peláez-Rodríguez et al. 2023). Techniques such as neural networks, fuzzy logic, and decision trees have been extensively utilized, enabling the integration of satellite, radar, and ground-based data for enhanced accuracy (Kim et al. 2023). ML models offer a computationally efficient alternative to NWP simulations but require careful training and validation to avoid overfitting (Lv et al. 2022). Recent research focuses on unifying data fusion, modelling, and mining frameworks for enhanced prediction performance (Dong et al. 2020; Galicia et al. 2019).

A review of key studies from recent years offers a list of important future directions for fog research, which are summarized and outlined below:

- Enhanced understanding of microphysics and boundary layer processes
- Integration of ensemble forecasting and land-atmosphere coupled models
- Advanced data assimilation techniques for improved accuracy
- Expanded use of machine learning for operational forecasting with efficient computational demands

By addressing these challenges, fog understanding and prediction will continue to advance, providing more accurate and reliable forecasts for diverse applications.

## 2.7 Conclusions

This chapter provides a comprehensive analysis of fog formation, development, and dissipation, highlighting its complex physical and dynamic processes. Fog types, radiation, advection, upslope, evaporation, ice, and precipitation fog are discussed in detail, focusing on key mechanisms such as radiative cooling, turbulence, synoptic conditions, and surface-atmosphere interactions. The study explores advancements in fog forecasting, including numerical weather prediction (NWP) models, data assimilation techniques, and the integration of machine learning to improve accuracy.

Challenges remain because of the micro-scale characteristics of fog, constraints in models, and uncertainties related to boundary layer processes and cloud microphysics. The chapter also analyses global trends in fog frequency and the possible role of climate warming, aerosol variability, and urbanization. In conclusion, the review highlights the need for a deeper comprehension of fog microphysics, better data assimilation methods, and the integration of new technologies to tackle forecasting difficulties amid evolving climate conditions.

**Acknowledgements** R.S. and J.E acknowledges the RYC2021 – 034044-I grant funded by Ministerio de Ciencia, Innovación y Universidades, Spain (MICIU/AEI/https://doi.org/10.13039/501100011033) and the European Union Next Generation EU/PRTR. M.S. and A.P-A. thank the support from the Xunta de Galicia (Consellería de Cultura, Educación, Formación Profesional e Universidades) under the Postdoctoral Grants No. ED481D–2024/017, and ED481B – 2023/016, respectively. R.S., M.S., A.P-A., and M.V. are also supported by the project Excelencia-ED431F-2024/03 funded by the Xunta de Galicia. EPhysLab members are supported by the Xunta de Galicia under Project ED431C2021/44 (Programa de Consolidación e Estructuración de Unidades de Investigación Competitivas (Grupos de Referencia Competitiva) and Consellería de Cultura, Educación e Universidade), and by the European Union "ERDF—A way of making Europe".

## References

Adhikari B, Wang L. The potential contribution of soil moisture to fog formation in the Namib Desert. J Hydrol. 2020;591:125326. https://doi.org/10.1016/j.jhydrol.2020.125326.

Alaoui B, Bari D, Bergot T, Ghabbar Y. Analog ensemble forecasting system for low-visibility conditions over the main airports of morocco. Atmosphere. 2022;13:1704. https://doi.org/10.3390/atmos13101704.

Andersen H, Cermak J, Douglas A, Myers TA, Nowack P, Stier P, Wall CJ, Wilson Kemsley S. Sensitivities of cloud radiative effects to large-scale meteorology and aerosols from global observations. Atmos Chem Phys. 2023;23:10775–94. https://doi.org/10.5194/acp-23-10775-2023.

Avotniece Z, Klavins M, Lizuma L. Fog climatology in Latvia. Theor Appl Climatol. 2015;122:97–109. https://doi.org/10.1007/s00704-014-1270-4.

Baldocchi D, Waller E. Winter fog is decreasing in the fruit-growing region of the Central Valley of California. Geophys Res Lett. 2014;41:3251–6. https://doi.org/10.1002/2014GL060018.

Ball L, Tzanopoulos J. Interplay between topography, fog and vegetation in the central south Arabian mountains revealed using a novel Landsat fog detection technique. Remote Sens Ecol. 2020;6:498–513. https://doi.org/10.1002/rse2.151.

Bari D. A preliminary impact study of wind on assimilation and forecast systems into the one-dimensional fog forecasting model COBEL-ISBA over Morocco. Atmos. 2019;10(10):615. https://doi.org/10.3390/atmos10100615.

Bari D, Bergot T, El Khlifi M. Numerical study of a coastal fog event over Casablanca, Morocco. Q J R Meteorol Soc. 2015;141(690):1894–905. https://doi.org/10.1002/qj.2494.

Bari D, Bergot T, Tardif R. Fog decision support systems: A review of the current perspectives. Atmosphere. 2023;14:1314. https://doi.org/10.3390/atmos14081314.

Belorid M, Lee CB, Kim JC, Cheon TH. Distribution and long-term trends in various fog types over South Korea. Theor Appl Climatol. 2015;122:699–710. https://doi.org/10.1007/s00704-014-1321-x.

Bell A, Martinet P, Caumont O, Vié B, Delanoë J, Dupont J-C, Borderies M. W-band radar observations for fog forecast improvement: an analysis of model and forward operator errors. Atmos Meas Tech. 2021;14:4929–46. https://doi.org/10.5194/amt-14-4929-2021.

Bell A, Martinet P, Caumont O, Burnet F, Delanoë J, Jorquera S, Seity Y, Unger V. An optimal estimation algorithm for the retrieval of fog and low cloud thermodynamic and micro-physical properties. Atmos Meas Tech. 2022;15:5415–38. https://doi.org/10.5194/amt-15-5415-2022.

Bergot T, Guedalia D. Numerical forecasting of radiation fog. Part I: numerical model and sensitivity tests. Mon Wea Rev. 1994;122:1218–30. https://doi.org/10.1175/1520-0493(1994)122<1218:NFORFP>2.0.CO;2.

Bergot T. Small-scale structure of radiation fog: a large-eddy simulation study. Q J R Meteorol. Soc. 2013;139:1099–1112. https://doi.org/10.1002/qj.2051.

Bergot T, Lestringant R. On the predictability of radiation fog formation in a mesoscale model: A case study in heterogeneous terrain. Atmosphere. 2019;10:165. https://doi.org/10.3390/atmos10040165.

Bergot T, Koracin D. Observation, simulation and predictability of fog: Review and perspectives. Atmosphere. 2021;12:235. https://doi.org/10.3390/atmos12020235.

Bokwa A, Wypych A, Hajto MJ. Impact of natural and anthropogenic factors on fog frequency and variability in Kraków, Poland, in the years 1966–2015. Aerosol Air Qual Res. 2018;18:165–77. https://doi.org/10.4209/aaqr.2016.12.0580.

Boutle I, et al. Demistify: a large-eddy simulation (LES) and single-column model (SCM) intercomparison of radiation fog. Atmos Chem Phys. 2022;22:319–33. https://doi.org/10.5194/acp-22-319-2022.

Boutle I, Price J, Kudzotsa I, Kokkola H, Romakkaniemi S. Aerosol–fog interaction and the transition to well-mixed radiation fog. Atmos Chem Phys. 2018;18:7827–7840. https://doi.org/10.5194/acp-18-7827-2018.

Boutle IA, Finnenkoetter A, Lock AP, Wells H. The London model: forecasting fog at 333 m resolution. Q J R Meteorol Soc. 2016;142:360–71. https://doi.org/10.1002/qj.2656.

Bowling SA, Ohtake T, Benson CS. Winter pressure systems and ice fog in Fairbanks, Alaska. J Appl Meteor Climatol. 1968;7:961–8. https://doi.org/10.1175/1520-0450(1968)007<0961:WPSAIF>2.0.CO;2.

Carswell C. Atmospheric science. California fogs are thinning. Science. 2015;347(6227):1184–5. https://doi.org/10.1126/science.347.6227.1184.

Chachere CN, Pu Z. Connections between cold air pools and mountain valley fog events in salt lake city. Pure Appl Geophys. 2016;173:3187–3196. https://doi.org/10.1007/s00024-016-1316-x.

Chaurasia S, Jenamani RK. Detection of fog using temporally consistent algorithm with INSAT-3D imager data over india. IEEE Journal of Selected Topics in Applied Earth Observations and Remote Sensing. 2017;10:5307–5313. https://doi.org/10.1109/JSTARS.2017.2759197.

Chen S, Shi Y, Wang L, Dong A. Impact of climate variation on fog in China. J Geogr Sci. 2006;16:430–8. https://doi.org/10.1007/s11442-006-0406-6.

Christensen, L.S., and Frost, W. Fog Dispersion. NASA Contract Report 3255.1980. Available at: https://ntrs.nasa.gov/api/citations/19800011422/downloads/19800011422.pdf. Last access: 26 October 2024.

Clark PA, Hopwood WP. One-dimensional site-specific forecasting of radiation fog. Part I: model formulation and idealized sensitivity studies. Meteorol Appl. 2001;8:279–86. https://doi.org/10.1017/S1350482701003036.

Coiffier J. Fundamentals of numerical weather prediction. New York: Cambridge University Press; 2011.

Cotton WR, Anthes RA. Fogs and stratocumulus clouds. In: Cotton WR, Anthes RA, editors. Storm and cloud dynamics. International geophysics series 44. San Diego: Academic Press; 1992. p. 303–67.

Cox RA. Predictions of fog formation due to a warm water lagoon proposed for power station cooling. Atmos Environ. 1973;7(3):363–368. https://doi.org/10.1016/0004-6981(73)90085-1.

Croft PJ. Fog. In: Holton JR, Curry JA, Pyle JA, editors. Encyclopedia of atmospheric sciences, vol. 2. Academic Press; 2003. p. 777–92.

Curry JA, Webster PJ. Thermodynamics of atmospheres and oceans. London: Academic Press; 1999. p. 467.

Cuxart J, Telisman Prtenjak M, Matjacic B. Pannonian Basin nocturnal boundary layer and fog formation: role of topography. Atmosphere. 2021;12:712. https://doi.org/10.3390/atmos12060712.

Didenko NK, Ivanov YN, Puzov YA, Smirnov VV. Experimental modelling of the arctic steam fogs. Atmos Res. 1994;32(1–4):195–201. https://doi.org/10.1016/0169-8095(94)90060-4.

Ding Y, Liu Y. Analysis of long-term variations of fog and haze in China in recent 50 years and their relations with atmospheric humidity. Science China Earth Sciences. 2014;57:36–46. https://doi.org/10.1007/s11430-013-4792-1.

Dione C, Haeffelin M, Burnet F, Lac C, Canut G, Delanoë J, Dupont J-C, Jorquera S, Martinet P, Ribaud J-F, Toledo F. Role of thermodynamic and turbulence processes on the fog life cycle during SOFOG3D experiment. Atmos Chem Phys. 2023;23:15711–31. https://doi.org/10.5194/acp-23-15711-2023.

Dong X, Yu Z, Cao W, Shi Y, Ma Q. A survey on ensemble learning. Front Comp Sci. 2020;14:241–58. https://doi.org/10.1007/s11704-019-8208-z.

Dorman CE, Mejia J, Koračin D, McEvoy D. World marine fog analysis based on 58-years of ship observations. Int J Climatol. 2019;40:145–68. https://doi.org/10.1002/joc.6200.

Dorman, CE, Mejia J, Koračin D, McEvoy D. World marine fog analysis based on 58-years of ship observations. Int J Climatol. 2020;40:145–68. https://doi.org/10.1002/joc.6200.

Dorman CE, Hoch SW, Gultepe I, Wang CQ, Yamaguchi RT, Fernando HJS, Krishnamurthy R. Large-scale synoptic systems and fog during the C-FOG field experiment. Bound-Layer Meteorol. 2021;181:171–202. https://doi.org/10.1007/s10546-021-00641-1.

Du P, Zeng Z, Zhang J, Liu L, Yang J, Qu C, Jiang L, Liu S. Fog season risk assessment for maritime transportation systems exploiting Himawari-8 data: a case study in Bohai Sea, China. Remote Sens. 2021;13:3530. https://doi.org/10.3390/rs13173530.

Ducongé L, Lac C, Vié B, Bergot T, Price JD. Fog in heterogeneous environments: the relative importance of local and non-local processes on radiative-advective fog formation. Q J R Meteorol Soc. 2020;146(731):2522–46. https://doi.org/10.1002/qj.3783.

Durán-Rosal AM, Fernández JC, Casanova-Mateo C, Sanz-Justo J, Salcedo-Sanz S, Hervás-Martínez C. Efficient fog prediction with multi-objective evolutionary neural networks. Appl Soft Comput. 2018;70:347–58. https://doi.org/10.1016/j.asoc.2018.05.035.

Dy CY, Fung JCH. Updated global soil map for the Weather Research and Forecasting model and soil moisture initialization for the Noah land surface model. J Geophys Res Atmos. 2016;121(15):8777–800. https://doi.org/10.1002/2015JD024558.

Egli S, Thies B, Drönner J, Cermak J, Bendix J. A 10 year fog and low stratus climatology for Europe based on Meteosat Second Generation data. Q J R Meteorol Soc. 2017;143:530–541. https://doi.org/10.1002/qj.2941.

Fernando HJS, et al. C-FOG: life of coastal fog. Bull. Amer. Meteor. Soc. 2021;102:E244–72. https://doi.org/10.1175/BAMS-D-19-0070.1.

Field PR, Hill A, Shipway B, Furtado K, Wilkinson J, Miltenberger A, et al. Implementation of a double moment cloud microphysics scheme in the UK met office regional numerical weather prediction model. Q J R Meteorol Soc. 2023;149(752):703–39. https://doi.org/10.1002/qj.4414.

Forthun G, Johnson M, Schmitz W, Blume J, Caldwell R. Trends in fog frequency and duration in the Southeast United States. PhysicalGeography. 2006;27:206–22. https://doi.org/10.2747/0272-3646.27.3.206.

Fu G, Guo J, Xie SP, Duan Y, Zhang M. Analysis and high-resolution modeling of a dense sea fog event over the Yellow Sea; Atmos. Res. 2006;81:293–303. https://doi.org/10.1016/j.atmosres.2006.01.005.

Fu GQ, Xu WY, Yang RF, Li JB, Zhao CS. The distribution and trends of fog and haze in the North China Plain over the past 30 years. Atmos Chem Phys. 2014;14:11949–11958. https://doi.org/10.5194/acp-14-11949-2014.

Gaffen DJ, Ross RJ. Climatology and trends of U.S. surface humidity and temperature. J Clim. 1999;12:811–28. https://doi.org/10.1175/1520-0442(1999)012<0811:CATOUS>2.0.CO;2.

Galicia A, Talavera-Llames R, Troncoso A, Koprinska I, Martínez-Álvarez F. Multi-step forecasting for big data time series based on ensemble learning. Knowl-Based Syst. 2019;163:830–41. https://doi.org/10.1016/j.knosys.2018.10.009.

Gandhi A, Geresdi I, Gyöngyösi AZ, et al. An observational case study of a radiation fog event. Pure Appl Geophys. 2024;181:2025–49. https://doi.org/10.1007/s00024-024-03498-w.

Ganeshi NG, Mujumdar M, Takaya Y, Goswami MM, Singh BB, Krishnan R, Terao T. Soil moisture revamps the temperature extremes in a warming climate over India. npj Climate and Atmospheric Science. 2023;6(1):12. https://doi.org/10.1038/s41612-023-00334-1.

Gao S, Lin H, Shen B, Fu G. A heavy sea fog event over the Yellow Sea in March 2005: analysis and numerical modeling. Adv Atmos Sci. 2007;24(1):65–81. https://doi.org/10.1007/s00376-007-0065-2.

Gao S, Qi Y, Zhang S, Fu G. Initial conditions improvement of sea fog numerical modeling over the Yellow Sea by using cycling 3DVAR. Part I: WRF numerical experiments (in Chinese). Journal of Ocean University of China. 2010;40:1–9. https://doi.org/10.16441/j.cnki.hdxb.2010.10.001.

Gao X, Bao X, Ma S, Chen Q, Wang B. An online assimilation method to improve the numerical forecast of sea fog using microwave radiometer-retrieved cloud water path. J Geophys Res Atmos. 2024;129(1):e2023JD040229. https://doi.org/10.1029/2023JD040229.

George JJ. Fog: its causes and forecasting with special reference to eastern and southern United States (I). Bull Amer Meteor Soc. 1940;21:135–48. https://doi.org/10.1175/1520-0477-21.4.135.

George A, Kelm S, Cheng X, Allelein HJ. Efficient CFD modelling of bulk condensation, fog transport and re-evaporation for application to containment scale. Nucl Eng Des. 2023;401:112067. https://doi.org/10.1016/j.nucengdes.2022.112067.

Gonçalves FLT, Silva Dias PL, Araújo GP. Climatological analysis of wintertime extreme low temperatures in São Paulo City, Brazil: impact of sea-surface temperature anomalies. Int J Climatol. 2002;22(12):1511–26. https://doi.org/10.1002/joc.764.

Gonçalves FLT, da Rocha RP, Fernandes GP, Petto S. Drizzle and fog analysis in the São Paulo Metropolitan Area: changes 1933-2005 and correlations with other climate factors. Erde. 2008;139:61–76. https://www.researchgate.net/publication/257823392_Drizzle_and_fog_analysis_at_Sao_Paulo_Metropolitan_area

González-Jardines P, Sierra-Lorenzo M, González-Ramírez CM. Data assimilation impact on fog/haze forecast applied to the short-range forecast system. Revista Cubana de Física. 2022;39(2):60–9.
Gray E, Gilardoni S, Baldocchi D, McDonald BC, Facchini MC, Goldstein AH. Impact of air pollution controls on radiation fog frequency in the Central Valley of California. Journal of Geophysical Research: Atmospheres. 2019;124:5889–5905. https://doi.org/10.1029/2018JD02941.
Gultepe I, Fernando HJS, Pardyjak ER, Hoch SW, Silver Z, Creegan E, Leo LS, Pu Z, De Wekker, SFJ, Hang C. An overview of the MATERHORN fog project: observations and predictability. Pure and Applied Geophysics. 2016;173:2983–3010. https://doi.org/10.1007/s00024-016-1374-0.
Gultepe I, Tardif R, Michaelides S, Cermak J, Bott A, Bendix J, Müller M, Pagowski M, Hansen B, Ellrod G, Jacobs W, Toth G, Cober SG. Fog research: a review of past achievements and future perspectives. Pure Appl Geophys. 2007;164:1121–59. https://doi.org/10.1007/s00024-007-0211-x.
Gultepe I, Zhou B, Milbrandt J, Bott A, Li Y, Heymsfield AJ, Ferrier B, Ware R, Pavolonis M, Kuhn T, Gurka J, Liu P, Cermak J. A review on ice fog measurements and modeling. Atmospheric Research. 2015;151:2–19. https://doi.org/10.1016/j.atmosres.2014.04.014.
Gultepe I, et al. Ice fog in Arctic during FRAM–ice fog project: aviation and nowcasting applications. Bull Amer Meteor Soc 2014;95:211–26. https://doi.org/10.1175/BAMS-D-11-00071.1.
Gultepe I, Heymsfield AJ, Gallagher M, Ickes L, Baumgardner D, D. Ice fog: the current state of knowledge and future challenges. Meteor Monogr. 2017a;58:4.1–4.24. https://doi.org/10.1175/AMSMONOGRAPHS-D-17-0002.1.
Gultepe I, Milbrandt JA, Zhou B. Marine fog: a review on microphysics and visibility prediction. In: Koračin D, Dorman C, editors. Marine fog: challenges and advancements in observations, modeling, and forecasting, Springer Atmospheric Sciences. Cham: Springer; 2017b. https://doi.org/10.1007/978-3-319-45229-6_7.
Gultepe I, Heymsfield AJ, Gallagher M. Arctic ice fog: its microphysics and prediction. In: Kokhanovsky A, Tomasi C, editors. Physics and chemistry of the Arctic atmosphere, Springer Polar Sciences. Cham: Springer; 2020. https://doi.org/10.1007/978-3-030-33566-3_6.
Guo J, Xu J, Xu X. An observational and modeling study of a sea fog event over the yellow and East China seas on 17 March 2014. Tropical Cyclone Research and Review. 2021;10(3):182–90. https://doi.org/10.1016/j.tcrr.2021.09.001.
Guy H, Brooks IM, Carslaw KS, Murray BJ, Walden VP, Shupe MD, Pettersen C, Turner DD, Cox CJ, Neff WD, Bennartz R, Neely RR III. Controls on surface aerosol particle number concentrations and aerosol-limited cloud regimes over the central Greenland Ice Sheet. Atmos Chem Phys. 2021;21:15351–74. https://doi.org/10.5194/acp-21-15351-2021.
Hardwick WC. Monthly fog frequency in the continental United States. Mon Weather Rev. 1973;101:763–6.
Heo KH, Ha KJ, Mahrt L, Shim JS. Comparison of advection and steam fogs: from direct observation over the sea. Atmos Res. 2010;98:426–37. https://doi.org/10.1016/j.atmosres.2010.08.004.
Herckes P, Marcotte AR, Wang Y, Collett JL. Fog composition in the Central Valley of California over three decades. Atmos Res. 2015;151:20–30. https://doi.org/10.1016/j.atmosres.2014.01.025.
Hoch SW, Whiteman CD, Mayer B. A systematic study of longwave radiative heating and cooling within valleys and basins using a three-dimensional radiative transfer model. J Appl Meteor Climatol. 2011;50:2473–89. https://doi.org/10.1175/JAMC-D-11-083.1.
Houssos EE, Lolis CJ, Gkikas A, Hatzianastassiou N, Bartzokas A. On the atmospheric circulation characteristics associated with fog in Ioannina, North-Western Greece. Int J Climatol. 2012;32:1847–62. https://doi.org/10.1002/joc.2399.
Hu H, Sun J, Zhang Q. Assessing the impact of surface and wind profiler data on fog forecasting using WRF 3DVAR: an OSSE study on a dense fog event over North China. J Appl Meteorol Clim. 2017a;56:1059–81. https://doi.org/10.1175/JAMC-D-16-0246.1.

Hu H, Zhang Q, Sun J, Ruan C, Huang F, Zhang S. Impact of high-frequency observations on fog forecasting: a case study of OSSE. Tellus A: Dynamic Meteorology and Oceanography. 2017b;69;1:1396182. https://doi.org/10.1080/16000870.2017.1396182.

Huang H, Liu H, Huang J, Mao W, Bi X. Atmospheric boundary layer structure and turbulence during sea fog on the southern China coast. Mon. Wea. Rev. 2015;143:1907–23. https://doi.org/10.1175/MWR-D-14-00207.1.

Hůnová I. Challenges in moving towards fog's contribution to spatial patterns of atmospheric deposition fluxes on a national scale. Sci Total Environ. 2024;946:174208. https://doi.org/10.1016/j.scitotenv.2024.174208.

Hůnová I, Brabec M, Malý M, Valeriánová A. Long-term trends in fog occurrence in the Czech Republic, Central Europe. Sci Total Environ. 2020;711:135018. https://doi.org/10.1016/j.scitotenv.2019.135018.

Hůnová I, Brabec M, Geletič J, Malý M, Dumitrescu A. Statistical analysis of the effects of forests on fog. Sci Total Environ. 2021;781:146675, https://doi.org/10.1016/j.scitotenv.2021.146675.

Jayakumar A, Rajagopal EN, Boutle IA, George JP, Mohandas S, Webster S, Aditi S. An operational fog prediction system for Delhi using the 330 m Unified Model. Atmos Sci Lett. 2018;19(1):e796. https://doi.org/10.1002/asl.796.

Jentsch H, Weidinger J. Spatio-temporal analysis of valley wind systems in the complex mountain topography of the Rolwaling Himal, Nepal. Atmosphere. 2022;2022(13):1138. https://doi.org/10.3390/atmos13071138.

Kawai H, Koshiro T, Endo H, Arakawa O, Hagihara Y. Changes in marine fog in a warmer climate. Atmos Sci Lett. 2016;17(10):548–55. https://doi.org/10.1002/asl.691.

Kawai H, Koshiro T, Endo H, Arakawa O. Changes in marine fog over the North Pacific under different climates in CMIP5 multimodel simulations. J Geophys Res Atmos. 2018;123(19):10911–24. https://doi.org/10.1029/2018JD028899.

Khoury D, Millet M, Jabali Y, Delhomme O. Fog water: a general review of its physical and chemical aspects. Environments. 2023;10:224. https://doi.org/10.3390/environments10120224.

Kim W, Yum SS, Hong J, Song JI. Improvement of fog simulation by the nudging of meteorological tower data in the WRF and PAFOG coupled model. Atmos. 2020;11(3):311. https://doi.org/10.3390/ATMOS11030311.

Kim D, Kim E, Kim E. Performance of decision-tree-based ensemble classifiers in predicting fog frequency in ungauged areas. Weather and Forecasting. 2023;38;11:2379–2393. https://doi.org/10.1175/WAF-D-23-0024.1.

Klemm O, Lin N. What causes observed fog trends: air quality or climate change? Aerosol Air Qual Res. 2016;16:1131–42. https://doi.org/10.4209/aaqr.2015.05.0353.

Koračin D. Modeling and forecasting marine fog. In: Koračin D, Dorman C, editors. Marine fog: challenges and advancements in observations, modeling, and forecasting, Springer Atmospheric Sciences. Cham: Springer; 2017. https://doi.org/10.1007/978-3-319-45229-6_9.

Koenig LR. Numerical modelling of ice deposition. J Almos Sci. 1971;28:226–237.

Krishnamurti TN, Bounoua L. An introduction to numerical weather prediction techniques. CRC press; 2018. p. 293. https://doi.org/10.1201/9781315137285.

Kutty SG, Dimri AP, Gultepe I. Climatic trends in fog occurrence over the Indo-Gangetic Plains. Int J Climatol. 2020;40:2048–61. https://doi.org/10.1002/joc.6317.

LaDochy S. The disappearance of dense fog in Los Angeles: another urban impact? Phys Geogr. 2005;26:177–91. https://doi.org/10.2747/0272-3646.26.3.177.

LaDochy, S., & Witiw, M. (2010). The continued reduction in dense fog in the Southern California region: Possible climate change influences. 5th International Conference on Fog, Fog Collection and Dew, Münster, Germany, 25–30 July 2010. https://meetings.copernicus.org/fog2010/fog-conference_2010_conference_book.pdf.

LaDochy S, Witiw M. The continued reduction in dense fog in the Southern California region: possible causes. Pure Appl Geophys. 2012;169:1157–63. https://doi.org/10.1007/s00024-011-0366-3.

Lakra K, Avishek K. A review on factors influencing fog formation, classification, forecasting, detection and impacts. Rend Fis Acc Lincei. 2022;33:319–53. https://doi.org/10.1007/s12210-022-01060-1.

Lavigne T, Liu C. Validity of global fog-day trends indicated by the Global Surface Summary of the Day (GSOD) data set. J Geophys Res Atmos. 2022;127:e2021JD035881. https://doi.org/10.1029/2021JD035881.

Lewis J, Koracin D, Rabin R, Businger J. Sea fog off the California coast: viewed in the context of transient weather systems. J Geophys Res. 2003;108:4457. https://doi.org/10.1029/2002JD002833. D15

Lewis JM, Koračin D, Redmond KT. Sea fog research in the United Kingdom and United States: a historical essay including outlook. Bull Am Meteorol Soc. 2004;85(3):395–408. https://doi.org/10.1175/BAMS-85-3-395.

Li X, Pu Z. Turbulence effects on the formation of cold fog over complex terrain with large-eddy simulation. Geophys Res Lett. 2022;49:e2022GL098792. https://doi.org/10.1029/2022GL098792.

Li X, Choi Y, Czader B, Roy A, Kim H, Lefer B, Pan S. The impact of observation nudging on simulated meteorology and ozone concentrations during DISCOVER-AQ 2013 Texas campaign. Atmos Chem Phys. 2016;16:3127–44. https://doi.org/10.5194/acp-16-3127-2016.

Li Y, Aemisegger F, Riedl A, Buchmann N, Eugster W. The role of dew and radiation fog inputs in the local water cycling of a temperate grassland during dry spells in Central Europe. Hydrol Earth Syst Sci. 2021;25:2617–48. https://doi.org/10.5194/hess-25-2617-2021.

Li Y, Eugster W, Riedl A, Westerhuis S, Buchmann N, Aemisegger F. Identifying key stages of radiation fog evolution using water vapor isotopes. Agric For Meteorol. 2023;334:109430. https://doi.org/10.1016/j.agrformet.2023.109430.

Lill E, Costa EJ, Barry K, Mirrielees JA, Mashkevich M, Wu J, et al. The abundance and sources of ice nucleating particles within Alaskan ice fog. J Geophys Res Atmos. 2024;129:e2024JD041170. https://doi.org/10.1029/2024JD041170.

Lin C, Zhang Z, Pu Z, Wang F. Numerical simulations of an advection fog event over Shanghai Pudong International Airport with the WRF model. J Meteorol Res. 2017;31(5):874–89. https://doi.org/10.1007/s13351-017-6187-2.

Liu Y, Ren J, Zhou X. The impact of assimilating sea surface wind aboard QuickSCAT on sea fog simulation (in Chinese). Journal of Applied Meteorological Science. 2011;22(04):472–81. https://doi.org/10.11898/1001-7313.20110410.

Liu DY, Yan WL, Yang J, et al. A study of the physical processes of an advection fog boundary layer. Boundary-Layer Meteorol. 2016;158:125–38. https://doi.org/10.1007/s10546-015-0076-y.

Liu J, Pu Z. Does soil moisture have an influence on near-surface temperature?. Journal of Geophysical Research: Atmospheres. 2019;124;12:6444–66. https://doi.org/10.1029/2018JD029750.

López A, Molina-Aiz FD, Valera DL, Peña A. Determining the emissivity of the leaves of nine horticultural crops by means of infrared thermography. Scientia Horticulturae. 2012;137:49–58. https://doi.org/10.1016/j.scienta.2012.01.022.

Lv SX, Peng L, Hu H, Wang L. Effective machine learning model combination based on selective ensemble strategy for time series forecasting. Inf Sci. 2022;612:994–1023. https://doi.org/10.1016/j.ins.2022.09.002.

Maalick Z, Kühn T, Korhonen H, Kokkola H, Laaksonen A, Romakkaniemi S. Effect of aerosol concentration and absorbing aerosol on the radiation fog life cycle. Atmos Environ. 2016;133:26–33. https://doi.org/10.1016/j.atmosenv.2016.03.018.

Mago PJ, Sherif SA. Frost formation and heat transfer on a cold surface in ice fog. International Journal of Refrigeration. 2005;28:538–546. https://doi.org/10.1016/j.ijrefrig.2004.10.004.

Manisalidis I, Stavropoulou E, Stavropoulos A, Bezirtzoglou E. Environmental and health impacts of air pollution: a review. Front Public Health. 2020;8:14. https://doi.org/10.3389/fpubh.2020.00014.

Maronga B, Bosveld F. Key parameters for the life cycle of nocturnal radiation fog: a comprehensive large-eddy simulation study. Q J R Meteorol Soc. 2017;143:2463–80. https://doi.org/10.1002/qj.3100.

Martinet P, Cimini D, Burnet F, Ménétrier B, Michel Y, Unger V. Improvement of numerical weather prediction model analysis during fog conditions through the assimilation of ground-based microwave radiometer observations: a 1D-Var study. Atmos Meas Tech. 2020;13(12):6593–611. https://doi.org/10.5194/amt-13-6593-2020.

Massey JD, Steenburgh WJ, Hoch SW, Knievel JC. Sensitivity of near-surface temperature forecasts to soil properties over a sparsely vegetated dryland region. J Appl Meteorol Climatol. 2014;53(8):1976–95. https://doi.org/10.1175/JAMC-D-13-0362.1.

Matveev LT. Models of fog formation. In: Cloud dynamics. Atmospheric sciences library, vol. 2. Dordrecht: Springer; 1984. https://doi.org/10.1007/978-94-009-6360-3_4.

Maurer M, Klemm O, Lokys HL, Lin NH. Trends of fog and visibility in Taiwan: climate change or air quality improvement? Aerosol Air Qual Res. 2019;19:896–910. https://doi.org/10.4209/aaqr.2018.04.0152.

Mazoyer M, Burnet F, Denjean C, Roberts GC, Haeffelin M, Dupont J-C, Elias T. Experimental study of the aerosol impact on fog microphysics. Atmos Chem Phys. 2019;19:4323–44. https://doi.org/10.5194/acp-19-4323-2019.

Mazoyer M, Burnet F, Denjean C. Experimental study on the evolution of droplet size distribution during the fog life cycle. Atmos Chem Phys. 2022;22:11305–21. https://doi.org/10.5194/acp-22-11305-2022.

McCabe A, Swinbank R, Tennant W, Lock A. Representing model uncertainty in the Met Office convection-permitting ensemble prediction system and its impact on fog forecasting. Q J R Meteorol Soc. 2016;142(700):2897–910. https://doi.org/10.1002/qj.2876.

Mohan TS, Temimi M, Ajayamohan RS, Nelli NR, Fonseca R, Weston M, Valappil V. On the investigation of the typology of fog events in an arid environment and the link with climate patterns. Mon. Wea. Rev. 2020;148:3181–202. https://doi.org/10.1175/MWR-D-20-0073.1.

Moore A, Zavala-Garay J, Arango HG, Edwards CA, Anderson J, Hoar T. Regional and basin scale applications of ensemble adjustment Kalman filter and 4D-Var Ocean data assimilation systems. Prog Oceanogr. 2020;189:102450. https://doi.org/10.1016/j.pocean.2020.102450.

Mühlig AC, Klemm O, Gonçalves FLT. Fog, temperature and air quality over the metropolitan area of São Paulo: a trend analysis from 1998 to 2018. Water Air Soil Pollut. 2020;231:535. https://doi.org/10.1007/s11270-020-04902-6.

National Weather Service. (NWS, n.d.). Fog Formation and Dissipation. Retrieved from https://www.weather.gov/media/zhu/ZHU_Training_Page/fog_stuff/fog_guide/fog.pdf. (Last access: Accessed on 25 October 2024).

Nelli N, Francis D, Abida R, Fonseca R, Masson O, Bosc E. In-situ measurements of fog microphysics: visibility parameterization and estimation of fog droplet sedimentation velocity. Atmos Res. 2024;309:107570. https://doi.org/10.1016/j.atmosres.2024.107570.

Neumann J. Forecasts of fine weather in the literature of classical antiquity. Bull. Amer. Meteor. Soc. 1989;70:46–9. https://doi.org/10.1175/1520-0477-70.1.46.

Nicholls ME, Pielke Sr RA. On the role of thermal expansion and compression in large-scale atmospheric energy and mass transports. Atmos Chem Phys. 2018;18:15975–6003. https://doi.org/10.5194/acp-18-15975-2018.

Niu F, Li Z, Li C, Lee KH, Wang M. Increase of wintertime fog in China: potential impacts of weakening of the eastern Asian monsoon circulation and increasing aerosol loading. J Geophys Res. 2010;115:D00K20. https://doi.org/10.1029/2009JD013484.

NOAA, 2024. https://www.noaa.gov/jetstream/synoptic/types-of-weather-phenomena.

Ohashi Y, Suido M. Numerical simulations of upslope fog observed at Beppu Bay in Oita Prefecture, Japan. Meteorological Applications. 2021;28(3):e2003. https://doi.org/10.1002/met.2003.

Oke TR. Boundary layer climates. 2nd ed. Methuen & Co. Ltd. 464 pp; 1987. https://doi.org/10.4324/9780203407219.

Parde AN, Ghude SD, Dhangar NG, Lonkar P, Wagh S, Govardhan G, et al. Operational probabilistic fog prediction based on ensemble forecast system: a decision support system for fog. Atmos. 2022;13(10):1608. https://doi.org/10.3390/atmos13101608.

Parde AN, Ghude SD, Dhangar NG, Bhautmage UP, Wagh S, Lonkar P, et al. Challenges in simulating prevailing fog types over urban region of Delhi. J Geophys Res Atmos. 2024;129:e2023JD039772. https://doi.org/10.1029/2023JD039772.

Peláez-Rodríguez C, Pérez-Aracil J, Casanova-Mateo C, Salcedo-Sanz S. Efficient prediction of fog-related low-visibility events with machine learning and evolutionary algorithms. Atmos Res. 2023;295:106991. https://doi.org/10.1016/j.asoc.2018.05.035.

Pérez-Díaz JL, Ivanov O, Peshev Z, Álvarez-Valenzuela MA, Valiente-Blanco I, Evgenieva T, Dreischuh T, Gueorguiev O, Todorov PV, Vaseashta A. Fogs: physical basis, characteristic properties, and impacts on the environment and human health. Water. 2017;9:807. https://doi.org/10.3390/w9100807.

Petterssen S. Weather analysis and forecasting. New York: McGraw-Hill. Book Co; 1940. 29p

Philip A, Bergot T, Bouteloup Y, Bouyssel F. The impact of vertical resolution on fog forecasting in the kilometric-scale model arome: a case study and statistics. Weather Forecast. 2016;31(5):1655–71. https://doi.org/10.1175/WAF-D-16-0074.1.

Pithani P, Ghude SD, Chennu VN, Kulkarni RG, Steeneveld GJ, Sharma A, et al. WRF model prediction of a dense fog event occurred during the winter fog experiment (WIFEX). Pure Appl Geophys. 2019;176:1827–46. https://doi.org/10.1007/s00024-018-2053-0.

Pithani P, Ghude SD, Jenamani RK, Biswas M, Naidu CV, Debnath S, Kulkarni R, Dhangar NG, Jena C, Hazra A, Phani R, Mukhopadhyay P, Prabhakaran T, Nanjundiah RS, Rajeevan M. Real-Time forecast of dense fog events over delhi: The performance of the WRF model during the WiFEX field campaign. Wea Forecasting. 2020;35:739–756. https://doi.org/10.1175/WAF-D-19-0104.1.

Price JD. On the formation and development of radiation fog: an observational study. Boundary-Layer Meteorol. 2019;172:167–97. https://doi.org/10.1007/s10546-019-00444-5.

Price J, Porson A, Lock A. An observational case study of persistent fog and comparison with an ensemble forecast model. Bound-Layer Meteorol. 2015;155:301–27. https://doi.org/10.1007/s10546-014-9995-2.

Pruppacher HR, Klett JD. Microphysics of clouds and precipitation. 2nd ed. Boston: Kluwer Pub. Inc.; 1997. p. 954.

Pu Z. Surface data assimilation and near-surface weather prediction over complex terrain. In: Park SK, Xu L, editors. Data assimilation for atmospheric, oceanic and hydrologic applications. 2017;III:219–40.

Pu Z, Chachere CN, Hoch SW, Pardyjak ER, Gultepe I. Numerical prediction of cold season fog events over complex terrain: the performance of the WRF model during MATERHORN-fog and early evaluation. Pure Appl Geophys. 2016;173:3165–86. https://doi.org/10.1007/s00024-016-1375-z.

Quan J, Zhang Q, He H, Liu J, Huang M, Jin H. Analysis of the formation of fog and haze in North China Plain (NCP). Atmos Chem Phys. 2011;11:8205–14. https://doi.org/10.5194/acp-11-8205-2011.

Rémy S, Pannekoucke O, Bergot T, Baehr C. Adaptation of a particle filtering method for data assimilation in a 1D numerical model used for fog forecasting. Q J R Meteorol Soc. 2012;138:536–51. https://doi.org/10.1002/qj.915.

Rita A, Bonanomi G, Allevato E, Borghetti M, Cesarano G, Mogavero V, Rossi S, Saulino L, Zotti M, Saracino A. Topography modulates near-ground microclimate in the Mediterranean Fagus sylvatica treeline. Sci Rep. 2021;14;11(1):8122. https://doi.org/10.1038/s41598-021-87661-6. PMID: 33854197; PMCID: PMC8046975

Robinson E, Thuman WC, Wiggins EJ. Ice fog as a problem of air pollution in the Arctic. Arctic. 1957;88–104. https://doi.org/10.14430/arctic3756.

Rodhe B. The effect of turbulence on fog formation. Tellus. 1962;14:49–86. https://doi.org/10.1111/j.2153-3490.1962.tb00119.x.

Román-Cascón C, Steeneveld GJ, Yague C, Sastre M, Arrillaga JA, Maqueda G. Forecasting radiation fog at climatologically contrasting sites: evaluation of statistical methods and WRF. Q J R Meteorol Soc. 2016;142:1048–63. https://doi.org/10.1002/qj.2708.

Sachweh M, Koepke P. Radiation fog and urban climate. Geophys Res Lett. 1995;22;9:0094–8276. https://doi.org/10.1029/95GL00907.

Schalkwyk LV, Dyson LL. Climatological characteristics of fog at Cape Town International Airport. Weather Forecast. 2013;28(3):631–46. https://doi.org/10.1175/WAF-D-12-00028.1.

Scholl, M.A., Gingerich, S.B., Tribble, G.W. (2002). The influence of microclimates and fog on stable isotope signatures used in interpretation of regional hydrology: East Maui, Hawaii, Journal of Hydrology, 264, 1–4,170–184, doi:https://doi.org/10.1016/S0022-1694(02)00073-2.

Sgoff C, Schomburg A, Schmidli J, Potthast R. Assimilating synthetic land surface temperature in a coupled land–atmosphere model. Q J R Meteorol Soc. 2020;146(733):3980–97. https://doi.org/10.1002/qj.3883.

Shao C, Nerger L. Assimilation of ground-based GNSS data using a local ensemble Kalman filter. Sci Rep. 2024;14(1):21682. https://doi.org/10.1038/s41598-024-72915-w.

Sim S, Im J, Jung S, Han D. Improving short-term prediction of ocean fog using numerical weather forecasts and geostationary satellite-derived ocean fog data based on AutoML. Remote Sens. 2024;16(13):2348. https://doi.org/10.3390/rs16132348.

Singh S, Bansal A, Sandhu R, Sidhu J. Fog computing and IoT based healthcare support service for dengue fever. International Journal of Pervasive Computing and Communications. 2018;14;2:197–207. https://doi.org/10.1108/IJPCC-D-18-0001.

Singh G, Harun. Multiple linear regression based analysis of weather data for precipitation and visibility prediction. In: Singh M, Tyagi V, Gupta P, Flusser J, Ören T, editors. Advances in computing and data sciences. ICACDS 2023. Communications in Computer and Information Science, vol. 1848. Cham: Springer; 2023. https://doi.org/10.1007/978-3-031-37940-6_6.

Smith DKE, Renfrew IA, Price JD, Dorling SR. Numerical modelling of the evolution of the boundary layer during a radiation fog event. Weather. 2018;73:310–6. https://doi.org/10.1002/wea.3305.

Smith PJ, Lawless AS, Nichols NK. The role of cross-domain error correlations in strongly coupled 4D-Var atmosphere–ocean data assimilation. Q J R Meteorol Soc. 2020;146:2450–2465. https://doi.org/10.1002/qj.3802.

Smith DK, Renfrew IA, Dorling SR, Price JD, Boutle IA. Sub-km scale numerical weather prediction model simulations of radiation fog. Q J R Meteorol Soc. 2021;147(735):746–63. https://doi.org/10.1002/qj.3943.

Smith DKE, Dorling SR, Renfrew IA, Ross AN, Poku C. Fog trends in India: relationships to fog type and western disturbances. Int J Climatol. 2023;43(2):818–36. https://doi.org/10.1002/joc.7832.

Smith DKE, Reka S, Dorling SR, Ross AN, Renfrew IA, Jayakumar A, Anurose TJ, Parde AN, Ghude SD, Rumbold H. Forecasts of fog events in northern India dramatically improve when weather prediction models include irrigation effects. Commun Earth Environ. 2024;5:141. https://doi.org/10.1038/s43247-024-01314-w.

Steeneveld GJ, de Bode M. Unravelling the relative roles of physical processes in modelling the life cycle of a warm radiation fog. Q J R Meteorol Soc. 2018;144:1539–54. https://doi.org/10.1002/qj.3300.

Steeneveld GJ, Ronda RJ, Holtslag AAM. The challenge of forecasting the onset and development of radiation fog using mesoscale atmospheric models. Bound-Layer Meteorol. 2015;154:265–89.

Stone RG. Fog in the United States and adjacent regions. Geogr Rev. 1936;26(1):111–34. https://doi.org/10.2307/209468.

Strajnar B, Žagar N, Berre L. Impact of new aircraft observations mode-S MRAR in a mesoscale NWP model. J Geophys Res Atmos. 2015;120(9):3920–38. https://doi.org/10.1002/2014JD022654.

Stull RB. Stable boundary layer Editor Stull R.B. (Ed.). An Introduction to Boundary Layer Meteorology. Springer, Dordrecht, Dordrecht. 1988:499–543. https://doi.org/10.1007/978-94-009-3027-8_12.
Sugimoto S, Sato T, Nakamura K. Effects of synoptic-scale control on long-term declining trends of summer fog frequency over the pacific side of hokkaido island. J Appl Meteor Climatol. 2013;52:2226–2242. https://doi.org/10.1175/JAMC-D-12-0192.1.
Syed FS, Körnich H, Tjernström M. On the fog variability over South Asia. Clim Dyn. 2012;39:2993–3005. https://doi.org/10.1007/s00382-012-1414-0.
Tav J, Masson O, Burnet F, Paulat P, Bourrianne T, Conil S, Pourcelot L. Determination of fog-droplet deposition velocity from a simple weighing method. Aerosol Air Qual Res. 2018;18(1):103–13. https://doi.org/10.4209/aaqr.2016.11.0519.
Taylor PA, Chen Z, Cheng L, Afsharian S, Weng W, Isaac GA, et al. Surface deposition of marine fog and its treatment in the Weather Research and Forecasting (WRF) model. Atmos Chem Phys. 2021;21(19):14687–702. https://doi.org/10.5194/acp-21-14687-2021.
Trigo IF, Boussetta S, Viterbo P, Balsamo G, Beljaars A, Sandu I. Comparison of model land skin temperature with remotely sensed estimates and assessment of surface-atmosphere coupling. J Geophys Res Atmos. 2015;120:12,096–12,111. https://doi.org/10.1002/2015JD023812.
van der Velde IR, Steeneveld GJ, Wichers Schreur BGJ, Holtslag AAM. Modeling and forecasting the onset and duration of severe radiation fog under frost conditions. Mon. Wea. Rev. 2010;138:4237–53. https://doi.org/10.1175/2010MWR3427.1.
Van Oldenborgh GJ, Yiou P, Vautard R. On the roles of circulation and aerosols in the decline of mist and dense fog in Europe over the last 30 years. Atmos Chem Phys. 2010;10:4597–609. https://doi.org/10.5194/acp-10-4597-2010.
Vautard R, Yiou P, van Oldenborgh GJ. Decline of fog, mist, and haze in Europe over the past 30 years. Nat Geosci. 2009;2:115–9. https://doi.org/10.1038/ngeo414.
Venne MG, Jasperson WH, Venne DE. Difficult weather: A review of thunderstorm, fog and stratus, and winter precipitation forecasting. Air Force Materiel Command. 1997.
Wærsted EG, Haeffelin M, Dupont J-C, Delanoë J, Dubuisson P. Radiation in fog: quantification of the impact on fog liquid water based on ground-based remote sensing. Atmos Chem Phys. 2017;17:10811–35. https://doi.org/10.5194/acp-17-10811-2017.
Wagh S, Kulkarni R, Lonkar P, Parde AN, Dhangar NG, Govardhan G, et al. Development of visibility equation based on fog microphysical observations and its verification using the WRF model. Modeling Earth Systems and Environment. 2023;9(1):195–211. https://doi.org/10.1007/s40808-022-01492-6.
Walker M. The science of weather: radiation fog and steam fog. Weather. 2003;58:196–7. https://doi.org/10.1256/wea.49.02.
Wang Y, Gao S, Fu G, Sun J, Zhang S. Assimilating MTSAT-derived humidity in nowcasting sea fog over the Yellow Sea. Weather Forecast. 2014;29(2):205–25. https://doi.org/10.1175/WAF-D-12-00123.1.
Wang JJ, Gao XY, Gao SH. Data assimilation experiments and formation mechanism study of a Yellow Sea fog event. J Mar Meteorol. 2017;37(1):42–53.
Wang X, Qiu X, Wu B, Dong Q, Liu L, Li Y, Tian M. Analysis of the different influence between initial/boundary and physical perturbation during ensemble forecast of fog. Meteorog Atmos Phys. 2023;135(5):44. https://doi.org/10.1007/s00703-023-00981-2.
Weedon GP, Osborne SR, Best MJ. Dew, frost, fog and lifted temperature minima: observations in southern England and implications for modelling. Q J R Meteorol Soc. 2024;150(761):2168–84. Available from:. https://doi.org/10.1002/qj.4702.
Westcott, N.E. (2002). Trends in dense fog frequency in the Midwestern United States. In 13th Conference on Applied Climatology, American Meteorological Society (Ed.), Boston, 173–176.
Westerhuis S, Fuhrer O, Cermak J, Eugster W. Identifying the key challenges for fog and low stratus forecasting in complex terrain. Q J R Meteorol Soc. 2020;146:3347–67. https://doi.org/10.1002/qj.3849.

Weston M, Chaouch N, Valappil V, Temimi M, Ek M, Zheng W. Assessment of the Sensitivity to the Thermal Roughness Length in Noah and Noah-MP Land Surface Model Using WRF in an Arid Region. Pure Appl. Geophys. 2019;176:2121–2137. https://doi.org/10.1007/s00024-018-1901-2.

Weston M, Temimi M. Application of a nighttime fog detection method using SEVIRI over an arid environment. Remote Sens. 2020;12(14):2281. https://doi.org/10.3390/rs12142281.

WMO, 2017. Fog. International Cloud Atlas: World Meteorological Organization (WMO). Accessed 20 November 2019.

Williams AP, Schwartz RE, Iacobellis S, Seager R, Cook BI, Still CJ, Husak G, Michaelsen J. Urbanization causes increased cloud base height and decreased fog in coastal Southern California. Geophys Res Lett. 2015;42:1527–1536. https://doi.org/10.1002/2015GL063266.

Witiw MR, LaDochy S. Trends in fog frequencies in the Los Angeles Basin. Atmospheric Research. 2008;87;3–4:293–300. https://doi.org/10.1016/j.atmosres.2007.11.010.

Wolfe SA. Cold-climate Aeolian environments. In: Shroder J(J)F, editor. Treatise on geomorphology. 2nd ed. Academic Press; 2022. p. 195–219. ISBN 9780128182352. https://doi.org/10.1016/B978-0-12-818234-5.00036-5.

Wu X, Li S. Automatic sea fog detection over Chinese adjacent oceans using Terra/MODIS data. Int J Remote Sens. 2014;35(21):7430–57. https://doi.org/10.1080/01431161.2014.968685.

Wu D, Lu B, Zhang T, Yan F. A method of detecting sea fogs using CALIOP data and its application to improve MODIS-Based Sea fog detection. J Quant Spectrosc Radiat Transf. 2015;153:88–94. https://doi.org/10.1016/j.jqsrt.2014.09.021.

Wu X, Zhu J, Wang X, Yang B. Sea fog simulation with assimilation of FY-3A microwave data (in Chinese). Chin J Atmos Sci. 2017;41(3):421–36. https://doi.org/10.3878/j.issn.1006-9895.1610.16105.

Xiao YF, Zhang J, Qin P. An algorithm for Daytime Sea fog detection over the Greenland Sea based on MODIS and CALIOP data. J Coast Res. 2019;90(SI):95–103.

Yadav P, Parde AN, Dhangar NG, Govardhan G, Lal DM, Wagh S, Prasad DSVVD, Ahmed R, Ghude SD. Understanding the genesis of a dense fog event over Delhi using observations and high-resolution model experiments. Modeling Earth Systems and Environment. 2022;8;4:5011–5022. https://doi.org/10.1007/s40808-022-01463-x/

Yang Y, Gao S. The impact of turbulent diffusion driven by fog-top cooling on sea fog development. J Geophys Res Atmos. 2020;125(4):e2019JD031562. https://doi.org/10.1029/2019JD031562.

Yang L, Liu J-W, Ren Z-P, Xie S-P, Zhang S-P, Gao S-H. Atmospheric conditions for advection-radiation fog over the western Yellow Sea. J Geophys Res Atmos. 2018;123:5455–68. https://doi.org/10.1029/2017JD028088.

Zhang F, Pu Z. Sensitivity of numerical simulations of near-surface atmospheric conditions to snow depth and surface albedo during an ice fog event over Heber Valley. J Appl Meteorol Climatol. 2019;58:797–811. https://doi.org/10.1175/JAMC-D-18-0064.1.

Zhang F, Yang Y, Wang C. The effects of assimilating conventional and ATOVS data on forecasted near-surface wind with WRF-3DVAR. Mon. Wea. Rev. 2015;143:153–64. https://doi.org/10.1175/MWR-D-14-00038.1.

Zhang FM, Ren BH, Dou CL, Wang CH. Numerical simulation of near-surface atmospheric conditions during a radiation fog over the complex terrain. IOP Conf Ser Earth Environ Sci. 2020;555:012093. https://doi.org/10.1088/1755-1315/555/1/012093.

Zhang G, et al. Fog/cloud processing of atmospheric aerosols from a single particle perspective: a review of field observations. Atmos Environ. 2024;329:120536. https://doi.org/10.1016/j.atmosenv.2024.120536.

Zhong S, Whiteman CD, Bian X, Shaw WJ, Hubbe JM. Meteorological processes affecting the evolution of a wintertime cold air Pool in the Columbia Basin. Mon Wea Rev. 2001;129:2600–13. https://doi.org/10.1175/1520-0493(2001)129<2600:MPATEO>2.0.CO;2.

Zhou X, Kollias P, Lewis ER. Clouds, precipitation, and marine boundary layer structure during the MAGIC Field campaign. J Clim. 2015;28:2420–42. https://doi.org/10.1175/JCLI-D-14-00320.1.

# Chapter 3
# Perspectives on Fog Chemistry: Characteristics, Variability, and Its Scavenging Efficiency

**Saurabh Sonwani**

## 3.1 Introduction

The fog is a suspension of microscopic water droplets in the air, reducing visibility at the Earth's surface (Fuzzi et al. 1996; Gultepe et al. 2007; Nieberding et al. 2018). Earlier it was considered as a collection of suspended water droplets or ice crystals near the Earth's surface that leads to a reduction of horizontal visibility below 1 km (5/8 of a statute mile) (NOAA 1995). Fog droplets are formed on aerosol particles as relative humidity (RH) increases and ambient temperature decreases (Lakhani et al. 2007). This phenomenon typically happens when the air cools to its dew point, often through radiative cooling, advection, or mixing of air masses with different temperatures and humidity levels. The process is influenced by various factors, including temperature, humidity, wind patterns, and the presence of condensation nuclei in the atmosphere. Fog supports the processing of aerosols by providing a site of interaction for the chemical species present in the atmosphere, involved in new particle formations, their incorporation into fog droplets, scavenging of soluble trace gases such as ammonia and nitric acid, and removal of gases and particles from the atmosphere by droplet deposition (Pandis et al. 1990; Lakhani et al. 2007; Saxena et al. 2019). The composition of fog includes water droplets as the primary component, but it also contains various other substances. These include atmospheric gases, particulate matter, and dissolved chemical compounds. The specific composition can vary depending on the local environment and pollution levels. In urban areas, fog may contain higher concentrations of pollutants such as sulfates, nitrates, and organic compounds from vehicle emissions and industrial activities. In contrast, rural fog might have a higher proportion of natural compounds from vegetation and

S. Sonwani (✉)
Department of Environmental Studies, Zakir Husain Delhi College, University of Delhi, New Delhi, India
e-mail: sonwani@zh.du.ac.in; sonwani.s19@gmail.com

 
P. Saxena et al. (eds.), *Fog: Formation, Chemistry, Dynamics, Impact and Dissipation*,
https://doi.org/10.1007/978-981-96-4422-3_3

soil. Fog droplets efficiently scavenge pollutants through various mechanisms, including activation, Brownian motion, and phoretic effects (Pérez-Díaz et al. 2017; Jeevan Kumar et al. 2023). Fog deposition can significantly enhance the removal of atmospheric pollutants several times in foggy periods than in non-foggy periods (Sonwani et al. 2022). Liquid-phase chemistry within fog droplets affects the size distribution and hygroscopicity of regenerated aerosol particles, potentially influencing subsequent fog events (Jeevan Kumar et al. 2023). The chemical composition of fog is dependent on the size of the fog droplets, airborne fine particle concentrations, and aqueous phase chemistry (Bator and Collett 1997). Several studies mentioned the aqueous phase photochemistry in fog drops (Erel et al. 1993; Anastasio et al. 1994; Zuo and Hoigne 1994). Fog water chemistry is also reported with the organic matter, its characterization, sources, and processing in several studies (Anastasio et al. 1994; Zuo and Hoigne 1994). Since that time, interest has increased, in particular, in the formation of secondary organic aerosol through fog processing of volatile organic precursors (Sonwani et al. 2022). Despite its importance, fog's chemistry on a national scale remains challenging due to the complex composition and its spatial and temporal variability (Hůnová 2024). Understanding fog chemistry is important to identify its impact on air quality, ecosystems, and human health (Pérez-Díaz et al. 2017). Fog is an important constituent of atmospheric deposition. Fog droplets are very efficient in scavenging boundary layer pollutants and removing them from the atmosphere. The deposition of acid and sulfate-bearing fog on buildings and structures can cause severe corrosion damage and chemical deterioration. Furthermore, the research indicates that the upper level of elevated fog layers tends to build up higher concentrations of pollutants and lower pH values, which could pose a special risk for forest damage, particularly at the upper level of fog layers during fall and winter. Pérez-Diaz et al. (2017) identified the elevated fog water ionic concentrations, particularly of $SO_4^{2-}$, $NO^{3-}$, $NH_4^+$, and $H^+$. Such atmospheric deposition facilitated by fog significantly enhances the influx of air pollutants into ecosystems, particularly observed in forested mountainous regions (Thalmann et al. 2002; Błaś et al. 2010). Several studies have suggested that fog can serve as a more critical mechanism for the deposition of air pollutants than precipitation, due to several factors: (1) the frequent prevalence of fog, (2) the typical liquid water content ranging from 0.2 to 0.6 $gm^{-3}$, (3) the higher concentration of pollutants found in fog compared to rain, and (4) the presence of effective fog/cloud receptors, such as mountainous forests (Błaś et al. 2002). Fog plays an essential role in ecosystems, and to evaluate any potential adverse effects of fog on the environment over time, it is important not only to analyze the chemical makeup of fog but also to examine trends in fog frequency and its deposition impact on ecosystems. In this regard, multiple studies have recently highlighted a significant decrease in fog frequency throughout Europe. An in-depth examination of daily horizontal visibility records from 1976 to 2006 indicated a marked reduction in the number of foggy days, with vast regions of Europe experiencing a decrease of up to 50% in fog occurrences (Vautard et al. 2009; van Oldenborgh et al. 2010).

In this review chapter, I have first summarized the chemistry behind fog formation and the factors influencing its chemistry. Further, the chemical characteristics of fog, and related recent observational studies about their ionic composition and concentrations, are highlighted. The fog chemistry in urban and rural settings and the role of fog in the scavenging of gas and particle-phase organics have also been reviewed.

## 3.2 Atmospheric Chemistry of Fog Formation

The chemistry of fog formation involves several complex processes and interactions between various atmospheric components, as mentioned in Fig. 3.1. Gases and particles present in the atmosphere participate in the formation of fog water. Sigg et al. (1987) reveal that fog droplets contain concentrations of ammonium, hydrogen, sulfate, bisulfite, nitrate, and chloride ions that are 10–50 times higher than those found in rainwater. The composition of fog water is influenced by various factors, including the absorption of gases ($SO_2$, $NO_x$, $CO_2$, HCl, and $NH_3$), oxidation processes, and the presence of aerosols. Furthermore, the acidity of aqueous atmospheric solutions is a key parameter driving both the partitioning of semi-volatile acidic and

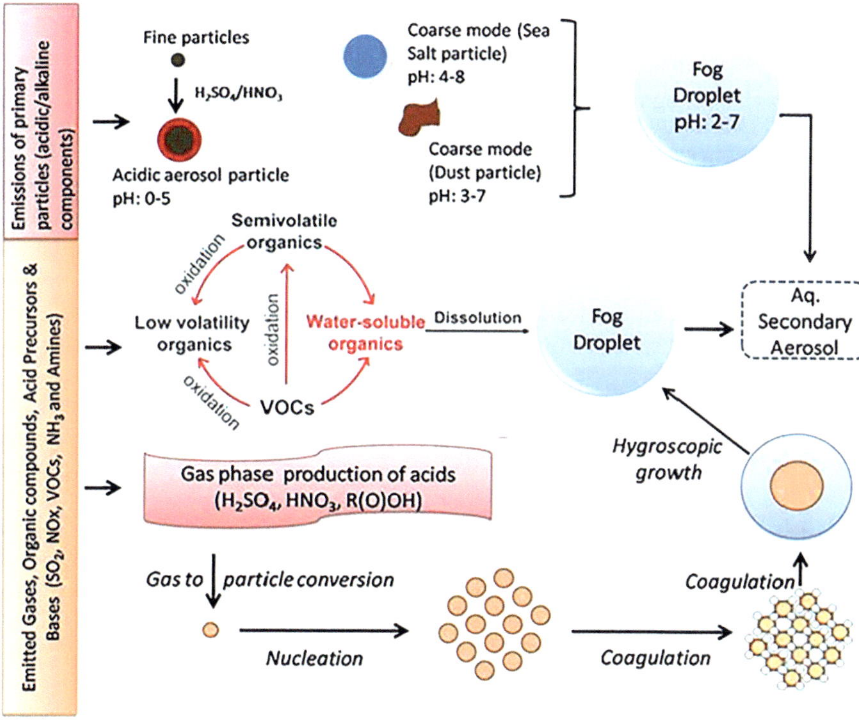


**Fig. 3.1** Showing the steps involved in the fog formation through gas phase and particle phase chemistry

basic trace gases and their aqueous-phase chemistry of fog and cloud formation (Tilgner et al. 2021). Studies mentioned two hypotheses to explain fog/cloud formation and mentioned about the gas-uptake and oxidation hypothesis, and the aerosol hypothesis. In case of $SO_2^-$ uptake hypothesis, it suggests that $SO_2$ is absorbed into fog water droplets in proportion to $NH_3$ due to pH- dependent solubility, followed by oxidation to $SO_4^{2-}$. The aerosol hypothesis proposes that $NH_4^+$, $SO_4^{2-}$, and $NO_3^-$ occur in relatively constant proportions in aerosols that act as nuclei for fog droplet growth (Tilgner et al. 2021). After emission, gases, organic compounds, acid precursors, and bases undergo a variety of changes such as ***condensation and nucleation*** (Fog droplets (5–50 μm in diameter) form when water vapor condenses on fine particles (aerosols) in the atmosphere. These aerosols act as nuclei for the growth of fog droplets), ***absorption of gases*** (Fog droplets absorb various gases from the atmosphere, including $SO_2$, $NO_x$, $CO_2$, HCl, and $NH_3$. These gases interact with $H_2O$ in an oxidative environment (containing $O_2$, $O_3$, OH•, $H_2O_2$) and in the presence of trace catalysts like metals (Cu, Fe, Mn) and organic materials), ***chemical reactions*** (Several chemical reactions occur within fog droplets: Oxidation of absorbed $SO_2$ to form sulfate ($SO_4^{2-}$); Absorption of $HNO_3$ to form nitrate ($NO_3^-$); and Acid-base reactions involving $NH_3$and other components), and ***aerosol formation*** (Interactions between gaseous components can lead to the formation of aerosols, such as ammonium sulfate and ammonium nitrate. These aerosols can then serve as nuclei for fog droplet formation or dissolve into existing droplets), ***pH Regulation*** (The pH of fog droplets is regulated by the balance of acidic components (like $H_2SO_4$, $HNO_3$, and HCl) and basic components (like $NH_3$)). The pH can vary widely, with some fogs being highly acidic (pH as low as 2) and ***concentration effects*** (The concentrations of solutes in fog water are typically 10–50 times higher than those in rainwater, making fog an efficient collector of pollutants close to the Earth's surface.). Apart from gases, fine/coarse primary particles (acidic/alkaline) emitted from the variety of sources participate in the acid aerosol/base aerosol formation that further participate in the fog formation process. This complex interplay of processes results in the formation of fog with varying chemical compositions, reflecting local atmospheric conditions and pollutant levels.

## 3.3 Factors Affecting Fog Chemistry

Several studies have been conduced to identify the effect of factors that influence the fog formation; the significant factors are listed below:

1. Global wind circulation: Large-scale atmospheric circulations can directly and indirectly influence fog formation by affecting relative humidity transport, cooling, and other meteorological conditions (Ghude et al. 2017). Synoptic-scale circulation significantly influences the intensity and magnitude of fog events, particularly marine fog. Studies also identified that advection fog formation is linked to synoptic-scale circulations, such as the South Atlantic high and the

to one-sixth and one-half of its original values, respectively. The pH of fog water collected in Delhi was recorded at 7.2 in 1985 (Khemani et al. 1987). This figure then decreased to 6.61 during the period of 2000–2003 (Ali et al. 2004) and further dropped to 5.38 in the years 2014–2015 (Nath and Yadav 2018). Over the span of 30 years, there has been a notable 35% reduction in the pH of fog water, indicating a trend toward increased acidity in the region. This decline in pH levels can be attributed to higher concentrations of sulfate and nitrate ions. The pH of fog water is approximately 5.6, largely due to the minimal contributions of Ca2+ and Mg2+ ions during the winter months of 2015–2016 in Delhi (Nath et al. 2023). According to Saikh et al. (2024), the pH of the fog water that was deposited varied between 6.4 and 6.8. The elevated levels of ammonium and calcium in the deposited fog water contributed to an increase in pH in the study area.

## 3.5 Comparative Analysis of Fog Water Chemistry

The chemical composition of fog water has been reported worldwide and found to be in different ionic compositions at different region as mentioned in Table 3.1. Studies have shown that the chemical composition of fog can vary considerably across different regions, with factors such as geographic location, local emissions, and meteorological conditions influencing the types and concentrations of chemical species present (Smith et al. 2023; Ferrero et al. 2019). Ions of calcium and magnesium are reported as predominant cations, however, the ions of chloride and sulfates are identified as predominant ions in the different Indian region (Saikh et al. 2024; Nath and Yadav 2018; Ali et al. 2004; Lakhani et al. 2007; Kulshrestha et al. 2005). Saikh et al. (2024) also found that the neutralizing factors, indicating the contribution of ions to neutralize acidity, were highest for $NH_4^+$ (1.13), followed by $Ca^{2+}$ (0.48) and $Mg^{2+}$ (0.07). Eastern Asian regions were identified with high ammonium and sulfate ions among all ions in fog water (Nieberding et al. 2018; Sun et al. 2015; Beiderwieden et al. 2007). Several studies have been conducted in the European region and reported a high concentration of ammonium and sodium ions among the cations, and a high concentration of nitrate ions among anions was observed in the European region (Giulianelli et al. 2014; Degefie et al. 2015; Wang et al. 2015; van Pinxteren et al. 2016). Studies in US regions were identified with a high concentration of nitrate and sulfate among anions and calcium and ammonium among cations (Wang et al. 2013; Aleksic et al. 2009). In a study of Beijing's winter haze events, researchers found that the aerosol chemistry during these events may be distinct and poorly understood, posing challenges for accurate air quality modeling (Liu et al. 2020). The chemical composition and formation processes of fog can vary considerably based on geographic location and environmental factors (Wang et al. 2020). For example, studies in China have shed light on the complex nature of winter haze events, indicating that the underlying atmospheric chemistry may not be fully understood (Liu et al. 2020). On the other hand, European research has explored methodological approaches to assess the potential for human intervention in cloud

**Table 3.1** Ionic concentration (mEqL-1) in fog water content in different parts of the world

| | | Anions | | | Cations | | | | | |
|---|---|---|---|---|---|---|---|---|---|---|
| Study site and period | | $Cl^-$ | $NO^-3$ | $_{SO}2\text{-}4$ | $Na^+$ | $_K+$ | $+Ca^2$ | $+Mg^2$ | $+NH_4$ | Reference |
| Central IGP, India | 2021 | 0.29 | 0.23 | **0.63** | 0.06 | 0.06 | **0.42** | 0.07 | **0.97** | Saikh et al. 2024 |
| New Delhi, India | 2014–2015 | 7.81 | 4.55 | **11.73** | 0.78 | 0.8 | **8.79** | 1.00 | **12.56** | Nath and Yadav 2018 |
| Delhi, India | 2000–2003 | 0.05 | 0.03 | 0.24 | 0.05 | 0.01 | 0.12 | 0.02 | 0.31 | Ali et al. (2004) |
| Agra, India | 1999–2000 | 0.38 | 0.49 | 0.69 | 0.18 | 0.02 | 0.34 | 0.15 | 1.70 | Lakhani et al. (2007) |
| Assam, India | 2002–2003 | 0.04 | 0.01 | 0.06 | 0.04 | 0.03 | 0.07 | 0.02 | 0.17 | Kulshrestha et al. 2005 |
| South Korea | 2002–2003 | 0.07 | 0.09 | 0.18 | 0.07 | 0.01 | 0.06 | 0.02 | 0.17 | Kim et al. (2006) |
| Ailaoshan Mountain, China | 2015 | 0.02 | 0.07 | **0.19** | 0.01 | 0.03 | 0.03 | 0.01 | **0.73** | Nieberding et al. (2018) |
| China | 2011–2012 | 0.03 | 0.18 | **0.34** | 0.02 | 0.01 | 0.06 | 0.01 | **0.28** | Shen et al. (2015) |
| Taiwan | 2006 | 0.04 | 0.18 | **0.40** | 0.06 | 0.01 | 0.03 | 0.02 | **0.24** | Beiderwieden et al. *(2007)* |
| Japan | 1993–2000 | 0.19 | **0.26** | **0.15** | 0.09 | 0.02 | 0.08 | 0.03 | **0.36** | Takeuchi et al. (2002) |
| Po-valley, Italy | 2011–2012 | 0.07 | **0.45** | 0.05 | 0.04 | 0.02 | 0.01 | 0.02 | **0.73** | Giulianelli et al. 2014 |
| Saclay plateau, South of Paris | 2012–2013 | 0.11 | **0.75** | 0.19 | 0.03 | 0.03 | **0.27** | **0.08** | **0.92** | Degefie et al. 2015 |
| Northern Poland | 2010 | 0.02 | 0.01 | 0.02 | 0.02 | 0.01 | 0.01 | 0.00 | 0.02 | Błaś et al. 2010 |
| Norway | 2011 | 0.05 | **0.06** | 0.04 | **0.07** | 0.01 | 0.03 | 0.01 | **0.09** | Wang et al. (2015) |
| Germany | 2010 | 0.03 | **0.16** | 0.04 | **0.04** | 0.01 | 0.01 | 0.01 | **0.22** | van Pinxteren et al. 2016 |
| USA 2Mangdang | 2009 | 0.01 | **0.11** | **0.06** | 0.03 | 0.01 | **0.09** | 0.01 | 0.06 | Huo et al. (2010) |
| Whiteface Mountain, USA | 1994–2006 | 0.01 | **0.08** | **0.20** | 0.00 | 0.00 | 0.02 | 0.01 | **0.12** | Aleksic et al. (2009) |

*IGP* Indo Gangetic plain.

and fog formation (Doronin et al. 2021). One study from China analyzed fog formation in Jinan, a city in the northern part of the country, during the cold season (Wang et al. 2020). The researchers found that the fog samples in Jinan exhibited high acidity, with pH values ranging from 4.4 to 5.7. The high acidity was attributed to the presence of sulfate and nitrate ions, which are typical indicators of secondary aerosol formation. In contrast, a study from South Korea's Cheju Island observed a different trend, with the concentrations of major ionic species, such as sulfate, nitrate, and ammonium, decreasing from fall to winter and spring (Kim et al. 2006). The chemical composition of fog water can also be influenced by proximity to the ocean and marine environments. (Nahar et al. 2022) analyzes the chemical composition of fog water in Bhola, Bangladesh, a coastal island located in the outflow region of the Indo-Gangetic Plain. The study found the dominance of water-soluble ions of $Ca^{2+}$, $NO_3^-$, $Cl^-$, $Na^+$, $SO_4^{2-}$, $NH_4^+$, $Mg^{2+}$, $K^+$, $F^-$, and $HCO_3^-$, in decreasing order of concentration, where $Ca^{2+}$ and $NH_4^+$ were identified as the primary neutralizing components in the fog water. Furthermore, enrichment factor analysis suggested anthropogenic sources for $NO_3^-$, $SO_4^{2-}$, Zn, Mn, and Cu.

## 3.6 Fog Chemistry in Urban and Rural Environment

### *3.6.1 Urban Fog Chemistry*

Atmospheric chemistry plays a crucial role in urban fog formation and composition. In urban environments, fog can act as a reservoir for pollutants, leading to complex chemical interactions and potential health impacts. The chemistry of urban fog is influenced by several factors:

1. Pollutant sources: Urban areas have numerous pollution sources, including vehicle emissions, industrial activities, and residential heating. These sources release various compounds into the atmosphere, such as nitrogen oxides (NOx), sulfur dioxide ($SO_2$), volatile organic compounds (VOCs), and particulate matter.
2. Aerosol composition: Urban fog droplets often form around aerosol particles, which serve as condensation nuclei. These particles can be composed of sulfates, nitrates, ammonium, and organic compounds, influencing the chemical composition of fog droplets.
3. Gas-to-particle conversion: Fog provides an ideal environment for gas-to-particle conversion processes. For example, sulfur dioxide can be oxidized to form sulfuric acid, which then contributes to the formation of sulfate particles in fog droplets.
4. Aqueous-phase chemistry: Within fog droplets, various chemical reactions can occur, leading to the formation of secondary pollutants. For instance, the oxidation of $SO_2$ to sulfate and NOx to nitrate can be enhanced in the aqueous phase.
5. pH levels: Urban fog tends to be more acidic due to the presence of pollutants like sulfuric acid and nitric acid. The pH of fog droplets can influence the solubility and reactivity of various chemical species.
6. Photochemical reactions: Although fog often occurs in low-light conditions, photochemical reactions can still play a role, especially during daytime fog

events. These reactions can lead to the formation of secondary organic aerosols and other complex compounds.

7. Vertical mixing and deposition: The stable atmospheric conditions associated with fog can trap pollutants near the ground, leading to higher concentrations. Fog droplets can also efficiently scavenge pollutants from the air through wet deposition processes.
8. Urban heat island effect: The higher temperatures in urban areas can affect fog formation and persistence, potentially altering the chemical composition and reactions within fog droplets.
9. Seasonal variations: The chemistry of urban fog can vary seasonally due to changes in emission patterns, meteorological conditions, and photochemical activity.
10. Interactions with urban surfaces: Fog droplets can interact with urban surfaces such as buildings and roads, potentially leading to the uptake or release of chemical species. Understanding the atmospheric chemistry of urban fog is crucial for assessing air quality, human health impacts, and ecosystem effects in urban environments. It requires interdisciplinary research combining atmospheric science, chemistry, and urban planning to develop effective strategies for managing and mitigating the impacts of urban fog pollution.

### *3.6.2 Rural Fog Chemistry*

While urban fog chemistry has been extensively studied, the atmospheric chemistry of rural fog presents unique characteristics and challenges. Rural environments typically have lower levels of anthropogenic pollutants, but they are influenced by different factors that shape fog composition and behavior:

1. Natural emission sources: Rural areas are often dominated by natural emission sources such as vegetation, soil, and agricultural activities. These sources release biogenic volatile organic compounds (BVOCs), ammonia, and other compounds that can interact with fog droplets.
2. Agricultural influences: In agricultural regions, fog chemistry can be affected by emissions from fertilizer use, livestock, and crop residue burning. These activities can introduce nitrogen-containing compounds, particulate matter, and organic substances into the atmosphere.
3. Lower aerosol concentrations: Rural areas generally have lower concentrations of anthropogenic aerosols compared to urban environments. This can affect fog formation processes and the chemical composition of fog droplets.
4. Biogenic secondary organic aerosols (SOA): The oxidation of BVOCs emitted by vegetation can lead to the formation of biogenic SOA, which can serve as condensation nuclei for fog droplets and influence their chemical properties.

5. Long-range transport: Rural fog can be influenced by pollutants transported from distant urban or industrial areas, potentially altering its chemical composition.
6. Dust and mineral particles: In some rural areas, especially those prone to dust storms or near mineral-rich soils, fog droplets may incorporate dust and mineral particles, affecting their pH and chemical reactivity.
7. Marine influences: Coastal rural areas may experience fog with chemical characteristics influenced by sea spray aerosols and marine biogenic emissions.
8. Reduced urban heat island effect: The absence of significant urban heat island effects in rural areas can lead to more frequent and persistent fog events, potentially affecting the duration of chemical processes within fog droplets.
9. Photochemistry in cleaner atmospheres: With generally cleaner air, rural fog may experience different photochemical processes compared to urban fog, potentially leading to unique chemical transformations.
10. Interactions with natural surfaces: Rural fog interacts with vegetation, soil, and water bodies, which can influence its chemical composition through processes such as deposition and re- emission of compounds.
11. Seasonal agricultural practices: The chemistry of rural fog may vary seasonally due to changes in agricultural activities, such as planting, harvesting, and fertilizer application.
12. Lower pH variability: Rural fog typically exhibits less acidic pH levels compared to urban fog, due to lower concentrations of acidic pollutants like sulfuric and nitric acids.
13. Organic matter composition: The organic matter in rural fog is likely to be dominated by biogenic sources, resulting in a different spectrum of organic compounds compared to urban fog.
14. Trace metal content: Rural fog may contain lower concentrations of trace metals associated with industrial and vehicular emissions, but could have higher levels of metals from natural sources or agricultural activities.
15. Nitrate to sulfate ratios: Rural fog often shows different nitrate to sulfate ratios compared to urban fog, reflecting the distinct emission sources in rural environments. Understanding the atmospheric chemistry of rural fog is essential for assessing its impacts on ecosystems, agriculture, and air quality in rural areas. It requires specialized research approaches that consider the unique characteristics of rural environments and their interactions with fog formation and composition.

## 3.7 Fog Scavenging of Atmospheric Chemicals

Fog deposition also plays a dominant role in the scavenging of organic materials from the atmosphere (Ducret and Cachier 1992). It was found that significant levels of elemental carbon (EC) and organic carbon (OC) are found in fog water (Saxena et al. 2020). However, the information on the carbonaceous content in fog water is still scarce as compared to the particulate phase. Several authors also mentioned the wet scavenging process and explained their significant impact in reducing atmospheric levels of important ions such as $Ca^{2+}$, $SO^{4-}$, $NH^{4+}$, and $NO^{3-}$ ions as well as several trace gases and particulate bound pollutants such as oxides of nitrogen, sulfur, and organic and elemental carbon (Kulshrestha et al. 2009a, b; Sonwani and Kulshrestha 2019; Saxena et al. 2020). The fog's ability to scavenge atmospheric particles and gaseous species plays a crucial role in determining the ionic composition of fog water (Sickles et al. 1999). This scavenging process varies significantly between different species, largely depending on their concentration distribution within fog droplets and aerosols. Additionally, when gaseous inorganic and organic species are absorbed by the fog droplets, they contribute to the formation of secondary aerosols through aqueous processing, subsequently released into the atmosphere (Chakraborty et al. 2016; Ervens et al. 2011; Herckes et al. 2007, 2013; Kaul et al. 2011). It is important to note that the predominance of one process over another can fluctuate, depending on the scavenging efficiency of the fog droplets. There are two prevalent methods for determining the overall mass scavenging efficiency (MSE) of fog or cloud. The first method involves conducting parallel sampling of both fog or cloud water and interstitial aerosol during foggy or cloudy events, followed by an assessment of the concentration and composition in each (Aikawa et al. 2007; Facchini et al. 1999; Kaul et al. 2011; Li et al. 2017; van Pinxteren et al. 2016). The second method entails examining the changes in aerosol composition before and after the formation of fog, typically over a sampling period of 3 to 4 hours (Collett et al. 2008; Gilardoni et al. 2014). The variations in the mass scavenging efficiency of chemical species are influenced by their solubility in water within an aqueous environment such as fog, clouds, and wet aerosols. Chemical species that exhibit greater hygroscopicity act as effective condensation nuclei, facilitating the formation of cloud or fog droplets. Furthermore, the dependence of mass scavenging efficiency on aerosol loading and liquid water content has been observed to differ based on the specific sampling site and the severity of fog conditions (Giulianelli et al. 2014; Herckes et al. 2013). According to Qian et al. (2023), the effectiveness of fog in scavenging particulate matter is influenced by the duration of the fog event. Initially, this effect is positive when the fog is forming, but it becomes negative as the fog begins to dissipate. Observations indicate that the concentration of PM2.5 can decrease from 159 μg m − 3 to38 μg m − 3 over a span of 11 hours of fog presence. The scavenging effect on PM2.5 is particularly pronounced in the vertical dimension, primarily due to the deposition of fog droplets on pollutant particles. In another study, it was found that during foggy nights in the Indo-Gangetic plain (IGP) of India, the concentrations of aerosols and their water-soluble components were notably lower compared to non-foggy nights, attributed to the effective scavenging action of fog droplets on aerosols (Izhar et al. 2020). The scavenging efficiency of wet deposition processes depends on several factors, such as wet deposition intensity, particle size, Brownian motion, reactivity of chemical reactivity of atmospheric species, hygroscopicity of particles and polarity, and water solubility of chemical species (Sonwani and Kulshrestha 2019). The scavenging ratio is a key method for assessing the efficiency of the scavenging process. This calculation is founded on the premise that the concentration of a substance in fog or rainwater correlates with its concentration in the atmosphere (Granat et al. 2002; Sonwani and Kulshrestha et al. 2019), as indicated in Eq. (3.1).

$$SR = \frac{Cr}{Ca}\rho_a \tag{3.1}$$

Where, Cr is the concentration of a carbonaceous species in rainwater (μgC/l), Ca is the concentration of carbonaceous species in air (μgC/m3) and ρa is the density of air (1.29 kg/$m^3$ at 0 °C). Another method to determining wet scavenging efficiency has been thoroughly examined by Latha et al. (2005) and Sonwani and Saxena (2021). According to their studies, the scavenging coefficient can be articulated by the following Eq. 3.2:

$$Ws = a(P)b \tag{3.2}$$

Where, Ws is the wet scavenging coefficient ($s^{-1}$), which varies with different particle sizes and rainfall rates. P is the rainfall intensity (mm $s^{-1}$), and the values for parameters 'a' and 'b' have been derived from the regression lot linking levels of different carbon fractions and rainfall intensity. Overall, the fog can scavenge trace gases, particulate matter, and associated chemical species through its scavenging process and can be analyzed through the above-mentioned scavenging efficiency concept.

## 3.8 Conclusion

This chapter identified fog as an important global atmospheric event that changes the atmospheric chemistry of a region. The atmospheric chemicals can change the chemical composition of fog water, as they participate in the fog formation process through heterogeneous and multiphase chemistry. Any change in the source of air pollution emission can change its chemistry; thus, the chemical composition of same region may be different from one site to another site. The chemical characterization of fog depends on the nearness of predominant sources of chemical emissions such as dissolved gases, particles, nutrients, metals, organic components, inorganic components, and salts. Thus, the chemical nature of fog at the valley side may be different from that in the coastal area, forest area, mountains, and city or rural area in the same region. The chemistry of fog in rural and urban areas exhibits distinct characteristics, reflecting the diverse sources and processes influencing atmospheric composition in these environments. Urban fog is generally characterized by higher concentrations of anthropogenic pollutants, lower pH values, and more complex chemical interactions, while rural fog shows a stronger influence of natural and agricultural sources. Understanding these differences is crucial for assessing the impacts of fog on air quality, ecosystem health, and human well-being.

As urbanization continues to expand globally, understanding the dynamics of fog chemistry in both urban and rural areas becomes increasingly important for sustainable development and environmental protection. It is also identified that the fog formation and deposition also play an active role in air pollution mitigation through its scavenging process.

Overall, this chapter provides a summary of fog chemistry by considering a chemistry of fog formation, chemical composition, and urban and rural fog chemistry and by scavenging efficiency of fog. On the basis of already published papers in this domain, it has been established that ongoing research in the area of fog chemistry will significantly enhance atmospheric models, pollution reduction strategies, and adaptations to climate change. By deepening our understanding of fog chemistry across various environments, we can create more effective methods for managing air quality, safeguarding ecosystems, and boosting the resilience of both urban and rural areas to changes in the atmosphere. Future studies should concentrate on clarifying the intricate interactions between fog and different environmental factors, as well as creating innovative technologies and management approaches that can minimize adverse effects and capitalize on the potential advantages associated with fog in diverse contexts.

## References

Aikawa M, Hiraki T, Suzuki M, Tamaki M, Kasahara M. Separate chemical characterizations of fog water, aerosol, and gas before, during, and after fog events near an industrialized area in Japan. Atmos Environ. 2007;41(9):1950–9.

Aleksic N, Roy K, Sistla G, Dukett J, Houck N, Casson P. Analysis of cloud and precipitation chemistry at Whiteface Mountain, NY. Atmos Environ. 2009;43(17):2709–16.

Ali K, Momin GA, Tiwari S, Safai PD, Chate DM, Rao PSP. Fog and precipitation chemistry at Delhi, North India. Atmos Environ. 2004;38(25):4215–22.

Anastasio C, Faust BC, Allen JM. Aqueous phase photochemical formation of hydrogen peroxide in authentic cloud waters. J Geophys Res Atmos. 1994;99(D4):8231–48.

Andersen H, Cermak J, Fuchs J, Knippertz P, Gaetani M, Quinting J, et al. Synoptic-scale controls of fog and low-cloud variability in the Namib Desert. Atmos Chem Phys. 2020;20(6):3415–38.

Anderson TL, Covert DS, Wheeler JD, Harris JM, Perry KD, Trost BE, et al. Aerosol backscatter fraction and single scattering albedo: Measured values and uncertainties at a coastal station in the Pacific Northwest. J Geophys Res Atmos. 1999;104(D21):26793–807.

Bator A, Collett Jr JL. Cloud chemistry varies with drop size. J Geophys Res Atmos. 1997;102(D23):28071–28078.

Beiderwieden E, Schmidt A, Hsia YJ, Chang SC, Wrzesinsky T, Klemm O. Nutrient input through occult and wet deposition into a subtropical montane cloud forest. Water Air Soil Pollut. 2007;186:273–88.

Błaś M, Sobik M, Quiel F, Netzel P. Temporal and spatial variations of fog in the Western Sudety Mts., Poland. Atmos Res. 2002;64(1–4):19–28.

Błaś M, Polkowska Ż, Sobik M, Klimaszewska K, Nowiński K, Namieśnik J. Fog water chemical composition in different geographic regions of Poland. Atmos Res. 2010;95(4):455–69.

Chakraborty A, Ervens B, Gupta T, Tripathi SN. Characterization of organic residues of size-resolved fog droplets and their atmospheric implications. J Geophys Res Atmos. 2016;121(8):4317–32.

Collett JL Jr, Herckes P, Youngster S, Lee T. Processing of atmospheric organic matter by California radiation fogs. Atmos Res. 2008;87(3–4):232–41.

Degefie DT, El-Madany TS, Held M, Hejkal J, Hammer E, Dupont JC, et al. Fog chemical composition and its feedback to fog water fluxes, water vapor fluxes, and microphysical evolution of two events near Paris. Atmos Res. 2015;164:328–38.

Doronin AP, Shchukin GG, Kozlova NA, Petrochenko VM. A methodological approach to assessing the current opportunities for changing the number of clouds in large are as in the interests of solving applied problems. In: IOP Conference Series: Earth and Environmental Science, vol. 840, No. 1. IOP Publishing; 2021. p. 012031.

Ducret J, Cachier H. Particulate carbon content in rain at various temperate and tropical locations. J Atmos Chem. 1992;15:55-67.

Erel Y, Pehkonen SO, Hoffmann MR. Redox chemistry of iron in fog and stratus clouds. J Geophys Res Atmos. 1993;98(D10):18423–34.

Ervens BTBWR, Turpin BJ, Weber RJ. Secondary organic aerosol formation in cloud droplets and aqueous particles (aqSOA): a review of laboratory, field and model studies. Atmos Chem Phys. 2011;11(21):11069–102.

Facchini MC, Fuzzi S, Zappoli S, Andracchio A, Gelencsér A, Kiss G, et al. Partitioning of the organic aerosol component between fog droplets and interstitial air. J Geophys Res Atmos. 1999;104(D21):26821–32.

Feng J, Huang B, Zhong M. Fabrication of super hydrophobic and heat-insulating antimony dopedtinoxide/polyurethane films by cast replica micro molding. J Colloid Interface Sci. 2009;336(1):268–72.

Ferrero L, Ritter C, Cappelletti D, Moroni B, Močnik G, Mazzola M, et al. Aerosol optical properties in the Arctic: The role of aerosol chemistry and dust. Composition in a closure experiment between Lidar and tethered balloon vertical profiles. Science of the total environment. 2019;686:452–67.

Fuzzi S, Facchini MC, Orsi G, Bonforte G, Martinotti W, Ziliani G, et al. The NEVALPA project: A regional network for fog chemical climatology over the Po Valley basin. Atmos Environ. 1996;30(2):201–13.

Gao Y, Zhang M, Liu Z, Wang L, Wang P, Xia X, et al. Modeling the feedback between aerosol and meteorological variables in the atmospheric boundary layer during a severe fog–haze event over the North China Plain. Atmos Chem Phys. 2015;15(8):4279–95.

Ghude SD, Bhat GS, Prabhakaran T, Jenamani RK, Chate DM, Safai PD, et al. Winter fog experiment over the Indo-Gangetic plains of India. Curr Sci. 2017;112:767–84.

Gilardoni S, Massoli P, Giulianelli L, Rinaldi M, Paglione M, Pollini F, et al. Fog scavenging of organic and inorganic aerosol in the PoValley. Atmos Chem Phys. 2014;14(13):6967–81.

Giulianelli L, Gilardoni S, Tarozzi L, Rinaldi M, Decesari S, Carbone C, et al. Fog occurrence and chemical composition in the Po valley over the last twenty years. Atmos Environ. 2014;98:394–401.

Granat L, Norman M, Leck C, Kulshrestha UC, Rodhe H. Wet scavenging of sulfur compounds and other constituents during the Indian Ocean Experiment (INDOEX). J Geophys Res Atmos. 2002;107(D19):INX2-24.

Gultepe I, Tardif R, Michaelides SC, Cermak J, Bott A, Bendix J, et al. Fog research: A review of past achievements and future perspectives. Pure Appl Geophys. 2007;164:1121–59.

Hay JS. Ming palace and tomb in early Qing Jiangning: dynastic memory and the openness of history. Late Imperial China. 1999;20(1):1–48.

Heald SM, Krupka KM, Brown CF. Incorporation of pertechnetate and perrhenate into corroded steel surfaces studied by X-ray absorption fine structure spectroscopy. Radiochimica Acta. 2012;100(4):243–53.

Herckes P, Chang H, Lee T, Collett JL. Air pollution processing by radiation fogs. Water Air Soil Pollut. 2007;181:65–75.

Herckes P, Valsaraj KT, Collett JL Jr. A review of observations of organic matter in fogs and clouds: origin, processing and fate. Atmos Res. 2013;132:434–49.

Hůnová, I. Challenges in moving towards fog's contribution to spatial patterns of atmospheric deposition fluxes on a national scale. Sci Total Environ. 2024;174208.

Huo M, Sun Q, Bai Y, Xie P, Liu Z, Li J, et al. Chemical character of precipitation and related particles and trace gases in the North and South of China. J Atmos Chem. 2010;67:29–43.

Izhar S, Gupta T, Panday AK. Scavenging efficiency of water soluble inorganic and organic aerosols by fog droplets in the Indo Gangetic Plain. Atmos Res. 2020;235:104767.

Jeevan Kumar B, Schmeller G. Assessment of WRF planetary boundary layer schemes in the simulation of fog events over Hungary. Időjárás/Quart J Hungarian Meteorolog Ser. 2023;127(1):1–22.

Kaul DS, Gupta T, Tripathi SN, Tare V, Collett JL Jr. Secondary organic aerosol: a comparison between foggy and nonfoggydays. Environ Sci Technol. 2011;45(17):7307–13.

Khemani LT, Momin GA, Prakasa Rao PS, Safai PD, Prakash P. Influence of alkaline particulates on the chemistry of fog water at Delhi, North India. Water Air Soil Pollut. 1987;34:183–9.

Kim MG, Lee BK, Kim HJ. Cloud/fog water chemistry at a high elevation site in South Korea. J Atmos Chem. 2006;55:13–29.

Kulshrestha MJ, Sekar R, Krishna D, Hazarika AK, Dey NC, Rao PG. Deposition fluxes of chemical components of fog water at a rural site in north-east India. Tellus B: Chem Phys Meteorol. 2005;57(5):436–9.

Kulshrestha UC, Reddy LAK, Satyanarayana J, Kulshrestha MJ. Real-time wet scavenging of major chemical constituents of aerosols and role of rain intensity in Indian region. Atmos Environ. 2009a;43(32):5123–7.

Kulshrestha UC, Sunder Raman R, Kulshrestha MJ, Rao TN, Hazarika PJ. Secondary aerosol formation and identification of regional source locations by PSCF analysis in the Indo-Gangetic region of India. J Atmos Chem. 2009b;63:33–47.

Lakhani A, Parmar RS, Satsangi GS, Prakash S. Chemistry of fogs at Agra, India: Influence of soil particulates and atmospheric gases. Environ Monit Assess. 2007;133:435–45.

Latha KM, Badarinath KVS, Reddy PM. Scavenging efficiency of rainfall on black carbon aerosols over an urban environment. Atmos Sci Lett. 2005;6(3):148–51.

Li S, Dardzinski BJ, Collins CM, Yang QX, Smith MB. Three-dimensional mapping of the static magnetic field inside the human head. Magn Reson Med. 1996;36(5):705–14.

Li R, Hu Y, Li L, Fu H, Chen J. Real-time aerosol optical properties, morphology and mixing states under clear, haze and fog episodes in the summer of urban Beijing. Atmos Chem Phys. 2017;17(8):5079–93.

Liu WJ, Zhang YP, Li HM, Meng FR, Liu YH, Wang CM. Fog-and Rainwater chemistry in the tropical seasonal rainforest of Xishuang banna, southwest China. Water Air Soil Pollut. 2005;167:295–309.

Liu XH, Wai KM, Wang Y, Zhou J, Li PH, Guo J, et al. Evaluation of trace elements contamination in cloud/fog water at an elevated mountain site in Northern China. Chemosphere. 2012;88(5):531–41.

Liu A, Wang H, Cui Y, Shen L, Yin Y, Wu Z, et al. Characteristics of aerosol during a severe haze-fog episode in the Yangtze River Delta: Particle size distribution, chemical composition, and optical properties. Atmos. 2020;11(1):56.

Nahar K, Nahian S, Jeba F, Islam MS, Rahman MS, Choudhury TR, et al. Characterization and source discovery of winter time fog on coastal Island, Bangladesh. Atmosphere. 2022;13(3):497.

Nath S, Yadav S. A comparative study on fog and dew water chemistry at New Delhi, India. Aerosol Air Qual Res. 2018;18(1):26–36.

Nath S, Chaudhary S, Rathi A, Yadav S. Chemistry and buffering capacity of fog water collected in and around New Delhi, India. Environ Sci Pollut Res. 2023;30(10):25635–49.

National Oceanic and Atmospheric Administration. Surface weather observations and reports, Federal Meteorological Handbook No. 1; 1995, 94 pp. [Available from Department of Commerce, NOAA, Office of the Federal Coordinator for Meteorological Services and Supporting Research, Silver Spring, MD, 20910.

Nieberding F, Breuer B, Braeckevelt E, Klemm O, Song Q, Zhang Y. Fog water chemical composition on Ailaoshan Mountain, Yunnan province, SW China. Aerosol Air Qual Res. 2018;18(1):37–48.

Pandis SN, Seinfeld JH, Pilinis C. Chemical composition differences in fog and cloud droplets of different sizes. Atmos Environ Part A. 1990;24(7):1957–69.

Pérez-Díaz JL, Ivanov O, Peshev Z, Álvarez-Valenzuela MA, Valiente-Blanco I, Evgenieva T, et al. Fogs: Physical basis, characteristic properties, and impacts on the environment and human health. Water. 2017;9(10):807.

Qian J, Liu D, Yan S, Cheng M, Liao R, Niu S, et al. Fog scavenging of particulate matters in air pollution events: observation and simulation in the Yangtze River Delta, China. Sci Total Environ. 2023;876:162728.

Saikh SR, Mushtaque MA, Das SK. A study on the understanding of chemical compositions of deposited fog water over Central Indo-Gangetic Plain in India. Aerosol Air Qual Res. 2024;24(3):230098.

Saxena S, Joosten A, Maze M. Brain fog: are clearer skies on the horizon? A review of perioperative neurocognitive disorders. Annu Update Intensi Care Emerg Med. 2019;2019:423–30.

Saxena P, Sonwani S, Sharma SK, Kumar P, Chandra N. Carbonaceous aerosol variations in foggy days: a critical analysis during the fireworks festival. Fresenius Environ Bull. 2020;29(8):6639–56.

Schemenauer RS, Banic CM, Urquizo N. High elevation fog and precipitation chemistry in southern Quebec, Canada. Atmos Environ. 1995;29(17):2235–52.

Shen J, Miao L, Guo Z. Research on the application and compensation for startup process of FOG based on RBF neural network. In: Proceedings of the 10th World Congress on Intelligent Control and Automation. IEEE; 2012. p. 3195–9.

Shen XJ, Sun JY, Zhang XY, Zhang YM, Zhang L, Che HC, et al. Characterization of submicron aerosols and effect on visibility during a severe haze-fog episode in Yangtze River Delta, China. Atmos Environ. 2015;120:307–16.

Sigg L, Stumm W, Zobrist J, Zürcher F. The chemistry of fog: factors regulating its composition. Chimia. 1987;41(5):159–65.

Smith DK, Dorling SR, Renfrew IA, Ross AN, Poku C. Fog trends in India: relationships to fog type and western disturbances. Int J Climatol. 2023;43(2):818–36.

Sonwani S, Kulshrestha UC. PM 10 carbonaceous aerosols and their real-time wet scavenging during monsoon and non-monsoon seasons at Delhi, India. J Atmos Chem. 2019;76:171–200.

Sonwani S, Saxena P. Water-insoluble carbonaceous components in rainwater over an urban background location in Northern India during pre-monsoon and monsoon seasons. Environ Sci Pollut Res. 2021;28(38):53058–73.

Sonwani S, Saxena P, Srivastava A. Carbonaceous aerosol variability and SOA formation during foggy days in Delhi, India. Front Sustain Cities. 2022;4:951340.

Sun L, Xia X, Wang P, Zhang R, Che H, Deng Z, et al. Surface and column-integrated aerosol properties of heavy haze events in January 2013 over the North China Plain. Aerosol Air Qual Res. 2015;15(4):1514–24.

Takeuchi M, Okochi H, Igawa M. Controlling factors of weak acid and base concentrations in urban Dew water—comparison of dew chemistry with rain and fog chemistry. Bull Chem Soc Jpn. 2002;75(4):757–64.

Thalmann E, Burkard R, Wrzesinsky T, Eugster W, Klemm O. Ion fluxes from fog and rain to an agricultural and a forest ecosystem in Europe. Atmos Res. 2002;64(1–4):147–58.

Tilgner A, Schaefer T, Alexander B, Barth M, CollettJr JL, Fahey KM, et al. Acidity and the multiphase chemistry of atmospheric aqueous particles and clouds. Atmos Chem Phys. 2021;21(17):13483–536.van Oldenborgh GJ, Yiou P, Vautard R. On the roles of circulation and aerosols in the decline of mist and dense fog in Europe over the last 30 years. Atmos Chem Phys. 2010;10(10):4597–609.

van Pinxteren D, Fomba KW, Mertes S, Müller K, Spindler G, Schneider J, et al. Cloud water composition during HCCT-2010: scavenging efficiencies, solute concentrations, and droplet size dependence of inorganic ions and dissolved organic carbon. Atmos Chem Phys. 2016;16(5):3185–205.

Vautard R, Yiou P, Van Oldenborgh GJ. Decline of fog, mist and haze in Europe over the past 30 years. Nat Geosci. 2009;2(2):115–9.

Wang T, Zeng J, Lai Y, Cai Y, Tian H, Chen Y, Wang B. Data collection from WSNs to the cloud based on mobile fog elements. Futur Gener Comput Syst. 2020;105:864–72.

Wang X, Wang W, Xue L, Gao X, Nie W, Yu Y, et al. Size-resolved aerosol ionic composition and secondary formation at Mount Heng in South Central China. FESE. 2013;7:815–826.

Wang Y, Zhang J, Marcotte AR, Karl M, Dye C, Herckes P. Fog chemistry at three sites in Norway. Atmos Res. 2015;151:72–81.

Wu D, Bi XY, Deng XJ, Li F, Tan H, Liao G, Huang J. Effect of atmospheric haze on the deterioration of visibility over the Pearl River Delta. Acta Meteorologica Sinica-English Edition. 2007;21(2), 215.

Zuo Y, Hoigné J. Photo chemical decomposition of oxalic, glyoxalic and pyruvic acid catalysed by iron in atmospheric waters. Atmos Environ. 1994;28(7):1231–9.

# Chapter 4
# Characterization of Chemical Constituents in Acidic Fog

**Merlyn Mathew, Anju Srivastava, Reena Jain, and Pallavi Saxena**

## 4.1 Introduction

In recent decades, concerns about air pollution and its detrimental effects on human health and the environment have escalated worldwide. Among the various forms of air pollution, fog is a weather phenomenon that has attracted significant attention due to its potential to exacerbate the negative impacts of air pollution. Air pollution incidents or disasters worldwide have often been closely linked to fog events. A prominent example is the devastating event that occurred in London in December 1952. On five days of dense fog, which severely limited visibility and light, the daily death rate surged from the typical average of 250 to approximately 1000 fatalities per day (Sheoran et al. 2021). As the fog dissipated, its harmful effects gradually waned. However, it was not until three months later that the death rate stabilized, revealing an alarming toll of 12,000 additional deaths attributable to the fog. Similar pollution-related occurrences of comparable magnitude have been observed in other regions, such as Donora, Pennsylvania, in the United States, and the Meuse Valley near Liege, Belgium. In both cases, the combination of air pollution and fog led to adverse impacts on public health, highlighting the urgent need for comprehensive research and effective measures to address the intricate relationship between air pollution and fog-related hazards (Sonwani et al. 2022). This knowledge is crucial for

M. Mathew
Environmental Studies and Resource Management, TERI School of Advanced Studies, Delhi, India

A. Srivastava · R. Jain
Department of Chemistry, Hindu College, University of Delhi, Delhi, India

P. Saxena (✉)
Department of Environmental Science, Hindu College, University of Delhi, Delhi, India
e-mail: pallavisaxena@hinducollege.ac.in; pallavienvironment@gmail.com

 
P. Saxena et al. (eds.), *Fog: Formation, Chemistry, Dynamics, Impact and Dissipation*,
https://doi.org/10.1007/978-981-96-4422-3_4

formulating policies and strategies aimed at mitigating the health risks associated with such events.

Fog, when mixed with certain pollutants, can transform into acidic fog, posing unique challenges to ecosystems and human populations living in affected regions. Understanding the chemical constituents of acidic fog and its sources is crucial for developing effective mitigation strategies and protecting vulnerable ecosystems. Certain studies that concentrated on analyzing the effect of acidic fog on respiratory function have suggested that the impact on the respiratory system is directly influenced by the degree to which the acid droplets that are inhaled can penetrate; or in other words, it depends on the size of the acid droplet (Hackney et al. 1989). Such studies require an understanding of the chemistry of acidic fog and the effects of prolonged exposure at ambient concentrations. Acidic fog has an acidic pH due to the presence of certain chemical components, primarily pollutants from anthropogenic activities. These pollutants can include sulfur dioxide ($SO_2$), nitrogen oxides (NOx), volatile organic compounds (VOCs), and particulate matter (PM). When these pollutants combine with water droplets in the atmosphere, they undergo complex chemical reactions, leading to the formation of acidic fog. Acidic fog has been linked to a wide range of adverse effects, including damage to vegetation, corrosion of infrastructure, and respiratory problems in humans. Several studies conducted in the United States, Europe, and Japan have highlighted the presence of highly acidic fog (e.g., Jacob et al. 1984, 1985; Igawa et al. 2001; Mrose 1966; Munger et al. 1983, 1990; Okita 1968; Waldman et al. 1982, 1985; Manabu et al. 2017; Bridges et al. 2002). According to observations made in Germany and North America (Georgii 1986; Bruck et al. 1989), a significant amount of the acidic cloud, fog, and haze that currently covers forests is caused by pollution from sulfate aerosols. Weathers et al., (2020), and Saxena and Lin (1990) have also reported on the decline of forests and the effects of acid cloud water on soil acidification and on the epicuticular waxes of red spruce needles. They contend that the harm to the environment from acidic haze and fog is at least as great as that caused by acid rain. In the northern parts of India, there is frequent heavy fog throughout the winter which has an impact on the economy, public transportation, aviation, and public life in general (Sheoran et al. 2021). The post-monsoon season (October to November) and the winter season (December to February) both see some fog over the Delhi region, but December and January see the most intense and continuous fog. The temperature drops to its lowest point during these months, and the frequency of the Western Disturbance—the sole weather event that brings rain to that area during the winter months—increases. The Delhi region receives moisture from this weather system to help with the creation of fog (Ali et al. 2004). Since 1984, research on the chemical makeup of Delhi's fog water has been ongoing. Published findings from the earlier investigations may be found (Khemani et al. 1987; Kapoor et al. 1993). According to which, soil-oriented cations like $Ca^2$+, $Na^+$, $K^+$, and $Mg^{2+}$ predominated over acidic ions like $SO^{2-}$ and NO, and the nature of the fog water was in fact alkaline.

Characterizing the chemical constituents in acidic fog is a multifaceted task that requires comprehensive monitoring and analytical techniques. By identifying and quantifying the specific compounds present in acidic fog, we can gain insights into the sources and transformations of pollutants in the atmosphere. This knowledge is essential for assessing the impact of acidic fog on different ecosystems and developing targeted strategies to mitigate its effects. Furthermore, understanding the chemical composition of acidic fog is crucial for evaluating the effectiveness of existing air pollution control measures and designing new mitigation strategies. By identifying the major sources of pollutants contributing to acidic fog formation, policymakers and environmental agencies can implement targeted regulations and emission reduction measures. These efforts can help minimize the occurrence and severity of acidic fog episodes and mitigate their environmental and health impacts.

The chapter aims to provide a comprehensive overview of the characterization of chemical constituents in acidic fog and its environmental implications. It will technically review the existing literature on the measurement techniques used for analyzing acidic fog samples, discuss the major chemical components found in acidic fog, and explore the sources and transformation mechanisms of these pollutants. Additionally, the review will examine the impacts of acidic fog on ecosystems, human health, and infrastructure. Finally, it will discuss potential mitigation strategies and future research directions to address the challenges posed by acidic fog. By advancing our understanding of the chemical constituents in acidic fog and their impacts, this chapter seeks to contribute to the development of effective measures to mitigate the adverse effects of acidic fog on ecosystems and human well-being. Ultimately, this knowledge can guide policymakers, scientists, and environmental agencies in making informed decisions and implementing targeted interventions to improve air quality and protect vulnerable regions from the detrimental effects of acidic fog.

## 4.2 Formation and Composition of Acidic Fog

Fog droplets, ranging in diameter from 5 to 50 μm, exhibit significantly smaller dimensions compared to rain droplets. Fog is characterized by a liquid water content of approximately $10^{-4}$ $Lm^{-3}$ of the atmosphere, typically around ten times lower than that of clouds. Fog droplets can arise through swift condensation on fine particles (aerosols) or dissipate via evaporation mechanisms. Due to its formation in lower atmospheric layers, fog droplets serve as efficient captors of pollutants in proximity to the Earth's surface. Consequently, their composition reflects the influence of local emissions to a greater extent than more remote clouds (Sigg et al. 1987). This suggests that the kinetics of fog droplet reactions should not be constrained by mass transfer considerations (Lange et al. 2006; Baboolal et al. 1981). Chemical reactions taking place within clouds and aqueous haze aerosols play a significant role in driving rapid transformations observed in both urban and rural atmospheres. These reactions are instrumental in accounting for the swift conversion of pollutants and other atmospheric constituents, leading to changes in air quality and environmental conditions. Understanding the intricate processes occurring within these atmospheric components is crucial for comprehending the dynamics of urban and rural atmospheric chemistry and its potential implications for public health and the environment.

### 4.2.1 Mechanism of Acidic Fog Formation

The atmosphere acts as an oxidizing environment, where numerous components arise from chemical reactions involving oxidation. Primarily, gaseous oxides such as $SO_2$, $SO_3$, $H_2SO_4$, NO, $N_2O$, $HNO_2$, and $HNO_3$ are generated through oxidative processes. Many of these reactions are influenced by catalysts in terms of kinetics and are frequently triggered by photochemical mechanisms. Ammonium-sulfate nitrate aerosols are formed due to the reactions between different gaseous elements, which usually have particles of soil, dust, charcoal, and traces of heavy metals. These aerosols in the atmosphere exhibit solid characteristics under low humidity conditions, transitioning into a liquid state through deliquescence at a critical relative humidity. Sulfuric acid, due to its exceptionally low vapor pressure ($<10^{-7}$ atm), predominantly exists as liquid particles (Sonwani et al. 2022). These particles can readily engage with water vapor and $NH_3$. As relative humidity rises, fog water forms on pre-existing aerosol nuclei. During this process, these aerosols dissolve as condensation occurs. The fog droplets not only incorporate gaseous components such as NO, $HNO_3$, $SO_2$, HCl, and $NH_3$ but also engage in oxidative and acid-base reactions within the aqueous phase. The concentration of substances within the fog water is intricately linked to the liquid water content, which is under the influence of water vapor condensation and evaporation processes. When the concentration of gaseous sulfuric acid becomes sufficiently high, homogeneous nucleation occurs, leading to the formation of pure sulfuric acid particles or droplets. Some of these sub-micron droplets of $H_2SO_4$ can be neutralized by ammonia gas, primarily originating from agricultural activities (Saxena et al. 2020). If the acidic particles attain a large enough size, they can settle to the Earth's surface through sedimentation. Otherwise, turbulent transport facilitates the dry deposition of these particles to the surface.

For the formation of cloud droplets or fog water droplets, the presence of condensation nuclei is crucial. These nuclei can include particles like soil dust, sea salt aerosols, fly ash, or secondary aerosols. The chemical composition of the fog water is, to some extent, influenced by the chemical composition of the condensation nuclei from which it originated. Some components of the nucleus are soluble, and these soluble chemicals will dissolve into the aqueous phase of the fog water. In the atmosphere, certain gases will also partition into the aqueous phase and undergo various chemical reactions (Aggarwal et al. 2013).

### 4.2.2 *Sources of Chemical Constituents in Acidic Fog*

Recent studies have elucidated the diverse anthropogenic and natural origins of major chemical constituents in acidic fog, focusing on sulfuric acid, nitric acid, ammonium, organic acids, trace metals, and other inorganic ions, shedding light on their environmental impacts. Sulfuric acid in acidic fog primarily arises from the oxidation of sulfur dioxide ($SO_2$), a byproduct of anthropogenic activities like fossil fuel combustion and industrial processes (Smith et al. 2020). Recent studies have reported increasing levels of sulfuric acid in urban areas, indicating a strong influence of industrial emissions (Zhang et al. 2020). Similarly, nitric acid is predominantly formed from the oxidation of nitrogen oxides ($NO_x$) released by combustion processes, with automobiles and power plants being significant sources (Seinfeld and Pandis 2016). A study in 2021 highlighted a rising trend of nitric acid concentrations in fog droplets, particularly in densely populated urban regions (Chen et al. 2021). On the other hand, Ammonium ions ($NH_4^+$) result from the neutralization of acidic species by ammonia ($NH_3$), originating from agricultural activities, industrial processes, and natural emissions (Aneja et al. 2019). Recent research have highlighted a notable increase in ammonium concentrations in fog water samples collected from regions with intensive agricultural practices (Zhang et al. 2022). Organic acids, including formic acid and acetic acid, have both biogenic and anthropogenic sources, with vehicular emissions and biomass burning contributing significantly (Ervens et al. 2011a, b). A study conducted by Zhu et al. 2020 demonstrated a surge in concentrations of organic acids in fog droplets in urban areas, indicating a direct link to anthropogenic activities. Trace metals such as aluminium, iron, and lead in acidic fog primarily originate from anthropogenic sources, including industrial emissions and vehicular activities (Barbante et al. 2017). Li et al. 2022 documented elevated levels of trace metals in fog water samples from industrial zones, underscoring their significant anthropogenic contribution. Other inorganic ions like chloride, fluoride, calcium, and magnesium are derived from various natural and anthropogenic sources, including sea spray, industrial emissions, and soil dust (Warneck 2000). Studies in 2019 revealed variations in the composition of inorganic ions in fog droplets, reflecting regional influences and anthropogenic activities (Wang et al. 2019). HCl is generated through the combustion and decomposition of organochlorine compounds, such as polyvinyl chloride, commonly found in refuse incineration. The acidification process is primarily initiated by the combustion of fossil fuels, resulting in the emission of sulfur dioxide ($SO_2$) and nitrogen oxides (NOx). Normally inert nitrogen and oxygen combine during high-temperature combustion to form nitrogen dioxide ($NO_2$), which is later oxidized to form nitric acid ($HNO_3$) (Table 4.1).

These gases are released into the atmosphere and dispersed through turbulent mixing and advection. While gas-phase reactions are not very fast under typical conditions, $SO_2$ and $NO_2$ can undergo oxidation to generate gaseous sulfuric acid ($H_2SO_4$) and nitrate radicals ($RNO_2$), respectively while nitric acid remains in the

**Table 4.1** Sources of chemical constituents in acidic fog

| S.No. | Sources | Major components | Nature of area |
|---|---|---|---|
| 1. | Industrial emissions | Sulfur dioxide ($SO_2$), Nitrogen oxides ($NO_x$) | Urban areas |
| 2. | Power plant discharges | Sulfur dioxide ($SO_2$), particulate matter | Industrial zones |
| 3. | Automobile exhaust | Nitrogen oxides (NOx), Carbon monoxide (CO) | High traffic areas |
| 4. | Agricultural activities | Ammonia ($NH_3$), pesticides | Rural and farming regions |
| 5. | Biomass burning | Organic compounds, particulate matter | Forested areas |
| 6. | Residential heating | Sulfur dioxide ($SO_2$), residential Biomass burning | Residential areas |
| 7. | Natural sources | Volcanic activity, forest emissions | Varied, depending on region |

gas phase. The swiftest production of sulfuric acid occurs within clouds. The moderately soluble $SO_2$, taken up by clouds, undergoes dissolution into the liquid phase, transforming into $H_2SO_4$ (comprising $H^+$, $HSO_4^-$, and $SO_4^{2-}$) due to chemical transformations within this assembly of aqueous droplets. Analogous chemical processes apply for converting nitrogen oxides into nitric acid, with clouds predominantly accumulating both acids. Naturally, these acids are reintroduced to the Earth's surface through precipitation events (Hoffmann 1984).

## 4.3 Sampling and Analytical Techniques in Acidic Fog

Methods for sampling and analyzing fog acidity involve several steps to accurately assess the concentration of acidic components within fog droplets. In the early days chemical makeup data were acquired by examining separate aerosol particles through analysis (Bigg et al. 1974; Mamane and de Pena 1978; Lodge 1986; Ganor and Pueschel 1988). pH determination involved the application of pH paper onto individual stages of a cascade impactor. Analysis of particle and droplet composition was conducted using localized chemical reactions and electron microscopy methods. For droplets, direct collection occurred on transmission electron microscope (TEM) grids positioned across each impact or stage. These grids were subjected to scrutiny for structural characteristics as well as levels of sulfate and nitrate content. There have been several developments in the collection and analysis techniques since that have paved the way for faster and more efficient experimentation.

### *4.3.1 Collection Methods for Acidic Fog Samples*

The most frequent method for studying chemical processes within fog droplets is to collect a sample of fog water and analyze its chemical composition. This method has several disadvantages, the most important of which is the averaging that happens during sample collecting (Ogren et al. 1989). Droplet size averaging is one sort of averaging in which the gathered sample incorporates contributions from droplets of widely varied sizes. The collection effectiveness as a function of droplet size varies greatly depending on the sampler principle and manufacturer design. In many circumstances, this information is unavailable. As a result, almost all of the cloud and fog water samplers now in use do not allow for the investigation of droplet-size-dependent physicochemical processes in the atmosphere. Nonetheless, a number of researchers have collected cloud and fog water using passive samplers (Unsworth and Fowler 1988).

Techniques for collecting acidic fog samples involve strategies to capture and preserve fog droplets while maintaining their original acidity (Fig. 4.1). In a study conducted by Michael Hoffmann (1984), to procure cloud water and fog water samples, a device termed the rotating arm collector (RAC) was innovated. This collector bears a resemblance to a helicopter blade or propeller, with sampling bottles securely affixed to its outer edges. Operating at approximately 1800 rotations per minute (rpm), the blades of the collector sweep through the air, entrapping fog water droplets via inertial impaction. While smaller particles or inactive condensation nuclei in the fog tend to follow the fluid dynamics around the revolving blade, larger fog water droplets, due to their momentum, collide with the blade's leading edge and are directed inward. Upon entry, these droplets encounter a tube or cylinder, propelled outward into the sampling bottles by centrifugal force. Another sort of

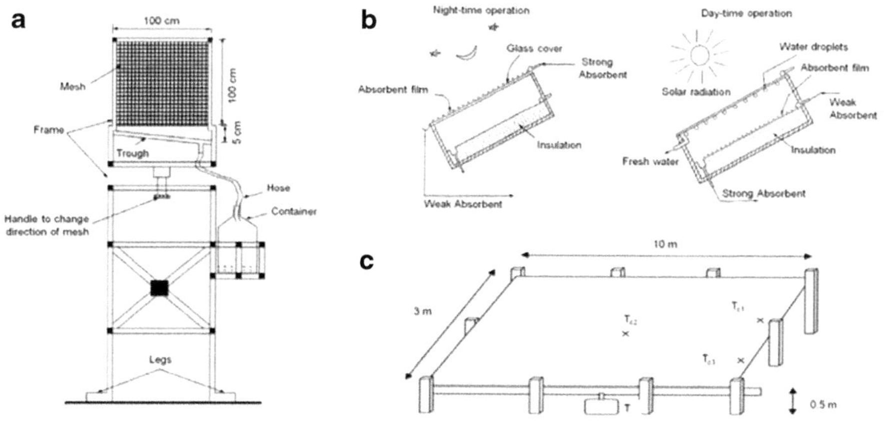


**Fig. 4.1** Apparatus for fog capturing. (**a**) Mesh Collector System and piping (**b**) Chemical Fog Collector (left: night collection, right: daytime release)) (**c**) Flat fog condenser (left: night collection, right: daytime release)

averaging is over time and space because all collectors require a set length of time to collect enough samples for chemical analysis. For a study conducted by Khemani et al. 1985 to study the influence of particulates on fog water in Delhi, a custom-designed apparatus was developed, comprising a stainless-steel cooling chamber and a collector that utilizes impaction methodology to gather fog droplets. This equipment was employed for the purpose of collecting water. Ambient fog-laden air was drawn into the system at a controlled flow rate. Within the collector, fog droplets were directed to impact onto a surface while being exposed to a temperature of −15 °C, maintained by the cooling chamber. Subsequently, the resulting melted fog droplets were amassed and subjected to analysis for both predominant ionic constituents and pH levels. Recent studies have witnessed significant progress in fog water collection systems, optimizing the efficiency of sample acquisition. Advanced fog water collectors, such as the Caltech Active Strand Cloud Water Collector (CASCC), offer improved collection efficiencies and reduced losses due to evaporation, thereby enhancing the accuracy of chemical analysis (Smith et al. 2020). Active fog collectors actively force air through a collection surface to promote droplet capture. These systems have gained popularity due to their ability to achieve higher collection efficiencies compared to passive collectors. Studies by Chen et al. (2021) and Khoury et al. (2023) utilized active fog collectors with specialized surfaces to enhance fog droplet capture, resulting in more representative samples. Determining the appropriate sampling duration is crucial for capturing a representative fog water sample. Recent research has emphasized the importance of extended sampling durations to ensure sufficient volume for chemical analysis. Studies by Li et al. (2021) conducted prolonged fog water sampling to obtain adequate quantities for accurate quantification of chemical constituents. Proper handling of fog water samples is also essential to prevent contamination and ensure data integrity. Recent studies have implemented meticulous protocols for sample preservation, storage, and transport. Zhang et al. (2022) emphasized the use of clean, acid-washed containers and immediate sample storage at low temperatures to preserve sample integrity. Proper handling of fog water samples is essential to prevent contamination and ensure data integrity. Recent studies have implemented meticulous protocols for sample preservation, storage, and transport. Zhang et al. (2022) emphasized the use of clean, acid-washed containers and immediate sample storage at low temperatures to preserve sample integrity. Certain studies that explore the potential of harvesting water from fog have devised simple, low-cost systems consisting of a fine mesh. In these fog collectors, water can be acquired by capturing fog carried by the wind through interception. The apparatus, resembling a sizable net, intercepts water droplets, causing them to adhere to its fibers. As water accumulates on the net, individual droplets coalesce, resulting in larger droplets that descend due to gravity toward a trough or channel positioned at the lower section of the structure (IDRC CIID). Such economical systems can also be employed in the collection of acidic fog samples that would reduce the cost of the entire experimentation.

### 4.3.2 Analytical Techniques for Characterizing Chemical Constituents

Acidic fog results from the interaction of human-made and natural emissions with atmospheric moisture, presenting complex challenges for the environment and human well-being. The intricate chemical composition of acidic fog, primarily stemming from acidic elements like sulfuric acid ($H_2SO_4$) and nitric acid ($HNO_3$), demands thorough analytical methods for precise characterization. Grasping the origins, changes, and potential impacts of these chemical components is essential for effective environmental management, policy development, and safeguarding public health. The intricate interplay of diverse atmospheric elements contributes to the unique chemistry of fog. While fog droplets encompass both inorganic and organic substances, discerning and quantifying these elements requires advanced analytical approaches that offer insights into their origins, quantities, and potential reactivity. A wide range of analytical tools, including traditional ones like ion chromatography (IC), as well as cutting-edge spectroscopic approaches, isotope analysis, and particle sizing techniques, contribute to the comprehensive understanding of the chemical contents found in acidic fog. The pH value and electric conductivity are key fog properties that represent both the total ionic content and the acid/base balance in the fog water. Fog is termed acidic when the pH is less than 5 (equilibrium with background $CO_2$ produces a pH of 5.6, and other naturally occurring acids often drop pH to 5 and basic when the pH rises to 6 or higher (Collett Jr et al. 2002). The balance of acidic and basic inputs governs fog pH (Jacob et al. 1985; Collett Jr et al. 1999). The presence of huge amounts of ammonia and calcium in the atmosphere raises the pH of fog water. In general, the collected fog samples are weighed and taken on-site for pH (pH meter) and electric conductivity (EC) measurements. After filtering the samples through a membrane, inorganic ions and organic acids are analyzed using ion chromatography, which includes a separation column, a guard column, and a conductivity detector. Prior to and after examination, samples are typically refrigerated.

Mass spectrometry has emerged as a cornerstone technique for fog droplet analysis due to its unparalleled sensitivity and versatility. Recent studies employing mass spectrometry have enabled the identification and quantification of various ions, including sulfuric acid, nitric acid, and organic acids in fog droplets (Chen et al. 2021). High-resolution mass spectrometry has further enhanced our ability to distinguish ions with similar mass-to-charge ratios, providing unprecedented insights into fog chemistry. Ion chromatography remains a pivotal technique for analyzing anions and cations in fog water samples. Recent advancements in ion chromatography have led to increased precision and accuracy in quantifying major ionic components. Study by Li et al. (2021) exemplify the application of ion chromatography in quantifying sulfate, nitrate, ammonium, and other ions in fog droplets, providing crucial data for understanding fog acidity. Another such analytical technique, NMR spectroscopy, is gaining prominence as a valuable technique for identifying and quantifying chemical compounds in fog droplets. Recent studies have utilized NMR to investigate molecular structures and compositions, providing insights into organic and inorganic species present in fog samples (Zhang et al. 2022). Recent research by Khoury et al. (2023) and Wang et al. (2019) showcased the applicability of capillary electrophoresis in characterizing fog water chemistry, particularly in urban environments. Capillary electrophoresis is emerging as a complementary technique for analyzing ions in fog water samples. Its high separation efficiency and low sample volume requirements make it a promising tool for studying ionic composition.

## 4.4 Major Chemical Constituents in Acidic Fog

### 4.4.1 *Sulfuric Acid*

Sulfuric acid ($H_2SO_4$) is a key component of acidic fog. It is produced as a byproduct of the oxidation of sulfur dioxide ($SO_2$) in the environment. Sulfur dioxide emissions are mostly caused by anthropogenic activities such as the combustion of fossil fuels and industrial processes. Sulfur dioxide conducts a sequence of chemical processes that result in the creation of sulfuric acid in the presence of sunshine and other oxidizing factors. This acid considerably contributes to the overall acidity of fog droplets, corroding flora, aquatic habitats, and materials (Smith et al. 2020).

### 4.4.2 *Nitric Acid*

Another important acidic chemical is found in fog droplets is nitric acid ($HNO_3$). It is mostly caused by nitrogen oxides ($NO_x$), which are formed during combustion processes in automobiles, power plants, and industrial sites. Nitrogen oxides are released as nitrogen dioxide ($NO_2$) and nitrogen monoxide (NO). These chemicals react with water vapor and other atmospheric elements to create nitric acid in the atmosphere. Its presence increases the acidity of fog, especially in urban regions with high amounts of nitrogen oxide emissions (Seinfeld & Pandis 2016). This exacerbates acidic fog's negative impacts on plants, aquatic habitats, and construction materials.

### 4.4.3 *Ammonium*

Acidic fog is made up entirely of ammonium ions. They are the result of ammonia ($NH_3$) neutralizing acidic species such as sulfuric and nitric acid, which is generated by agricultural operations, industrial processes, and natural sources. Ammonium is important in buffering the pH of fog droplets, which influences their overall acidity and reactivity with the environment (Aneja et al. 2019).

### 4.4.4 Organic Acids

Organic acids, such as formic acid (HCOOH) and acetic acid ($CH_3COOH$), contribute significantly to acidic fog. They are produced by both biogenic and anthropogenic sources, including vegetative emissions and incomplete combustion of fossil fuels (Ervens et al. 2011a, b). Organic acids add to the overall acidity of fog droplets and can increase trace metal mobilization, influencing aquatic ecosystems and soil pH.

### 4.4.5 Metal Traces

Acidic fog can carry trace metals from the atmosphere to terrestrial and aquatic habitats, including aluminium (Al), iron (Fe), and lead (Pb). These metals, which are frequently related to anthropogenic activity, can harm plants, aquatic creatures, and soil quality. These metals' enhanced solubility in acidic settings exacerbates their environmental impact.

### 4.4.6 Other Inorganic Ions

In addition to the aforementioned primary ingredients, acidic fog can contain a variety of additional inorganic ions such as chloride ($Cl^-$), fluoride ($F^-$), and other cations such as calcium ($Ca^{2+}$) and magnesium ($Mg^{2+}$). These ions help to maintain the ion balance and electrical conductivity of fog droplets.

## 4.5 Factors Influencing the Chemical Constituents of Acidic Fog

### 4.5.1 Meteorological Factors

#### 4.5.1.1 Temperature

Temperature is a fundamental determinant of the chemical constituents within acidic fog. Its influence on chemical reaction rates, solubility, vapor pressure, chemical equilibria, and deposition processes imparts significant variability to the composition of fog droplets. One of the foremost ways in which temperature influences the composition of acidic fog is through its effect on chemical reaction rates. Higher temperatures accelerate chemical reactions, leading to an increased rate of formation of acidic compounds like sulfuric acid ($H_2SO_4$) and nitric acid ($HNO_3$) (Wang et al. 2018). This phenomenon is particularly pronounced in regions characterized by elevated temperatures, where reactions proceed at a heightened pace (Andreae et al. 1986). Conversely, lower temperatures impede chemical reactions, resulting in a decelerated formation of acidic compounds within fog droplets (Liljestrand and Morgan 1978). Temperature exerts a significant influence on the solubility of gases and the reactivity of dissolved species within fog droplets. In warmer conditions, the enhanced solubility of gases and heightened reactivity of dissolved species promote increased dissolution of pollutants. This leads to the formation of more complex chemical species within the fog droplets (Jaeschke et al. 2002). In contrast, colder temperatures reduce the solubility of certain gases and ions in fog droplets, potentially resulting in a decrease in the concentration of specific chemical constituents (Waldman et al. 1982).

The vapor pressure of volatile compounds is directly influenced by temperature. Higher temperatures elevate the vapor pressure of these compounds, facilitating their volatilization from fog droplets back into the gas phase. This process has the potential to alter the concentration of certain chemicals in the fog (Andreae et al. 1986). Conversely, lower temperatures diminish the vapor pressure of volatile compounds, diminishing their propensity to evaporate from fog droplets. Temperature-induced changes in chemical equilibria further contribute to the dynamic composition of acidic fog. Alterations in temperature can shift equilibria in chemical reactions involving acids and bases. For instance, higher temperatures may favor the formation of specific products over others, potentially leading to shifts in pH levels and modifications in the chemical makeup of the fog. Conversely, lower temperatures may induce equilibria shifts in the opposite direction, influencing the distribution of chemical species within the fog. Temperature impacts the rates of deposition and adsorption of chemical species onto surfaces. Elevated temperatures accelerate these processes, potentially influencing the availability of certain compounds in the atmosphere and their potential incorporation into fog droplets. Conversely, colder temperatures slowdown the rates of deposition and adsorption, potentially leading to alterations in the composition of fog droplets (Bikkina et al. 2017).

#### 4.5.1.2 Humidity

Humidity levels play a critical role in the formation and size of fog droplets. In conditions of high humidity, ample water vapor is available for condensation. This fosters the creation of fog droplets, acting as carriers for dissolved gases and particles (Warneck 2000). Moreover, higher humidity can lead to the formation of smaller fog droplets due to heightened condensation rates. Conversely, lower humidity levels restrict the availability of water vapor for condensation. This may impede the formation of fog droplets or result in larger droplets, potentially influencing the concentration of chemical constituents in the fog. Humidity exerts a significant influence on the solubility of gases in water. High humidity augments the solubility of gases, allowing more atmospheric gases to dissolve in fog droplets. This results in higher concentrations of dissolved species, a phenomenon particularly pertinent for gases like sulfur dioxide ($SO_2$) and ammonia ($NH_3$), both of which contribute to the acidity of fog. Conversely, lower humidity diminishes the solubility of gases in water, limiting the number of certain gases that can dissolve in fog droplets. This may lead to lower concentrations of specific chemical constituents.

Humidity levels impact the occurrence of chemical reactions within fog droplets. In environments characterized by increased humidity, chemical reactions involving acidic and basic species occur more readily. This can potentially lead to alterations in the composition of the fog. Conversely, lower humidity may slowdown or even limit chemical reactions within fog droplets, thereby influencing the concentration and composition of chemical species. Humidity levels govern the interactions between aerosol particles and fog droplets (Aiken et al. 2008). High humidity fosters increased interactions, resulting in the absorption of aerosol constituents into the fog droplets. This has the potential to alter the chemical composition of the fog. In contrast, lower humidity levels may reduce the interactions between aerosol particles and fog droplets, potentially impacting the incorporation of aerosol constituents into the fog. Humidity levels affect the rates of deposition and evaporation of chemical species in fog droplets. Elevated humidity levels reduce the rates of evaporation, allowing more chemical constituents to remain in the liquid phase. This potentially leads to higher concentrations of dissolved species. Conversely, lower humidity may increase the rates of evaporation, leading to a concentration of chemical constituents as the droplets shrink. This can influence the overall composition of the fog (Saxena et al. 2020).

#### 4.5.1.3 Wind Patterns

Wind patterns dictate the trajectory of air masses, acting as conduits for the movement of pollutants from their source regions to other areas. This means that air masses carrying pollutants from regions with high emissions, such as industrial zones, can be transported to areas where fog formation occurs. As a result, pollutants are introduced into fog-forming regions, potentially altering the composition of the resulting fog. In addition to local transport, strong winds have the capacity to facilitate the long-distance movement of pollutants. This can lead to the infusion of a diverse array of chemical species into fog-forming regions. The introduction of such a varied chemical mixture has the potential to significantly impact the composition of acidic fog (Aiken et al. 2008). Areas situated downwind from major emission sources are at a higher risk of experiencing fog with elevated concentrations of specific pollutants. Urban centers or regions near industrial facilities are particularly susceptible to fog containing higher levels of sulfur dioxide ($SO_2$) and nitrogen oxides ($NO_x$). The proximity to emission sources directly influences the composition of fog in these areas. Higher wind speeds play a crucial role in the dilution of pollutants in the atmosphere. This means that the concentration of certain chemical constituents in fog droplets may be reduced. The increased dispersion of pollutants contributes to a more uniform distribution of chemical species in fog. Wind patterns have a substantial impact on the dispersion of pollutants in the atmosphere. Strong winds can disperse pollutants over a larger area, potentially leading to a more even spread of chemical species in fog (Zhang et al. 2024). This dispersion effect has significant implications for the overall composition of fog. Strong winds enhance the process of dry deposition, whereby pollutants are directly deposited onto surfaces. This can result in a reduction in the concentration of certain chemical constituents available for incorporation into fog droplets. The process of dry deposition alters the availability of pollutants for inclusion in fog (Ervens et al. 2011a, b). Wind patterns influence the uptake of gases by vegetation. Vegetation acts as a sink for certain pollutants, potentially altering their availability for inclusion in fog droplets. This interaction between vegetation and pollutants is a significant factor in determining the composition of acidic fog. In coastal areas, wind patterns significantly influence the incorporation of sea spray and marine emissions into fog droplets. This introduces additional chemical constituents, such as sea salts and marine-derived organic compounds, into the fog. These unique constituents contribute to the distinct composition of coastal acidic fog. Wind patterns play a critical role in facilitating the transport of pollution from one region to another. This introduces chemical constituents from distant sources, potentially influencing the composition of fog in receiving regions. The regional influence of wind patterns on fog composition underscores the interconnectedness of atmospheric systems (Seinfeld and Pandis 2016).

## *4.5.2 Anthropogenic Factors*

### 4.5.2.1 Industrial Emissions

Industrial activities have long been recognized as major contributors to air pollution, emitting a complex mixture of pollutants into the atmosphere. These emissions play a significant role in shaping the chemical constituents of acidic fog. One of the most prominent contributions of industrial emissions to acidic fog composition is the release of primary pollutants. Sulfur dioxide ($SO_2$) emissions, emanating primarily from processes like fossil fuel combustion and metals melting, serve as crucial precursors to sulfuric acid ($H_2SO_4$), a cornerstone of acidic fog chemistry (Seinfeld and Pandis 2016). Upon release into the atmosphere, $SO_2$ undergoes chemical transformations, culminating in the formation of sulfuric acid in fog droplets. Similarly, nitrogen oxides ($NO_x$), comprising nitrogen dioxide ($NO_2$) and nitric oxide (NO), are released by industrial activities. These compounds are instrumental in the formation of nitric acid ($HNO_3$), which can be incorporated into fog droplets, intensifying the acidity of the fog (Finlayson-Pitts and Pitts 2000). Particulate matter (PM) emissions, a common byproduct of industrial processes, may include sulfate aerosols. These fine particles act as nuclei for fog droplet formation and contribute significantly to the chemical makeup of fog (Clegg et al. 2016). Industrial processes release an array of VOCs into the atmosphere. These compounds, comprising hydrocarbons and other organic substances, undergo reactions in the presence of sunlight and other pollutants, leading to the formation of secondary organic aerosols (SOA) (Hallquist et al. 2009). SOA plays a pivotal role in influencing fog composition when incorporated into fog droplets.

Certain industrial processes, notably metal smelting and the combustion of fossil fuels, release heavy metals into the atmosphere. These metals, including lead (Pb), cadmium (Cd), and mercury (Hg), among others, have the potential to be incorporated into fog droplets, thereby contributing to the chemical composition of acidic fog (Huang et al. 2012). Industrial emissions can introduce acidic gases like hydrochloric acid (HCl) and hydrofluoric acid (HF) into the atmosphere. These acidic species can directly augment the acidity of fog when incorporated into fog droplets (McHenry and Dennis, 1994). Industrial activities related to fertilizer production and combustion processes release ammonia ($NH_3$) into the atmosphere. $NH_3$ can act as a neutralizing agent for acidic species in fog droplets, exerting a notable influence on the pH and chemical composition of the fog (Seinfeld and Pandis 2006). Sulfate aerosols, released through various industrial processes, serve as cloud condensation nuclei and wield influence over fog droplet formation. These aerosols, when integrated into fog, contribute to the overall chemical composition of the fog (Ramanathan et al. 2001). Certain industrial processes release organic acids and aldehydes into the atmosphere. These compounds undergo chemical reactions in the atmosphere and may become components of fog droplets, further influencing the chemical makeup of acidic fog (Ervens et al. 2011a, b). Industrial emissions are integral to the complex array of chemical reactions transpiring in the atmosphere. These reactions lead to the formation of secondary pollutants, including sulfuric acid, nitric acid, and organic compounds, which are subsequently incorporated into fog droplets (Seinfeld and Pandis 2016).

#### 4.5.2.2 Agricultural Practice

Agriculture is a vital sector of the global economy, providing food, fiber, and other essential resources. However, agricultural practices can have unintended environmental consequences, including effects on atmospheric chemistry. One such consequence is the influence of agricultural activities on the chemical constituents of acidic fog. Agricultural practices are synonymous with the application of fertilizers, which serve as a significant source of ammonia ($NH_3$) emissions. $NH_3$, when released into the atmosphere, has the remarkable capacity to neutralize acidic species in fog droplets, thereby influencing both the pH and chemical composition of the fog (Dentener et al. 2006). Additionally, the management of livestock, integral to agriculture, produces manure that releases ammonia into the atmosphere, further contributing to the neutralization of acidic species in fog droplets.

Tillage and plowing, essential components of agricultural activities, often generate substantial amounts of dust and particulate matter. These minute particles can act as nuclei for fog droplet formation, playing a pivotal role in influencing the chemical composition of acidic fog. The inclusion of these particles can lead to alterations in the fog's chemical makeup. Agriculture is not limited to the cultivation of crops alone; it encompasses an entire ecosystem. Some plants naturally emit volatile organic compounds (VOCs) as part of their metabolic processes. These VOCs, once released into the atmosphere, can undergo a series of complex reactions. Under the influence of sunlight and other pollutants, they can transform into secondary organic aerosols (SOA) (Hallquist et al. 2009). These SOA particles can become integral components of fog droplets, thereby influencing the composition of acidic fog. Modern agricultural practices frequently involve the application of pesticides and herbicides to manage pests and ensure crop yields. However, these chemical agents can leave residues in the environment. Over time, these residues may be transported into the atmosphere and potentially become components of fog droplets, affecting the chemical makeup of acidic fog.

Fertilizer application is a cornerstone of agriculture, significantly enhancing soil fertility. However, the application of fertilizers can alter the pH of the soil. Changes in soil pH can have cascading effects on the composition of chemicals released into the atmosphere. These altered chemical compositions may eventually find their way into fog droplets, further influencing the chemical constituents of acidic fog. Agricultural landscapes often feature substantial vegetation, and certain plants emit biogenic volatile organic compounds (VOCs). These emissions contribute to the pool of organic compounds present in the atmosphere. Consequently, these compounds have the potential to be incorporated into fog droplets, impacting the chemical composition of acidic fog. Agricultural operations often include livestock, which can release ammonia and methane into the atmosphere. Ammonia, in particular, can act as a potent neutralizing agent for acidic species in fog droplets. Consequently, the presence of livestock can influence the pH and chemical composition of acidic fog (Reay et al. 2012). In some agricultural practices, such as the burning of crop residues, various compounds are released into the atmosphere. These compounds may include carbonaceous particles, which can potentially become components of fog droplets and contribute to the chemical composition of acidic fog.

## 4.6 Health and Environmental Impacts of Acidic Fog

### *4.6.1 Effects on Human Health*

The presence of acidic fog has been associated with a range of adverse health effects on humans. Inhaling acidic fog can lead to irritation of the respiratory tract, resulting in symptoms such as coughing, sore throat, and shortness of breath. Individuals with pre-existing respiratory conditions, like asthma or bronchitis, are particularly susceptible to these effects (Bell et al. 2019). Furthermore, people with chronic respiratory diseases may experience exacerbation of their conditions when exposed to acidic fog. The acidic particles in the fog can intensify inflammation, leading to increased difficulty in breathing. Contact with acidic fog can cause eye irritation, including redness and discomfort. Prolonged exposure may lead to more serious eye conditions. Additionally, direct contact with acidic fog or rain can cause skin irritation, especially in individuals with sensitive skin. Acidic fog can exacerbate cardiovascular conditions. Fine acidic particles can enter the bloodstream through inhalation, potentially worsening heart conditions and increasing the risk of heart attacks. This is of significant concern for individuals with pre-existing cardiovascular diseases. Prolonged exposure to acidic fog may weaken the immune system, making individuals more susceptible to infections and illnesses (Chakraborty et al. 2023). The weakened immune response can lead to a higher incidence of respiratory infections and other health complications. While studies on the neurological effects of acidic fog are limited, some research suggests potential impacts on the nervous system. Exposure to acidic fog may lead to cognitive impairments and other nervous system issues. However, further research is needed to fully understand the extent of these effects. Prolonged and continuous exposure to acidic fog over extended periods may contribute to the development of chronic health conditions. This includes respiratory and cardiovascular diseases, which can have lasting implications on an individual's overall health and well-being.

### *4.6.2 Impacts on Ecosystems and Vegetation*

Acidic fog, characterized by heightened concentrations of acid aerosols and droplets, exerts profound effects on ecosystems and vegetation. These impacts stem from the deposition of acidifying substances onto soils and plant surfaces, subsequently altering soil chemistry, nutrient availability, and overall plant health. Acidic fog contributes to soil acidification by depositing acids such as sulfuric acid ($H_2SO_4$) and nitric acid ($HNO_3$). This leads to a reduction in soil pH levels, which can adversely affect soil microorganisms and nutrient availability (Gundersen et al. 1998). The subsequent leaching of essential nutrients like calcium, magnesium, and potassium from the soil due to acidic conditions results in nutrient imbalances and deficiencies in plants (Johnson and Lindberg 1992). Altered nutrient uptake and availability, a consequence of changed soil chemistry, can lead to varying accessibility of essential nutrients. Some nutrients become less available, while others, like aluminium and manganese, may become more accessible, potentially leading to toxic effects (Likens et al. 1996). Nutrient deficiencies in plants, a direct consequence of acidic conditions, significantly impact their growth, development, and overall health (Eastaugh et al. 1998). Acidic conditions can mobilize toxic elements like aluminium and heavy metals from the soil. This results in potential harm as these elements can be taken up by plants and accumulates in their tissues (Isil et al. 2022). Acidic conditions impair root functionality, leading to diminished growth. This limitation affects the plant's ability to access water and nutrients from the soil (Fenn et al. 2006). Acidic fog directly injures plant leaves, causing symptoms like necrosis, discoloration, and reduced photosynthetic efficiency (McLaughlin et al. 2007). Acidic fog-induced shifts in vegetation composition favor acid-tolerant species over sensitive ones. This can have cascading effects on ecosystem structure and function (Padgett et al. 1997). Acidic conditions influence the composition and activity of soil microorganisms, potentially disrupting crucial nutrient cycling processes (Gundersen et al. 1998). Sensitive ecosystems, like wetlands and high-elevation areas, are particularly vulnerable to the impacts of acidic fog due to their lower natural buffering capacities (Sullivan et al. 2011). Acidic fog contributes to the acidification of water bodies when deposited acids are washed in. This can have detrimental effects on aquatic life, including fish, amphibians, and invertebrates (Nieminen et al. 2012). Changes in vegetation and aquatic ecosystems due to acidic fog have downstream effects on wildlife, disrupting food sources and shelter for various species (Mitchell et al. 1999). The impacts of acidic fog on ecosystems and vegetation are far-reaching and multifaceted. The severity of these effects depends on various factors, including the duration and frequency of exposure, soil buffering capacity, and specific ecosystem characteristics. Monitoring and mitigating the effects of acidic fog are imperative to preserving the health and functionality of these vital components of our environment.

## 4.7 Mitigation Strategies and Future Directions

### *4.7.1 Pollution Control Measures*

One of the foundational pillars in mitigating the impacts of acidic fog lies in the implementation of stringent emission reduction policies. This necessitates setting and enforcing rigorous standards for industrial and vehicular emissions, with a particular focus on curbing pollutants like sulfur dioxide ($SO_2$) and nitrogen oxides ($NO_x$). Industrial facilities and power plants must adhere to prescribed limits, while the promotion of cleaner transportation technologies is crucial for reducing mobile source emissions (Nault et al. 2020). Continued research and development in pollution abatement technologies are pivotal in the battle against acidic fog. Innovations in flue gas desulfurization (FGD) systems, catalytic converters, and cleaner energy production methods hold the potential to significantly curtail the emission of precursor pollutants responsible for the formation of acidic fog (Nath and Yadav 2018). Shifting from fossil fuel-dependent energy generation to renewable sources, including solar, wind, and hydroelectric power, presents a promising avenue for mitigating the impacts of acidic fog. This transition not only reduces emissions associated with combustion processes but also promotes sustainability and resilience in the face of climate change (Cheng et al. 2020). Effective urban planning plays a critical role in minimizing pollution hotspots and mitigating the concentration of pollutants contributing to acidic fog. This involves the strategic placement of industrial zones, the creation of green spaces, and the implementation of sustainable urban design practices, which collectively disperse emissions and alleviate their localized impacts (Chen et al. 2019).

Agricultural practices are significant contributors to the formation of acidic fog, particularly through ammonia ($NH_3$) emissions. Implementing precision agriculture techniques and optimizing fertilizer application practices can substantially reduce these emissions, thereby mitigating their contribution to acidic fog (Chen et al. 2018). Raising public awareness about the causes and consequences of acidic fog is fundamental in garnering support for emission reduction policies. Educating communities about individual actions that can reduce pollution and advocating for policy changes are essential steps in fostering a collective commitment to mitigating the impacts of acidic fog (Wang et al. 2019). Continued refinement of atmospheric models and monitoring networks is indispensable for accurately predicting and characterizing acidic fog events. High-resolution models enable the identification of pollution hotspots and provide invaluable insights for refining mitigation efforts (Li et al. 2011). Comprehensive studies on the biogeochemical cycling of key elements in ecosystems affected by acidic fog offer crucial insights into its long-term impacts. By understanding how acidic fog influences nutrient availability, soil health, and plant physiology, targeted mitigation strategies can be developed to protect vulnerable ecosystems (Ali et al. 2004). A rigorous assessment of implemented mitigation measures is imperative to gauge their effectiveness. This involves regular monitoring of air quality, deposition rates, and ecosystem health. Such assessments provide crucial feedback for refining interventions and adapting strategies as needed (Yu et al. 2017). Given the transboundary nature of acidic fog, international cooperation and policy harmonization are essential. Collaborative efforts to establish consistent emission reduction policies prevent the movement of pollutants across borders and promote a collective commitment to mitigating this environmental challenge (Wang et al. 2018). Efforts to combat climate change offer an additional avenue for indirectly reducing the formation of acidic fog. Transitioning to low-carbon economies and reducing greenhouse gas emissions not only addresses global climate concerns but also leads to co-benefits in air quality improvement (Chen et al. 2017).

## 4.8 Conclusion

This chapter offers a thorough summary of the approaches used to describe the chemical components of acidic fog. The variety of analytical methods demonstrates the field's multidisciplinary nature, with developments in mass spectrometry, ion chromatography, and spectroscopic techniques driving the development. Geographic differences in fog composition demonstrate the important role that local emissions play in determining the chemical composition of fog droplets, with industrial areas displaying different signs of anthropogenic contaminants. The importance of these species in the development and composition of acidic fog is highlighted by the dominance of sulfate ions as well as the constant presence of nitrate and ammonium ions. The complicated interactions between climatic factors, emission patterns, and atmospheric chemistry are further highlighted by seasonal and temporal fluctuations in fog composition. In addition to consolidating current knowledge, this chapter also outlines research needs, particularly when it comes to new analytical techniques and the requirement for extensive, long-term monitoring studies. The knowledge gathered from this review contributes to a deeper comprehension of the chemical components of acidic fog and their effects on the environment and public health, serving as a crucial starting point for future investigations in this important area.

## References

Aggarwal SG, Kawamura K, Umarji GS, Tachibana E, Patil RS, Gupta PK. Organic and inorganic markers and stable C-, N-isotopic compositions of tropical coastal aerosols from megacity Mumbai: sources of organic aerosols and atmospheric processing. Atmos Chem Phys. 2013;13:4667–80. https://doi.org/10.5194/acp-13-4667-2013.

Aiken AC, DeCarlo PF, Kroll JH, Worsnop DR, Huffman JA, Cocherty K, et al. O/C and OM/OC ratios of primary, secondary, and ambient organic aerosols with high resolution time-of-flight aerosol mass spectrometry. Environ Sci Technol. 2008;42:4478–85. https://doi.org/10.1021/es703009q.

Ali K, Momin GA, Tiwari S, Safai PD, Chate DM, Rao PSP. Fog and precipitation chemistry at Delhi, North India. Atmos Environ. 2004;38:4215–222.

Ali K, Momin GA, Tiwari S, Safai PD, Chate DM, Rao PSP. Fog and precipitation chemistry at Delhi, North India. Atmospheric Environment, (2004);38:4215–22.

Andreae MO, Roehl A, Slinn WG. The chemical composition of fog and cloudwater. Atmos Environ. 1986;20(6):1103–17.

Aneja VP, et al. Atmospheric ammonia (NH3) in the southeastern United States: part I. A review of air concentrations and dry deposition fluxes. Curr Pollut Rep. 2019;5(2):98–124.

Baboolal LB, Pruppacher HR, Topalian JH. A sensitivity study of a theoretical model of SO2 scavenging by water drops in air. Journal of Atmospheric Sciences. 1981;38(4):856–870.

Barbante C, et al. Atmospheric chemistry and physics of airborne particulate matter at an urban site in northern Italy. J Environ Monit. 2017;9(5):495–504.

Baboolal LB, Pruppacher HR, Topalian JH. A sensitivity study of a theoretical model of SO2 scavenging by water drops in air. Journal of Atmospheric Sciences. 1981;38(4):856–70.

Bell ML, et al. Climate change, air quality, and human health. Am J Prev Med. 2019;57(6):779–85.

Bigg EK, Ono A, Williams JA. Chemical tests for individual submicron aerosol particles. Atmospheric Environment 1974;8(1):1–13.

Bikkina S, Andersson A, Ram K, Sarin MM, Sheesley RJ, Kirillova EN, et al. Carbon isotope-constrained seasonality of carbonaceous aerosol sources from an urban location (Kanpur) in the Indo-Gangetic Plain. J Geophys Res Atmos. 2017;122:4903–23. https://doi.org/10.1002/2016JD025634.

Bruck RI, Robarge WP, McDaniel A. Forest decline in the boreal montane ecosystems of the southern Appalachian Mountains. Water, Air, and Soil Pollution. 1989;48:161–80.

Bridges KS, Jickells TD, Davies TD, Zeman Z, Hunova I. Aerosol, precipitation and cloud water chemistry observations on the Czech Krusne Hory plateau adjacent to a heavily industrialized valley. In: Atmospheric Environment 2002;36:353–360.

Chakraborty A, Gupta T, Mandariya AK, Tripathi S. Trace elements in ambient aerosols and size-resolved fog droplets: trends, enrichment, and risk assessment. Heliyon. 2023;9(6):e16400.

Chen L, et al. Nitric acid in fog droplets: concentrations, influencing factors, and implications for atmospheric acidity. Atmos Res. 2021;257:105647.

Chen C, Wang Y, Jia J, Mao L, Meurk CD. Ecosystem services mapping in practice: A Pasteur's quadrant perspective. Ecosystem Services. 2019;40:101042.

Cheng Y, Voisin N, Yearsley JR, Nijssen B. Reservoirs modify river thermal regime sensitivity to climate change: A case study in the southeastern United States. Water Resources Research. 2020;56(6):e2019WR025784.

Chen, Q, Jiang, J.C, Wu, F. and Zou, M.Y, 2018. Performance evaluation of water mist with mixed surfactant additives based on absorption property. Procedia Engineering, 211, pp.85-93.

Chen H, Kang JN, Liao H, Tang BJ, Wei YM. Costs and potentials of energy conservation in China's coal-fired power industry: a bottom-up approach considering price uncertainties. Energy Policy. 2017;104:23–32.

Collett Jr JL, Bator A, Sherman DE, Moore KF, Hoag KJ, Demoz BB, Rao X, Reilly JE. The chemical composition of fogs and intercepted clouds in the United States. Atmos Res. 2002;64:29–40.

Collett Jr JL, Hoag KJ, Rao X, Pandis SN. Internal acid buffering in San Joaquin Valley fog drops and its influence on aerosol processing. Atmospheric Environment. 1999;33(29):4833–47.

Clegg SL, et al. An improved model for the composition of multiphase hygroscopic aerosols. Environ Sci Technol. 2016;50(13):6925–34.

Dentener FJ, et al. Nitrogen and sulfur deposition on regional and global scales: a multimodel evaluation. Glob Biogeochem Cycles. 2006;20(4)

Eastaugh CS, Gaydon DS, Walker TW. Acid fog and forest nutrient relations. Environ Sci Technol. 1998;32(17):276A–9A.

Ervens B, et al. Secondary Organic Aerosol Formation in CloudDroplets and Aqueous Particles (aqSOA): a review of laboratory, field, and model studies. Atmos Chem Phys. 2011a;11(21):11069–102.

Ervens B, et al. Secondary organic aerosol yields from cloud-processing of isoprene oxidation products. Geophys Res Lett. 2011b;38(14)

Fenn ME, Poth MA, Aber JD. Nitrogen excess in north American ecosystems: predisposing factors, ecosystem responses, and management strategies. Ecol Appl. 2006;6(3):706–33.

Finlayson-Pitts BJ, Pitts JN. Chemistry of the upper and lower atmosphere: theory, experiments, and applications. San Diego: Academic Press; 2000.

Ganor E, Pueschel RF. Composition of individual nitrate-containing particles in non-urban atmospheres of Colorado, USA and Tel Aviv, Israel. Water, Air, and Soil Pollution. 1988;42:169–81.

Georgii HW, Perseke C, Rohbock E. Investigation of the wet and dry deposition of acidic and heavymetal components. In Chemistry of Multiphase Atmospheric Systems. 1986;(pp. 227-248). Berlin, Heidelberg: Springer Berlin Heidelberg.

Gundersen P, Callesen I, de Vries W, Hertel O. Nitrate leaching in forest ecosystems is related to forest floor C/N ratios. Environ Pollut. 1998;102(1):403–7.

Hackney JD, Linn WS, Avol EL. Acid fog: effects on respiratory function and symptoms in healthy and asthmatic volunteers. Environ Health Perspect. 1989;79: Guangdong. China Journal of Rare 159–62.

Hallquist M, et al. The formation, properties and impact of secondary organic aerosol: current and emerging issues. Atmos Chem Phys. 2009;9(14):5155–236.

Hoffmann M. Acid fog. Engineering and Science. 1984;48(1):5–11.

Huang D, et al. Distribution, sources and assessment of heavy metal contamination in soils around an industrial area in Shaoguan, Guangdong. China Journal of Rare Earths. 2012;30(5):544–51.

Igawa M, Matsumura K, Okochi H. Fog water chemistry at Mt. Oyama and its dominant factors. Water, Air, and Soil Pollution. 2001;130:607–12.

Isil S, Collett Jr J, Lynch J, Weiss-Penzias P, Rogers CM. Cloud and fog deposition: Monitoring in high elevation and coastal ecosystems. The past, present, and future. Atmospheric Environment. 2022;274:118997.

Jacob DJ, Waldman JM, Munger JW, Hoffmann MR. A field investigation of physical and chemical mechanisms affecting pollutant concentrations in fog droplets. Tellus B, 1984;36(4):272–85.

Jacob DJ, Waldman JM, Munger JW, Hoffmann MR. Chemical composition of fogwater collected along the California coast. Environmental Science & Technology. 1985;19(8):730–36.

Jaeschke P, Andreae MO, Prospero JM, Pandis SN. The global distribution of organic acids in cloudwater. J Geophys Res Atmos. 2002;107(D19):4629.

Jaeschke P, Andreae MO, Prospero JM, Pandis SN. The global distribution of organic acids in cloudwater. Journal of Geophysical Research: Atmospheres. 2002:107(D19):4629.

Johnson DW, Lindberg SE. Atmospheric deposition and forest nutrient cycling: a synthesis of the integrated forest study. Ecological Studies. 1992;91:349–68.

Kapoor RK, Tiwari S, Ali K, Singh G. Chemical analysis of fog water at Delhi, North India. Atmos Environ. 1993;27A:2453–55.

Khemani LT, Momin GA, Rao PSP, Safai PD, Prakash P. Influence of alkaline particulates on the chemistry of fog water at Delhi, North India. Water Air Soil Pollut. 1987;34:183–189.

Khemani LT, Momin GA, Naik MS, Prakasa Rao PS, Kumar R, Ramana Murty Bh. V. Trace elements and sea salt aerosols over the sea areas around the Indian sub-continent. Atmospheric Environment 1985;19:277–84.

Khoury D, Millet M, Jabali Y, Delhomme O. Fog water: A general review of its physical and chemical aspects. Environments. 2023;10(12):224.

Lange OL, Green TGA, Meyer A, Zellner H. Water relations and CO2 exchange of the terrestrial lichen Teloschistes capensis in the Namib fog desert. Measurements during two seasons in the field and under controlled conditions. Flora 2006;201:268–80.

Li P, Li X, Yang C, Wang X, Chen J, Collett Jr JL. Fog water chemistry in Shanghai. Atmos Environ. 2011;45:4034–41.

Li Y, Aemisegger F, Riedl A, Buchmann N, Eugster W. The role of dew and radiation fog inputs in the local water cycling of a temperate grassland during dry spells in central Europe. Hydrology and Earth System Sciences, 2021;25(5):2617–48.

Li M, et al. Elevated concentrations of trace metals in fog water of a megacity: seasonal variation and source apportionment. Environ Sci Technol. 2022;56(4):2167–75.

Li J, Zhou Y, Wang W, Du F, Ren L. A bio-inspired superhydrophobic surface for fog collection and directional water transport. Journal of Alloys and Compounds. 2020;819:152968.

Likens GE, Driscoll CT, Buso DC. Long-term effects of acid rain: response and recovery of a forest ecosystem. Science. 1996;272(5269):244–6.

Liljestrand HM, Morgan JJ. The chemistry of fog water in the northeastern United States. Atmos Environ. 1978;12(12):1271–7.

Lodge JP. A rational basis for national ambient air standards for particulate matter. Aerosol science and technology, 1986;5(4):459–68.

Mamane Y, De Pena RG. A quantitative method for the detection of individual submicrometer size sulfate particles. In Sulfur in the Atmosphere 1978;(pp. 69–82). Pergamon.

Manabu I, Kosuke K, Ben N, Kiyoshi M. Chemical composition of polluted mist droplets. Atmospheric Environment. 2017;171:230–6.

McHenry JN, Dennis RL. The relative importance of oxidation pathways and clouds to atmospheric ambient sulfate production as predicted by the regional acid deposition model. Journal of Applied Meteorology and Climatology. 1994;33(7):890–905.

McLaughlin SB, Nosal M, Wullschleger SD. Interactive effects of atmospheric CO2 enrichment and air temperature on gross photosynthesis and dark respiration in sweet gum seedlings. Environ Exp Bot. 2007;59(2):213–23.

Mitchell MJ, Driscoll CT, Kahl JS. Effects of watershed geomorphology on stream chemistry and hydrology in the Catskill Mountains. Water Resour Res. 1999;35(11):3471–81.

Mrose H. Measurements of pH, and chemical analyses of rain-, snow-, and fog-water. Tellus. 1966;18(2-3):266–70.

Munger JW, Jacob DJ, Waldman JM, Hoffmann MR. Fogwater chemistry in an urban atmosphere. Journal of Geophysical Research: Oceans, 1983;88(C9):5109–21.

Munger JW, Collett Jr J, Daube Jr B, Hoffmann MR. Fogwater chemistry at riverside, California. Atmospheric Environment. Part B. Urban Atmosphere. 1990;24(2):185–205.

Nath S, Yadav S. A comparative study on fog and dew water chemistry at New Delhi, India. Aerosol and Air Quality Research. 2018;18(1):26–36.

Nault BA, Campuzano-Jost P, Day DA, Guo H, Jo DS, Handschy AV, Pagonis D, Schroder JC, Schueneman MK, Cubison MJ, Dibb JE. Interferences with aerosol acidity quantification due to gas-phase ammonia uptake onto acidic sulfate filter samples. Atmospheric Measurement Techniques. 2020;13(11):6193–213.

Nieminen TM, Piirainen S, Kiikkilä O, Poikonen A. Chemical composition of fog water in the southern Finland, 1987-1992. Boreal Environ Res. 2012;17(2):153–63.

Ogren JA, Heintzenberg J, Zuber A, Noone K, Charlson RJ. Measurements of the sizedependence of solute concentrations in cloud droplets. Tellus B: Chemical and Physical Meteorology. 1989;41(1):24–31.

Okita T. Concentration of sulfate and other inorganic materials in fog and cloud water and in aerosol. Journal of the Meteorological Society of Japan. Ser. II, 1968;46(2):120–127.

Padgett PE, Allen EB, Bytnerowicz A, Minnich RA, Carroll-Wood A. A gradient analysis of southern California conifers. Madrono. 1997;44(2):203–25.

Ramanathan V, Crutzen PJ, Lelieveld J, Mitra AP, Althausen D, Anderson J, Andreae MO, Cantrell W, Cass GR, Chung CE, Clarke AD. Indian Ocean Experiment: An integrated analysis of the climate forcing and effects of the great Indo-Asian haze. Journal of Geophysical Research: Atmospheres, 2001;106(D22):28371–98.

Reay DS, et al. Global agriculture and nitrous oxide emissions. Nat Clim Chang. 2012;2(6):410–6.

Saxena P, Sonwani S, Sharma SK, Kumar P, Chandra N. Carbonaceous aerosol variations in foggy days: a critical analysis during the fireworks festival. Fresenius Environ Bull. 2020;29(8):6639–56.

Saxena VK, Lin NH. Cloud chemistry measurements and estimates of acidic deposition on an abovecloud base coniferous forest. Atmospheric Environment. Part A. General Topics. 1990;24(2):329–52.

Seinfeld JH, Pandis SN. Atmospheric Chemistry and Physics: From Air Pollution to Climate Change. John Wiley & Sons, Hoboken, New Jersey, USA. 2006.

Seinfeld JH, Pandis SN. Atmospheric chemistry and physics: from air pollution to climate change. Hoboken: Wiley; 2016.

Sigg L, Stumm W, Zobrist J, Zurcher F. The chemistry of fog: Factors regulating its composition. Chimia. 1987;41(5):159–65.

Sheoran, R., Dumka, U. C., Sowmya, H. N., Bisht, D. S., Srivastava, A.K., Tiwari, S., Attri, S. D., & Hopke, P. K. (2021). Changes in inorganic chemical species in fog water over Delhi

Smith SJ, et al. Future sulfur dioxide emissions. Proc Natl Acad Sci. 2020;117(32):18757–64.

Smith P, Davis SJ, Creutzig F, Fuss S, Minx J, Gabrielle B, Kato E, Jackson RB, Cowie A, Kriegler E, Van Vuuren DP. Biophysical and economic limits to negative CO2 emissions. Nature Climate Change. 2016;6(1):42–50.

Sonwani S, Saxena P, Srivastava A. Carbonaceous aerosol variability and SOA formation during foggy days in Delhi, India. Frontiers in Sustainable Cities. 2022;4:951340.

Sullivan TJ, McDonnell TC, McPherson GT. Sulfur and nitrogen deposition on forests in the northeastern United States. Environ Sci Technol. 2011;45(16):6886–93.

Unsworth MH, Fowler D. Deposition of pollutants on plants and soils; principles and pathways. In Air Pollution and Ecosystems: Proceedings of an International Symposium held in Grenoble. 1988; France, 18–22 May 1987 (pp. 68-84). Springer Netherlands.

Waldman JM, Munger JW, Jacob DJ, Flagan RC, Morgan JJ, Hoffmann MR. Chemical composition of acid fog. Science. 1982;218(4573):677–9.

Wang X, Wei W, Cheng S, Li J, Zhang H, Lv Z. Characteristics and classification of PM2. 5 pollution episodes in Beijing from 2013 to 2015. Science of the Total Environment. 2018;612:170–79.

Wang Z, et al. Inorganic ions in fog water: temporal variation, size distribution, and sources in a megacity of eastern China. Atmos Res. 2019;220:42–52.

Warneck P. Chemistry of the natural atmosphere. Academic Press; 2000.

Weathers KC, Ponette-González AG, Dawson TE. Medium, vector, and connector: Fog and the maintenance of ecosystems. Ecosystems, 2020;23(1);217–229.

Yu H, He N, Wang Q, Zhu J, Gao Y, Zhang Y, Jia Y, Yu G. Development of atmospheric acid deposition in China from the 1990s to the 2010s. Environmental Pollution. 2017;231:182–90.

Zhang G, Peng X, Sun W, Fu Y, Yang Y, Liu D, Shi Z, Tang M, Wang X, Bi X. Fog/cloud processing of atmospheric aerosols from a single particle perspective: a review of field observations. Atmos Environ. 2024;329:120536.

Zhang R, et al. Diurnal and seasonal variations in concentration and size distribution of sulfate aerosols in Nanjing. China Environmental Pollution. 2020;267:115494.

Zhang Y, et al. Elevated concentrations and potential sources of nitrogen-containing compounds in the fog water of a subtropical urban area. South China Atmospheric Research. 2022;277:105994.

Zhu L, et al. Organic and inorganic species in aerosols and clouds in the 2019 Los Angeles supersite study: sources and processes affecting secondary aerosol formation. Environmental Science &Technology. 2020;54(9):5417–27.

# Chapter 5
# Urban Fog and its Impact on Atmospheric Visibility Degradation

**Saumya G. Kutty and Madhavi Jain**

## 5.1 Introduction

For the majority of the global population, cities and the urban environment have become the norm. To put things in perspective, the urban population has increased from roughly 3% of the total population in the year 1800 to 55% in the year 2018, and is projected to reach 68% by the year 2050 (Alberti 2005; United Nations 2013, 2019; Oke et al. 2017; Jain 2023). Apart from the rapid increase in the urban population growth rate, a considerable percentage of the rural population is also continuously migrating to cities (Davis 1955; Cohen 2006; United Nations 2014; Jain et al. 2016). With its built environment's ability to interact and change the local urban climate and weather patterns (Oke 1973; Collier 2006; Lindberg et al. 2013; Yang et al. 2016; Jain et al. 2017; Niyogi et al. 2017; Jain 2022a, b, 2023), cities present us with a unique ecosystem, unlike any other on this planet. The uniqueness of cities and larger urban metropolitan areas is also due to their nature as concentrated hubs of people, infrastructure, technology and scientific advancements, economic wealth, and cultural heritage, which leaves them more vulnerable to human and economic losses under various extreme events and natural disasters (Baccini et al. 2011; Lankao and Qin 2011; Baklanov et al. 2018; Bokwa et al. 2018a, b; Jain 2022b, 2024). In this context, it becomes increasingly important to identify and understand the processes that govern climate modifications in the urban environment.

Some of the urbanization-induced local weather changes include the creation of urban heat islands (UHIs) (Oke 1973; Kim 1992; Rizwan et al. 2008; Zhao et al. 2014; Jain et al. 2017; Jain 2023), and the modification of cloud formation (Braham

S. G. Kutty
School of Environmental Sciences, Jawaharlal Nehru University, New Delhi, India

M. Jain (✉)
Urban Systems Lab, The New School, New York, NY, USA
e-mail: jainm@newschool.edu; madhavijain.90@gmail.com



P. Saxena et al. (eds.), *Fog: Formation, Chemistry, Dynamics, Impact and Dissipation*,
https://doi.org/10.1007/978-981-96-4422-3_5

Jr 1974; Changnon 1992; Romanov 1999; Vo et al. 2023), thunderstorm cells (Shem and Shepherd 2009; Haberlie et al. 2015; Jain 2019), precipitation (Huff and Changnon Jr 1973; Shepherd 2005; Schmid and Niyogi 2017; McLeod et al. 2024), lightning activity (Shepherd et al. 2010; Jain 2019; Stallins and Rose 2008; Shi et al. 2024), flood patterns (Ahmed et al. 2013; Miller and Hutchins 2017; Zhou et al. 2017), as well as, fog (LaDochy 2005; Shi et al. 2008; Bokwa et al. 2018a; Kutty et al. 2019, 2020) and haze events (Zhong et al. 2018; Yang et al. 2021; Han and Cao 2022). This book chapter is broadly focused on fog in the built environments, and how it impacts the visibility conditions specifically in cities.

Fog is a significant atmospheric hazard and can be defined as a ground-level (2 m height from the surface) cloud consisting of minute (diameter between 2 and 50 microns) water droplets or ice crystals (in concentrations of 20–500 droplets per cu.cm) suspended in the air that reduces horizontal visibility to a distance of 1000 m or less (Myers 1968; Gultepe et al. 2007; Román-Cascón et al. 2016). In urban settings, the conditions favoring or opposing fog formation, and its characteristics, are heavily influenced by several anthropogenic factors, including increased heat retention, altered surface properties, altered radiation fluxes and water budgets, and enhanced aerosol concentrations (Oke et al. 2017). This distinctive form of fog, often referred in literature as **urban fog**, is not only more complex in its formation but also significantly different in its behavior, persistence, and structure as compared to rural fog. As urban areas expand, the unique meteorological conditions created by dense infrastructure, city morphology type, increased pollution, presence or absence of blue-green infrastructure (BGI), and existence of different microclimates within the same city, have a significant impact on the urban fog dynamics (Shi et al. 2008; Bokwa et al. 2018a; Kutty et al. 2020, 2021). Since these elements vary drastically among cities, the overall impact on whether it would create more conducive or antagonistic conditions for fog formation is unique to every city. Over time the fog dynamics of any city are also influenced by the changing land use characteristics, and its urbanization patterns, such as urban sprawl or urban verticalization. Urban fog can drastically reduce visibility, cause huge financial losses by disrupting all modes (air, road, rail, and waterway) of passenger and goods transportation, cause stoppage of construction activities, and contribute to safety hazards for millions of people (Jacobson 2002). Flight delays, diversions, and cancellations, due to recurrent foggy conditions at airports in some cities, amount to huge financial losses every winter season (Román-Cascón et al. 2016). Besides these, dense persistent fog is a concern for military operations and defense systems, cold and frost-sensitive crops, efficiency of solar panels, and many other sectors (Myers 1968). Fog thickness and fog-top height, which in turn are impacted by atmospheric stability conditions in the urban ecosphere, are critical in determining the extent to which the life and movement of the cities are brought to a halt (Liu et al. 2021).

Within the urban areas, six major different types of fog are observed which are characterized on the basis of conditions that lead to their formation. These are radiation fog, advection fog, evaporation fog, upslope fog, valley fog, and precipitation fog.

(a) **Radiation fog** forms when the ground loses heat via nighttime radiation, cooling the air just above the surface. As the air temperature reaches the dew point, water vapor condenses into small droplets, leading to supersaturation of the lower atmosphere (Gultepe et al. 2007). Radiation fog is typically shallow and dissipates after sunrise as the ground warms up again. This is the most widely observed fog type in both rural and urban areas.
(b) **Advection fog** occurs when warm, moist air moves horizontally over a cooler surface, causing the air to cool and reach its dew point. This type is prevalent in coastal regions, often observed along coastlines of San Francisco and London, where warm air from the ocean meets cooler land or water temperatures.
(c) **Evaporation fog**, or steam fog, forms when cold air passes over warm water, causing the water to evaporate and mix with the cooler air. The sudden cooling of the humid air leads to condensation and fog formation. This type of fog is commonly seen in lakes and rivers during the autumn months when the water temperature is significantly warmer than the surrounding air. Cities like Minneapolis and Toronto often experience this phenomenon in the fall and winter.
(d) **Upslope fog** forms when moist air is forced to ascend a mountain or a slope. The consequent adiabatic cooling of air is responsible for forming this type of fog, and is therefore very common in mountainous regions like the Appalachian Mountains in the eastern United States and the Rocky Mountains.
(e) **Valley fog** forms in low-lying areas, particularly in valleys, where cool air becomes trapped by surrounding higher terrain. The trapped air cools and condenses, forming fog. Valley fog can persist for days, especially in winter months.
(f) **Precipitation fog** forms when rain falls into cooler air at frontal zones and the rain evaporates, increasing the humidity in the air until it reaches saturation and forms fog. It is also called frontal fog and usually occurs during or after the passage of a warm front, often accompanying storm systems.

These fog types are modulated by a myriad of factors in the urban environment. By understanding these different factors and their role in modulating the characteristics and dynamics of urban fog, we can better address the challenges it poses to urban planning, transportation, and public health in increasingly urbanized environments. In Sect. 5.2 we explore different aspects of the urban environment and their role in modifying the urban boundary layer and creating urban microclimates. In Sect. 5.3 we describe the processes governing fog lifecycle and its impact on visibility degradation while emphasizing the role of urban boundary layer, microclimates, anthropogenic emissions, and their combined effects on atmospheric conditions conducive or detrimental for urban fog. Lastly, in Sect. 5.4 we summarize the chapter highlighting the importance of urban planning in preventing and mitigating the losses due to urban fog.

## 5.2 Factors Impacting Formation of Urban Fog

Fog in the urban settings is typically due to complex feedbacks between multiple anthropogenic and natural factors. One of the key factors that influence how urban fog is shaped is the urban morphology of the city. The other major factor to consider is the presence of aerosols and air pollution in cities. We begin with explaining these urban characteristics, and then move on to how they impact urban fog formation, duration, and thickness.

### *5.2.1 Urban Morphology*

The fabric that the city is composed of, i.e., manmade construction materials, such as reinforced concrete, brick, asphalt, steel, glass, etc., or natural materials, such as wood, urban vegetation, bare soil, urban water bodies, etc., essentially determines the **thermal properties**—albedo, thermal conductivity, thermal admittance, thermal diffusivity, specific heat, and heat capacity of the surfaces (Oke 1973; Taha 1997; Arnfield 2003; Stathopoulou et al. 2009; Stewart and Oke 2012; Mohajerani et al. 2017; Jain 2019, 2022a, 2023). The proportion of these urban fabric elements within a city block or neighborhood, most notably the impervious area fraction and the green area fraction, plays an important role in determining the way the fluxes, evaporation rates, radiation, and moisture budgets would be altered (Arnfield 2003; Ching et al. 2009; Bowler et al. 2010; Stewart and Oke 2012; Paschalis et al. 2021). At a micro-scale, a higher fraction of urban fabric and impervious surfaces trap more heat from the incoming solar radiation during the day, and releases this heat later during the night in the form of longwave radiation, leading to elevated nocturnal temperatures (Oke 1973; Arnfield 2003). At the city-scale, the **urban heat island (UHI) effect**, one of the most well-documented phenomena across cities in the world, wherein urban centers experience significantly warmer air temperatures than their rural surroundings, is observed. This difference in the urban–rural air temperatures is termed as the urban heat island intensity (UHII), and can range from a few degrees to over 10 °C, and is most noticeable on calm winter nights (Oke 1973, 1982; Li et al. 2012; Jain 2023). Additional sources of anthropogenic heat that are not as prevalent in rural areas, and not present in unaltered natural landscapes, are the heat emissions generated from the use of air conditioning systems, vehicles, industries, power plants, airports, and, in some circumstances, very large concentrations of people themselves (Shahmohamadi et al. 2011; Li et al. 2018; Raj et al. 2020). Anthropogenic heat emissions also follow metabolic cycles that vary at daily, weekly, and seasonal scales and are an important factor when assessing urban climate modifications (Allen et al. 2011; Earl et al. 2016; Oke et al. 2017; Li et al. 2021). The role of air pollutants and aerosol emissions is discussed in more detail in Sect. 5.2.2.

One of the most critical impacts of urbanization is on the planetary boundary layer (PBL). The PBL is considered as the lowest layer of the troposphere, and forms a dynamic link between the lithosphere and the atmosphere via land-air interactions and feedback mechanisms (Oke et al. 2017; Jain et al. 2017; Zhang et al. 2020). The height of the PBL determines vertical mixing of the air mass and pollutants in the troposphere, a key aspect impacting fog formation and persistence (Zhang et al. 2020). The height of the PBL is significantly influenced by surface and anthropogenic heating (and therefore UHI) as warmer turbulent air tends to deepen the PBL height. Radiative cooling in urban areas during the night leads to a thermal inversion and a significantly shallower PBL than during the day is noted. These variations in PBL height are more pronounced in urban areas. The **three-dimensional urban structure or urban geometry** determines the heat transport and flux exchange via building walls, the modifications of air flow patterns and eddy formations due to varying street widths and urban canyon depths, and urban canopy layer (UCL) and PBL height modulations because of roughness length (Ching et al. 2009; Salamanca et al. 2011; Gutiérrez et al. 2015; Demuzere et al. 2017; Oke et al. 2017; Mansouri and Zarghami 2025). The presence of high-rise buildings and other large structures can channel winds, creating wind tunnels and turbulent wakes that affect fog distribution and thickness (Fernando 2010). In some areas, **turbulent mixing** caused by buildings can also limit the duration of fog. Tall structures may increase wind speed at street level, dispersing fog more quickly and leading to localized clearings, particularly near skyscrapers or along wide urban boulevards. On the other hand, densely packed buildings can create sheltered areas where fog can linger longer, especially in low-lying zones where cool air accumulates (Grimmond and Oke 1999).

Impervious surfaces do not allow water permeability, and built-up surfaces such as pavements, roads, buildings, and construction therefore typically show an enhanced surface water runoff, a reduced infiltration rate, and significantly low evapotranspiration due to lack of vegetation and tree cover (Taha 1997). This adversely impacts the moisture fluxes and develops **urban dry islands (UDIs)** within the urban core. In such a case a stronger UHII would be likely to form. An important note to add is the growing interest in incorporating nature-based solutions (NbS) and blue-green infrastructure (BGI) within the urban centers as a way to increase thermal comfort (Bowler et al. 2010; Jain 2022a). Urban parks, green roofs, vertical gardens, urban ponds, lakes, and rivers create cooler local microclimates within the cities, which may offset or reverse the UHI effects. As cities increasingly adopt NbS, these **urban cool islands (UCIs)** are likely to become more prevalent worldwide. Therefore, at a surface level, microclimates created by hotspots of either UHIs or UCIs define if fog formation would be supported or suppressed, and this can considerably differ from one city to another.

The energy balance equation, important for understanding urban fog dynamics in the context of vegetation, is expressed as:

$$Q^* = Q_{\mathrm{H}} + Q_{\mathrm{E}} + Q_{\mathrm{G}} \tag{5.1}$$

where,

$Q^*$ is the net radiation, $Q_H$ is sensible heat flux, $Q_E$ is latent heat flux, and $Q_G$ is ground heat flux.

Increased nighttime temperatures in UHI hotspots can inhibit fog formation, as the warm ground and air may not cool sufficiently during the night to reach dew point. The presence of BGI significantly influences urban fog formation via increased moisture availability for evaporation, temperature regulation by increasing the latent heat flux, and the increased availability of condensation nuclei. Croplands on the outskirts of urban areas, seasonal plantations, and green complexes within the city can influence seasonal patterns of fog. Localized variations in fog density and persistence can also occur based on how different crops transpire and affect soil moisture retention. For instance, in cities in the Indo-Gangetic plains (IGP) in India (e.g., New Delhi, Agra, Varanasi, etc.), dense fog often occurs in the early morning due to the combination of surface moisture from the rivers, vast cultivated wheat fields on the river floodplains, and the cool, still air (Badarinath et al. 2009; Kutty et al. 2020). Tree canopies help regulate temperature by providing shade during the day and reducing heat loss at night. The formation of urban fog is thus uneven across a city. While it may be thick in one area, it may be virtually absent just a few blocks away due to UHI and UCI microclimate characteristics.

The occurrence and characteristics of urban fog also vary **seasonally** and **diurnally,** largely due to changes in temperature and humidity. For instance, coastal cities like San Francisco experience thick marine fog during summer due to the interaction of warm inland air with cooler ocean waters, while in winter, radiation fog becomes more common as clear skies at night allow for rapid cooling of the surface (Tardif and Rasmussen 2007). In polar regions, fog can form in summer when warm air masses move over cold surfaces, but it is generally less common during the extreme cold of winter due to lower humidity levels (Tjernström et al. 2015). The most common types of fog—namely radiation and advection fog—are more likely to form in winter when **temperature inversions** are common, trapping cooler air near the surface under a layer of warmer air, which suppresses vertical mixing. In urban areas, this effect can be intensified by heat released from buildings and vehicles, leading to localized inversions that support the formation of longer duration fogs (Myers 1968; Sachweh and Koepke 1995; Bokwa et al. 2018a) as the warmer air above acts as a cap, preventing the fog from dissipating (Seinfeld and Pandis 2016). **Diurnal variations** also play a role, with urban fog often forming during the early morning hours when temperatures are at their lowest and relative humidity is highest. As the sun rises, differential heating between urban surfaces and green spaces often leads to rapid dissipation of fog in some areas while allowing it to persist in shaded zones or in areas shielded from direct sunlight (Myers 1968). In contrast, urban areas near water bodies, such as rivers or coastal regions, can experience fog more frequently due to the advection of warm urban air over cooler water surfaces (Voogt and Oke 2003).

### 5.2.2 Aerosols and Gaseous Air Pollutants

The presence of **aerosols and gaseous air pollutants** plays a crucial role in fog formation, especially those easily available in urban environments. When the right atmospheric conditions in urban areas permit fog formation, the high concentrations of aerosols and air pollutants present in the polluted urban air result in a dangerous mixture called smoke-fog, or more commonly known as smog. Although not meteorologically classified as fog, it assumes significance due to its negative impact on visibility and health, especially in cities where aerosol concentrations are extremely high. This combination of fog and pollution can exacerbate visibility reduction and cause adverse health issues, especially respiratory and cardiovascular problems (Wang et al. 2015). The infamous **London smog of 1952**, which killed over 4000 people, remains a stark example of how fog, when combined with urban pollution, can lead to severe visibility degradation and health impacts (Bell and Davis 2001). On December 5, 1952, a dense fog had settled over London, and at the same time, due to a strong inversion layer formation and very light winds, any chance of pollutant dispersion to either upper layers of the troposphere or to downwind of the city was blocked. At the time, tons of chemical fumes, smoke, and dust emitted from the factories, vehicles, furnaces, and industries turned the fog from hazardous brown to toxic black, causing deaths of over 4000 people, and several deaths were caused due to drowning or other accidents as visibility had degraded to just a few inches (Myers 1968).

Vehicular pollutants come from the exhaust and other emissions related to vehicles, such as cars, trucks, buses, motorcycles, and construction equipment. The most common vehicular pollutants are carbon monoxide (CO) produced due to incomplete combustion of fuel in engines, nitrogen oxides ($NO_x$, which is a mixture of NO and $NO_2$) produced due to high-temperature combustion in engines, particulate matter (PM) including soot, black carbon (BC), and other particles that are primarily emitted from diesel engines, volatile organic compounds (VOCs) emitted from unburned fuel and evaporation of fuel, and carbon dioxide ($CO_2$) which is a byproduct of complete combustion of fossil fuels. Industrial pollutants are emitted from various industrial processes and include much larger quantities of CO, $NO_x$, VOCs, fine and coarse PM, as well as, sulfur dioxide ($SO_2$) which is primarily emitted from coal-based thermal power plants, industrial facilities, and refineries. The difference between the two pollutants is the source of emissions—vehicular emissions are not stationary (mobile) and dispersed, which are mostly concentrated around ground transportation networks and traffic node junctions, while industrial emissions are stationary points. In cities, primary air pollutants—$NO_x$ and VOCs in the presence of sunlight—are also known to produce secondary air pollutants. These photochemical reactions produce tropospheric ozone ($O_3$), an important ground-level air pollutant and a major component of smog as well as peroxyacyl nitrates (PANs) in urban and industrial areas in high-temperature conditions.

Both aerosols and gaseous air pollutant particles impact the physics and the chemistry of the atmosphere, and act as condensation nuclei. Urban areas, due to their high levels of air pollution, often have a significantly larger quantity of condensation nuclei compared to rural regions. More surfaces for condensation allow more water vapor to condense, forming the tiny droplets that make up fog. Many air pollutants form smaller particles, which can lead to smaller fog droplets. This can create denser and dirtier fog (as more condensation nuclei are available to initiate droplet formation) with longer persistence, as smaller droplets are less likely to coalesce and fall out of the fog layer (Myers 1968; Seinfeld and Pandis 2016). On the other hand, larger droplets may lead to reduced fog density and visibility. Commonly, smog can occur near industrial and polluted urban areas, during both dusk and dawn, because of the formation of a nocturnal inversion layer (Umar et al. 2021). Daily and weekly cycles of vehicular and anthropogenic emissions are influenced by peak traffic movement hours, school and office open days, road cleaning and construction activities, etc. (Saxena et al. 2021; Jain et al. 2024) which can significantly influence smog formation.

Based on its composition, smog can be categorized into acidic and photochemical smog.

(a) **Acidic smog** (e.g., London smog of 1952) forms in cool, damp climates, primarily due to the burning of coal or other fossil fuels. Exhaust gasses including $SO_2$, $NO_x$, CO, $CO_2$, etc. emitted from the combustion process are trapped close to the ground by a stable inversion layer (Czerwińska et al. 2019).
(b) **Photochemical smog** (e.g., Los Angeles smog of 1943) which forms in warm, sunny climates through chemical reactions between sunlight and pollutants like vehicle emissions. VOCs, $SO_2$, $NO_x$, etc., primarily from car exhaust fumes, go through several photochemical reactions forming ozone and aldehydes, particularly formaldehyde which causes lung diseases and severe ailments.

Industrialized and developed nations have over the last few decades put increasingly tough measures to curb air pollution, while the less developed nations, mostly in Asia and Africa, are still rapidly urbanizing, growing their economies, and as a result are dealing with significantly high rates of air pollution (Biswas et al. 2008). Some of the prominent examples include the Asian brown haze (Szidat 2009) and the recurrent smog episodes every winter in northern India and the IGP (Ghude et al. 2017; Singh and Chauhan 2022; Saxena et al. 2021; Kutty et al. 2020). The steady transport of regional and transcontinental sources of dust, for example, the Sahara Desert, the Gobi Desert, the Taklimakan Desert, the Arabian Desert, the Great Basin, the Thar Desert, etc., can also significantly increase the PM pollution and supply condensation nuclei for fog development (Filonchyk et al. 2022; Jain 2022a, b; Jain et al. 2024). Besides this, fugitive emissions of dust, soot and larger particulates from cleaning of roads, unpaved pathways, barren lands, mining and stockpiles, and solid waste burning contribute to increased larger PM that may promote the formation of larger fog droplets (Jain et al. 2024). Depending on the characteristics of the dust, it may contribute to quicker droplet growth and potential precipitation, which can shorten the duration of fog.

Unregulated agricultural practices such as crop residue burning and slash-and-burn agriculture are important sources of significant seasonal aerosols and air pollution, and cause a severe worsening of the local and regional air quality of the area, including cities and urban centers. These practices are mostly witnessed in the less developed nations, which typically face weak policy enforcements of air pollution laws, gaps in environmental awareness, and lesser economic incentives. Crop residue burning is a practice most prevalent in South Asia (particularly in India and Pakistan) and Southeast Asia (particularly in Thailand, Vietnam, and Cambodia), where the leftover crop stubble (mostly rice or wheat) biomass is burned away. This way the farmers can prepare their agricultural lands easier, faster, and cheaper for the next sowing season, ensuring multiple crops can be harvested in a year instead of just one or two (Umar et al. 2021). The IGP is one of the most fertile, populated, and polluted regions in the entire world and shows high background aerosol optical depth due to high $PM_{2.5}$, $PM_{10}$, $NO_x$, and $SO_x$ (Saxena et al. 2021). Air quality is poor due to high dust influx from local and transcontinental sources, high vehicular density, and industrial emissions that remain largely trapped due to the Himalayas blocking wind flow in the north (Ghude et al. 2017; Singh and Chauhan 2022). During winters meteorological conditions inhibit mixing and dispersion of air pollutants, and they remain trapped in a dense low-level thermal inversion layer. On top of this, the prevalence of crop residue burning after the harvest of rice and paddy and large-scale pollutant emissions from firecracker bursting (due to celebration of the Diwali festival) have resulted in smog being a seasonal occurrence every year in the IGP (Saxena et al. 2021).

The other agricultural practice of slash-and-burn is particularly prevalent in forested regions of Amazonia, Southeast Asia, South Asia, and Africa. This differs from crop residue burning where the farmers clear away patches of forest and burn vegetation to enrich soil with nutrients, farm until the soil loses its fertility, and then move on to the next patch of forest (Kleinman et al. 1995; Yadav 2013). As many farmers are driven by limited water, time, financial, or labor resources for practicing crop residue burning or slash-and-burn agriculture, it poses a major challenge for the worsening air quality, deforestation, and other ill effects on the environment, including smog formation. Additionally, smoke from wildfires can travel long distances, degrade the air quality of cities far away from the wildfire source, and cause smog formation under the right circumstances (Filonchyk et al. 2022).

## 5.3 Impact on Visibility

Fog is characterized by visibility <1000 m and can be further classified into (a) **light fog** (<1000 m and ≥ 500 m), (b) **moderate fog** (<500 m and ≥200 m), (c) **thick fog** (<200 m and ≥50 m), and (d) **very thick fog** (<50 m) (Kutty et al. 2019). Mist and haze are similar phenomena but associated visibilities are 1–2 km and 2–5 km, respectively. A significant trend in fog frequency has been observed in several cities (Table 5.1) with a corresponding change in visibility.

**Table 5.1** Fog frequency trends in urban centers across the world. Radiation fog is the most commonly observed fog type. Positive (negative) fog trend corresponds to increase (decrease) in fog frequency

| Region | Country | City/urban center | Location | Dominant fog types | Fog trend | Period | Reference |
|---|---|---|---|---|---|---|---|
| Asia | Indo-Gangetic Plains (IGP) in India | Amritsar | Inland | Radiation | Positive | 1979–2014 | Kutty et al. (2019) Hingmire et al. (2019) |
| | | New Delhi | | | | | |
| | | Lucknow | | | | | |
| | | Varanasi | | | | | |
| | China | Shanghai | Coast | Radiation, advection | Negative | 1989–2017 | Gu et al. (2019) |
| | | East-Central China (Yangtze corridor) | Inland, coast | Radiation | Positive | 1976–2007 | Niu et al. (2010) |
| | | South Eastern China (Sichuan basin) | | | Negative | | |
| | Japan | Asahikawa | Inland | Radiation | Negative | 1966–2005 | Akimoto and Kusaka (2015) |
| | | Miyako | Coast | Advection | Positive | | |
| | South Korea | Wonju | Inland | Radiation | Negative | 1989–2013 | Belorid et al. (2015) |
| | | Ulleongdo | Island | Frontal | Positive | | |
| America | USA | Los Angeles | Coast | Advection-radiation | Negative | 1948–2004 | Witiw and LaDochy (2008) |
| | Brazil | Sau Paulo | Coast | Radiation, Advection | | 1933–2005 | Gonçalves et al. (2008) |
| Europe | Germany | Munich | Inland | Radiation | Negative | 1949–1990 | Sachweh and Koepke (1995) |
| | | Potsdam | Inland | – | | 1976–2006 | van Oldenborgh et al. (2010) |
| | Switzerland | Zurich | | | | | |
| | Netherlands | De Bilt | | | | | |

A decreasing trend in fog frequency is observed in 9 out of 16 cases, associated with the increasing rate of urbanization in these cities leading to visibility degradation. Visibility depends on the fog intensity and is commonly parameterized as a function of the liquid water content (LWC) or the number of fog droplets ($N_d$) and can be expressed as Eq. (5.2) by Kunkel (1984) or Eq. (5.3) by Gultepe et al. (2006):

$$\mathrm{Vis}(\mathrm{m}) = 27 * \mathrm{LWC}\left(\mathrm{gm}^{-3}\right)^{-0.88} \tag{5.2}$$

$$\mathrm{Vis}(\mathrm{m}) = 1002\left[\mathrm{LWC}\left(\mathrm{gm}^{-3}\right) * N_d\left(\mathrm{cm}^{-3}\right)\right]^{-0.6473} \tag{5.3}$$

In cities, where aerosol concentrations are typically higher due to anthropogenic activities, visibility can be parameterized based on the Mie theory using the rule proposed by Koschmieder (1924):

$$\mathrm{Vis}(\mathrm{m}) = \frac{3.912}{\beta(\mathrm{km}^{-1})} \times 1000 \tag{5.4}$$

where $\beta$ is the atmospheric extinction coefficient.

In the case of fog, $\beta$ is inversely proportional to wavelength (Liu et al. 2017). Anthropogenic forcings have a greater influence on fog as opposed to other types of clouds (Yan et al. 2019) due to its proximity to the surface. Fog formation and dissipation in cities mainly rely on a balance between the components of the built environment, primarily governed by the urban morphology, its land use and land cover, and the aerosol and pollutant concentrations.

## 5.3.1 *Role of the Built Environment in Modulating the Fog Life Cycle*

Fog macrophysics is significantly modified in the urban environment due to the multiple factors involved and their complex feedback processes. The modulation of surface fluxes and the PBL by the combined influence of urban elements and morphology as well as anthropogenic emissions determines the process of fog evolution. The spatiotemporal evolution of fog may be categorized into two main stages: (a) formation and development; (b) maturity and dissipation, and are discussed in more detail.

### 5.3.1.1 Formation and Development

Fog forms under saturated conditions, with the near-surface temperature approaching the dew point temperature such that the dew point depression ($T_{dep}$) = 0 and relative humidity ($RH_{2m}$) = 100%. UHIs weaken the nocturnal radiative cooling, a critical factor that is responsible for cooling the moist air near the ground to its dew point (Gu et al. 2019). **Radiative cooling** results in the establishment of a **stable inversion layer**, which in conjunction with conducive factors like temperature, humidity, and turbulence, leads to fog formation. As explained in Sect. 5.2.1, the urban surfaces, including buildings, roads, and pavements, are prone to heat retention owing to their high thermal conductivity. The reduced nighttime radiative cooling in UHI hotspots has an adverse impact on the stability of the urban inversion layer and, as a result, can inhibit fog onset and maturation (Fitzjarrald and Lala 1989). Further, once fog is formed, an instability near the surface is observed prior to the fog layer reaching the ground, which may be attributed to the combined effect of urban heat and increased forced convection at night, commonly associated with the UCL (Ye et al. 2015; Oke 1973, 1982). In coastal areas, the raised nighttime near-surface dew point depression results in a lifting of the altitude of condensation and **cloud base height** (Williams et al. 2015).

A lower albedo combined with higher longwave radiation also has a significant impact on the urban radiation budget. Bowen ratio is a term that describes the ratio of sensible heat flux to latent heat flux and is used to explain energy partitioning. A higher Bowen ratio is typically observed in cities, and combined with the higher amounts of anthropogenic heat generated, this results in a significant storage heat flux term observed, alongside strong upward sensible heat flux and a weak evapotranspiration during the day. At night, the release of this stored heat leads to upward directed turbulent heat fluxes, which prevents saturation via increasing air temperature and limits the stability of the inversion layer, thereby suppressing fog formation (Lietzke et al. 2014). Both fog formation and development are impacted by the air-surface exchange of heat and vapor (Liu et al. 2011). In UDI hotspots within the cities, low moisture availability and therefore low specific humidity levels lead to delayed onset of fog. Once formed, the fog's evolution depends on the variability of surface humidity conditions ($RH_0$). The **limited moisture availability** in urban areas inhibits condensational growth and vertical development, thereby resulting in shallow fog with lower fog-top height and greater visibility compared to their rural counterparts.

**Turbulence and wind** also play significant roles in fog formation and development. As explained in Sect. 5.2.1, because of the three-dimensional urban geometry, several factors including greater surface roughness, higher urban and anthropogenic heat, greater vertical wind shear, and a possible wind channeling (or wind tunneling) result in enhanced turbulence parameters, including **turbulent kinetic energy** (TKE) and **friction velocity** (u*). Low values of TKE ($\sim 10^{-1}$ $m^2s^{-1}$), u* ($<0.3$ $ms^{-1}$) and wind speed ($WS < 0.2$ $ms^{-1}$) were observed close to surface (~40 m) prior to fog onset in the Tianjin area of China (Ye et al. 2015), with large values of TKE and u* at upper levels due to shear production. During a fog episode, a strengthening of both these parameters indicates that moderate turbulence is favorable for fog development (Zhou and Ferrier 2008). Also, a low vertical velocity variance near the ground (at 2 m) ~0.002–0.005 $m^2$ $s^{-2}$ or less plays a crucial role in the formation of thermally stable radiation fog (Price 2019). Moreover, these have implications for visibility. Ye et al. (2015) found several instances of reduced visibility during fog to be associated with fluctuations in the wind direction and attributed it to the mesoscale motions linked to weak large-scale flows (Mahrt 2007).

#### 5.3.1.2 Maturation and Dissipation

During the fog transition phase, as the fog thickness increases, various processes taking place at the top of the fog layer assume greater significance (Duynkerke 1999). Radiative cooling at the fog top can enhance turbulent mixing and promote the **vertical thickening of the fog layer**. The rates of fog-top entrainment and fog condensational growth determine if fog persists or dissipates (Shao et al. 2023). While moderate turbulence is beneficial for fog development, high turbulence can suppress or deepen the fog-top height depending on the **critical turbulence coefficient (ε)** (Zhou and Ferrier 2008). Fog persists (dissipates) when the turbulence intensity is $<\varepsilon$ ($>\varepsilon$). However, due to high turbulence parameters and vertical wind velocity in urban centers, the strong updrafts can contribute to fog dissipation.

Moreover, high aerosol loading in urban areas promotes early onset of fog while delaying fog dissipation (Shao et al. 2023; Quan et al. 2021). Shao et al. (2023) found that in an urban center like Nanjing in China, the fog area, fog-top height, and duration increase by, respectively, 23.1%, 109.6%, and 20.0% under polluted (high emission) conditions compared to clean (low emission) conditions. A higher PBL height can also allow for more surface heating and further turbulent mixing, reinforcing the process. However, in polluted urban areas, the high aerosol loading can, depending on the chemical composition and size, either reflect or absorb incoming shortwave radiation and, as a result, shallow or deepen the daytime urban boundary layer. Due to this feedback between PBL height and aerosol-cloud interactions (ACI), a lower PBL height induced by high aerosol loading promotes higher humidity and stability in foggy conditions. Higher aerosol and gaseous air pollutant concentrations in cities increase the **fog-top height**, promoting turbulence and horizontal distribution while also reducing visibility (Jia et al. 2019). As fog evolves vertically, aerosols enhance longwave radiative cooling in the middle-upper part of fog alongside a warming in the lower part. The vertical temperature gradient thus further weakens temperature inversion, leading to turbulence, which is then further strengthened via condensation and latent heat release. Excessive turbulence on the other hand leads to fog dissipation.

An interesting phenomenon called **urban fog hole** is increasingly being observed over major cities in the world (Gautam and Singh 2018; Yan et al. 2020; Anurose et al. 2024). The UHI-induced sharp urban–rural temperature gradient, and the consequent reduction in fog thickness, as well as, early fog dissipation in built environments compared to surrounding rural areas, shows as apparent holes in the blanket of fog spread across a larger region. These fog holes are limited to the urban areas and can be easily detected with the help of satellite images (Fig. 5.1).

Though high aerosol loading is conducive to fog formation due to its role as condensation nuclei, the limited moisture availability results in a large number of droplets ($N_d$) with smaller radius ($R_e$). The shallow, patchy fog thus formed is more prone to dissipate quickly post sunrise as compared to its denser, more persistent rural counterpart (Anurose et al. 2024). Urban fog holes are more prominent and prolonged over major cities with a larger urban and anthropogenic footprint (e.g., New Delhi: Delhi-National Capital Region (NCR) in Fig. 5.1) compared to smaller cities and towns. An increase in fog frequency in the IGP (Table 5.1) in the recent years (Kutty et al. 2019; Hingmire et al. 2019) can be attributed to enhanced aerosol loading mainly due to crop residue burning commonly observed during the winter season in the north western part of IGP (Saxena et al. 2021).

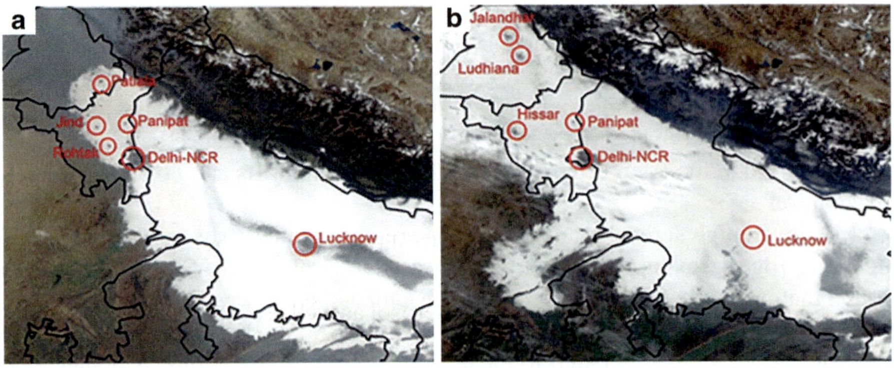


**Fig. 5.1** Urban fog holes over cities located in the Indo-Gangetic plains (IGP) in India on (**a**) 07 December 2016, and (**b**) 25 January 2024. Images captured by MODIS onboard NASA's Terra satellite and accessed via NASA Worldview Snapshots. Average daily visibility at Indira Gandhi International (IGI) Airport situated in Delhi-NCR, was 322 m (moderate fog), and 965 m (light fog) for (**a**) and (**b**) respectively

### 5.3.2 Fog Microphysics

The micro-scale evolution of fog is influenced by its macro-scale dynamics, low-level wind fluctuations, and thermodynamic structures in the boundary layer (Liu et al. 2011), which are all in turn modified by urbanization. The UHI and UDI effects limit radiative cooling such that the high saturation levels required for dense fog formation are not attained. The limited extent of **condensation growth** results in smaller fog droplets with low LWC (Liu et al. 2017). In contrast, high saturation temperature of water vapor in the suburbs gives way to an enhanced condensation growth, resulting in larger droplets with high LWC. Elevated aerosol levels commonly observed in urban centers, with values almost twice as those in suburbs and city outskirts (Liu et al. 2017), influence fog microphysics significantly (Quan et al. 2011; Wang et al. 2021). The greater concentration of aerosols and gaseous air pollutants in cities serves as condensation nuclei, resulting in higher fog droplet number concentrations ($N_d$ > 1000 $cm^{-3}$) with smaller effective radius ($R_e$ ~ 7 μm) and low visibility compared to smaller $N_d$ (<100 $cm^{-3}$) with larger $R_e$ (~9 μm) in clean fog cases. Consequently, a decreasing trend in LWC is observed coinciding with a decrease in the number of fog days and an increase in temperature. As fog develops, the available moisture condenses over the large number of aerosols acting as condensation nuclei. The fog is shallow at this point with only a slight reduction in visibility. As the fog droplets grow via condensational growth, collision, and later coalescence, a significant drop in visibility is observed. This process depends on the extent of turbulent mixing and also competes with fog-top entrainment. Dry air entrainment or high turbulence can inhibit droplet growth by favoring evaporation and consequently result in a reduction in visibility and fog thickness. Yan et al. (2020) observed horizontal wind convergence and enhanced updraughts due to greater surface roughness in urban areas which increased LWC in the upper level

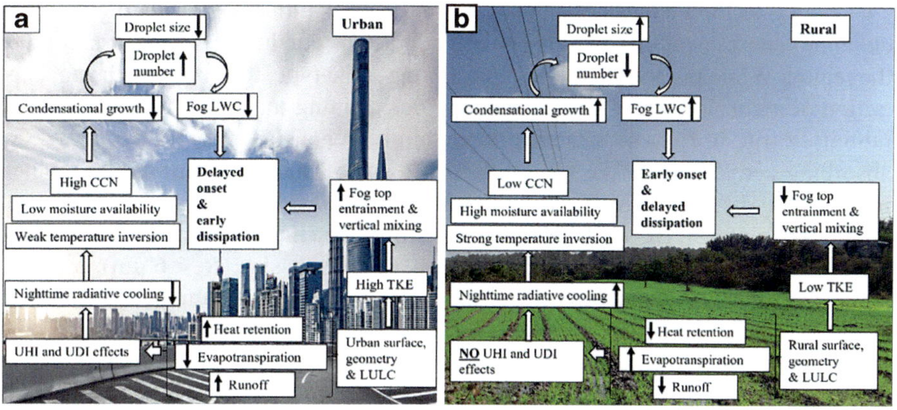


**Fig. 5.2** Schematic diagram summarizing the role of urbanization in modulating the fog lifecycle and the urban-rural contrast. Opposing processes affect fog formation and dissipation in (**a**) urban and (**b**) rural settings resulting in delayed (early) onset, and early (delayed) dissipation respectively. Note that only the aerosol concentration (high CCN in (**a**)) is shown here

and a drop in LWC at low levels due to vapor flux divergence. Toward the end of the mature stage, drop settlement and auto-conversion lead to reduced visibility. Figure 5.2 summarizes the fundamental processes involved in the fog lifecycle. The chemical composition, size, and affinity of CCN to water vapor all play an important role in ACI, and thereby the feedback processes that exacerbate the impact of urban fog on visibility degradation.

### *5.3.3 Rate of Urban Expansion and Implications for Fog and Induced Visibility Degradation*

Both urbanization and aerosol loading have a significant impact on the fog life cycle and influence visibility degradation. Urbanization delays fog onset and expedites its dissipation while aerosol promotes fog formation, persistence, and thickness through its role as CCN as well as via ACI. However, the negative impact of urbanization is stronger and takes precedence over the positive aerosol effect (Yan et al. 2020). The extent to which these factors modulate fog and visibility depends heavily on the stage of urban development as well as the rate of expansion. The urban impact on fog is significant when the natural land cover is reduced to a small fraction (<10%, Zhao-Ming et al. 2012). In relatively new cities at the early stage of urbanization, aerosols played a dominant role in promoting fog whereas in older cities at a mature development stage of rapid urbanization, the negative impact of UHI and UDI dominated fog inhibition (Yan et al. 2019; Shi et al. 2008). In China, Niu et al. (2010) observed the highest significant positive trend in fog frequency over the most heavily populated east-central region of the country (Table 5.1) and attributed it to the increase in aerosol loading and its influence on the regional circulation pattern, resulting in atmospheric conditions conducive to fog formation in the region. While predominantly fog promoting—the type, the size, and the concentration of aerosols—are critical factors for determining the extent of promoting or inhibitory role in fog and visibility modulation. For instance, presence of BC, an absorbing aerosol above the fog layer, is shown to cause a **"dome effect"** which prevents fog dissipation by blocking the shortwave radiation (Ding et al. 2016). But within the fog layer, BC leads to rapid dissipation. The fine mode aerosols dominant in case of haze fog days (Yu et al. 2011), owing to their size ($R_e < 6$ μm), promote the early onset of fog by initiating condensation under less saturated conditions. These effects are further complicated due to the urban structures, land use, and BGI. Industrial areas within a city are more prone to fog due to high aerosol concentrations. The presence of water bodies such as lakes, ponds, and rivers acts as localized moisture hotspots and encourages fog formation in spite of the UHI effect. Urban structures can give rise to zones that favor fog formation, including areas of shade like parks, plantations, urban canyons, etc., where the influence of turbulence and UHI-induced warming is limited. Climate change and large-scale forcings also play a significant role in fog modulation in cities. For instance, the coastal city of San Francisco is prone to frequent fog due to the influence of the Pacific high, which causes a shift in the wind pattern and coastal upwelling. An increasing trend in fog frequency was observed over the IGP in India which was characterized by a regime shift coinciding with a strong El Niño year (Kutty et al. 2020). Shifts in the general circulation patterns and associated changes in wind speed, direction, and relative humidity viz. convergence (divergence) of water vapor resulted in significant positive (negative) fog trends over east-central (south-eastern) China (Table 5.1, Niu et al. 2010). However, Yan et al. (2020) found that the combined effect of urbanization and aerosol on fog is 1.6 times that of climate change. The geographic location and topography of a city also determine the extent of fog frequency and thickness. Declining (increasing) fog trends have been observed in inland (coastal and island) areas (Table 2) due to the limited (enhanced) influence of regional advection.

### 5.3.4 *Adverse Impact of Fog and Induced Visibility Degradation*

Reduced visibility due to urban fog directly impacts transportation networks, including roadways, airports, and railways due to an increase in accidents, delays, and economic costs. The aviation sector is particularly vulnerable to fog, with airports often forced to delay or cancel flights when visibility drops below safe thresholds (Gultepe et al. 2009). Between 2011 and 2016, dense fog occurrences at IGI Airport, Delhi-NCR (India), caused economic losses of approximately \$3.9 million (Kulkarni et al. 2019). Similarly, fog-induced visibility reduction on highways increases the risk of multi-vehicle accidents, particularly in high-traffic urban corridors. Fog and smoke-related crashes involve more vehicles and a greater extent of injuries compared to those occurring in clear-visibility conditions (Abdel-Aty et al. 2011). The phenomenon of **"fog bursting"** which refers to a sudden drop in visibility below 500 m in a duration of 30 min is associated with the polluted urban environment (Liu et al. 2021) and is a major contributor toward fog-related accidents. In the United States alone, 13,720 people lost their lives in fog-related accidents between 1995 and 2004 (Forthun et al. 2006). Supercooled evaporation fog or "black frost" observed on the North Atlantic can result in loss of ships due to the heavy ice accretion (Wessels 1979). The financial losses can range between $10,000 and $25,000 per day for each ship, potentially amounting to millions for a moderately busy harbor (Pérez-Díaz et al. 2017). While reduced visibility associated with fog is detrimental to the economy through hampering air and road transportation, **smog** has severe implications on the health of city inhabitants. Fine aerosols present in smog increase the risk of heart attack and stroke while pollutants such as $SO_2$, $NO_x$, and $PM_{2.5}$ exacerbate health-related issues including asthma, bronchitis, lung emphysema, eye irritation, breathlessness, etc. (Kim et al. 2012). Development of several chronic diseases as well as lung cancers has been attributed to toxic carcinogens and chemicals, such as benzene, $O_3$, VOCs, heavy metals, etc. found in the urban fog (Pérez-Díaz et al. 2017).

Fog has both positive and negative impacts on agriculture. Acidic fog leads to reduced yields, low root weight, and delayed maturity among different crop species (Musselman and Sterrett 1988). High humidity levels during fog make crops susceptible to fungal diseases along with hindered pollination (Kumar et al. 2015). On the bright side, pest activity and frost damage are inhibited. Moreover, in drought-prone areas, fog benefits agriculture and plays a significant role in maintaining the water balance via fog catching. Fog catching is the interception of fog water via throughfall or collecting fog droplets suspended in the air using nets or other methods, especially in coastal regions (Sawaske and Freyberg 2015). For instance, during the dry summer in Northern California, non-rainfall water in the form of fog and dew is solely responsible for mitigating the severe water stress (Chung et al. 2017). A reduction in water deficit by 25% was attributed to inputs from fog water throughfall (10–30 mm/year) and fog suppression of evapotranspiration (125 mm/year), where evaporation suppression played a more significant role. Another positive impact of fog is the reduction in UHI effect by ~1 K under particular conditions (Bokwa et al. 2018b) which can considerably modify the urban microclimate. Additionally, fog can act as both source and sink of aerosols. Various chemical reactions due to the collision and coagulation of fog droplets can give rise to new aerosol particles, and following fog dissipation, the aerosol concentration can increase (Yang et al. 2012). In contrast, fog eliminates particulate matter ($PM_{10}$), among others, through the process of wet deposition, resulting in lower concentrations of $PM_{10}$ during the dissipation stage (Liu et al. 2017).

Several efforts have been made by the scientific modeling community to capture the underlying processes and associated feedbacks that modulate the urban fog (Zhao-Ming et al. 2012; Gu et al. 2019; Yan et al. 2020; Kutty et al. 2021), but the complex mechanisms and rapidly changing urban environment serve as limitations.

Incorporating urban materials, geometries, anthropogenic emissions into high resolution weather models (e.g., Weather Research and Forecasting, WRF) as well as suitable modification of visibility parameterizations is key to capturing these complex dynamical processes. More recently, Bhautmage et al. (2024) found that the WRF-Urban Asymmetric Convective Model (U-ACM) is able to better simulate fog conditions compared to the baseline ACM. However, fog modeling faces several challenges including limited spatiotemporal data availability, resolution and scale limitations, and high computational costs. The rapid pace of urban development also makes it challenging for weather and climate models to stay current, as they require frequent updates to reflect the latest urban layouts and land use. Real time fog and visibility advisories are limited due to a lack of interdisciplinary integration as data from multiple, different sectors including weather, public health, and transportation. This, combined with the lack of cohesive inputs from meteorologists, urban planners, health experts, and urban policy makers limit the accuracy of fog models and their effectiveness in mitigating losses due to visibility degradation.

## 5.4 Conclusion

In summary, fog in the urban environments is a weather hazard, and presents a distinct challenge, especially in terms of horizontal visibility. The visibility degradation associated with urban fog not only poses a significant risk to the transportation sector, but also has much serious implications for urban air quality and public health. The ongoing rapid spatial and vertical expansion of cities, which leads to continuous evolution of the urban dynamics, is an important determinant of urban fog behavior. Moreover, the major role played by anthropogenic inputs, which is based on human decisions, remains critical in understanding urban meteorology, especially urban fog. The interaction between heat, moisture, air pollution, and urban morphology creates complex land-air interactions and local-scale climate dynamics that can both enhance and inhibit fog formation, in different parts of the same city.

While urban fog is a disruptive hazard for most cities, it can prove to be beneficial for others. Urban planning and design should leverage the fog modification capabilities of the urban ecosystem to either promote suppression or enhancement of fog. For effective control over urban fog formation, cities need to strategize what materials the urban fabric should be composed of, what the three-dimensional urban geometry should be, which areas of the city need NbS implementation, and which BGI approaches to opt for. As an example, wind tunnels, an outcome of urban geometry (combination of high-rise buildings and long narrow gridded streets) greatly reduces urban fog formation, but more urban vegetation or water bodies (i.e., BGI) on the other hand can provide humidity necessary in aiding fog occurrence. A shift toward sustainable energy consumption can cap anthropogenic heat and pollution emissions, therefore improving visibility in foggy conditions. Some other techniques to improve visibility include fog seeding using dry ice to condense particles into ice particles, ground-based heating or infrared systems to disperse fog, use of wind turbines to hinder fog persistence, or the use of electric fields to coalesce fog into larger water droplets. Simultaneously, cities situated in semi-arid or arid regions can explore the possibility of creating artificial fogs and fog water harvesting to overcome water scarcity.

Besides strategic urban planning and formulating fog dispersal solutions, fog prediction accuracy needs to improve. Accurate short-range, medium-range, and long-range fog forecasts are crucial in mitigating financial, economic, societal, and public health costs due to urban fog. However, effective fog predictions are still limited by lack of scientific capacity and experts who understand the urban fog dynamics, very high computational costs needed to run the sophisticated urban physics-based weather prediction models, and the insufficient resolution of climate inputs to the models which is crucial to resolve the urban physics of fog at either micro- or local-scales.

**Acknowledgments** We acknowledge the use of imagery from the Worldview Snapshots application (https://wvs.earthdata.nasa.gov), part of the Earth Observing System Data and Information System (EOSDIS).

## References

Abdel-Aty M, Ekram AA, Huang H, Choi K. A study on crashes related to visibility obstruction due to fog and smoke. Accid Anal Prev. 2011;43(5):1730–7.

Ahmed Z, Rao DRM, Reddy KRM, Raj YE. Urban flooding–case study of Hyderabad. Global J Eng Design Technol. 2013;2(4):63–6.

Akimoto Y, Kusaka H. A climatological study of fog in Japan based on event data. Atmos Res. 2015;151:200–11.

Alberti M. The effects of urban patterns on ecosystem function. Int Reg Sci Rev. 2005;28(2):168–92.

Allen L, Lindberg F, Grimmond CSB. Global to city scale urban anthropogenic heat flux: model and variability. Int J Climatol. 2011;31(13):1990.

Anurose TJ, Jayakumar A, Sandhya M, Gordon H, Aryasree S, Mohandas S, Bhati S, Prasad VS. Unraveling the mechanism of the holes in the blanket of fog over the Indo-Gangetic Plains: are they driven by urban Heat Islands or aerosol? Geophys Res Lett. 2024;51(10):e2023GL107252.

Arnfield AJ. Two decades of urban climate research: a review of turbulence, exchanges of energy and water, and the urban heat Island. Int J Climatol. 2003;23(1):1–26.

Baccini M, Kosatsky T, Analitis A, Anderson HR, D'Ovidio M, Menne B, Michelozzi P, Biggeri A, PHEWE Collaborative Group. Impact of heat on mortality in 15 European cities: attributable deaths under different weather scenarios. J Epidemiol Community Health. 2011;65(1):64–70.

Badarinath KVS, Kharol SK, Sharma AR, Roy PS. Fog over indo-Gangetic plains—a study using multisatellite data and ground observations. IEEE J Sel Top Appl Earth Obs Remote Sens. 2009;2(3):185–95.

Baklanov A, Grimmond CSB, Carlson D, Terblanche D, Tang X, Bouchet V, Lee B, Langendijk G, Kolli RK, Hovsepyan A. From urban meteorology, climate and environment research to integrated city services. Urban Clim. 2018;23:330–41.

Bell ML, Davis DL. Reassessment of the lethal London fog of 1952: novel indicators of acute and chronic consequences of acute exposure to air pollution. Environ Health Perspect. 2001;109(suppl 3):389–94.

Belorid M, Lee CB, Kim JC, Cheon TH. Distribution and long-term trends in various fog types over South Korea. Theor Appl Climatol. 2015;122:699–710.

Bhautmage UP, Ghude SD, Parde AN, Kamath HG, Gokul Dhangar N, Pleim J, Mau Fung Wong M, Wagh S, Kumar R, Niyogi D, Rajeevan M. Implementation of WRF-urban asymmetric convective model (UACM) for simulating urban fog over Delhi, India. J Geophys Res Atmos. 2024;129(20):e2024JD040760.

Biswas KF, Ghauri BM, Husain L. Gaseous and aerosol pollutants during fog and clear episodes in South Asian urban atmosphere. Atmos Environ. 2008;42(33):7775–85.

Bokwa A, Wypych A, Hajto MJ. Impact of natural and anthropogenic factors on fog frequency and variability in Kraków, Poland in the years 1966–2015. Aerosol Air Qual Res. 2018a;18(1):165–77.

Bokwa A, Wypych A, Hajto MJ. Role of fog in urban heat Island modification in Kraków, Poland. Aerosol Air Qual Res. 2018b;18(1):178–87.

Bowler DE, Buyung-Ali L, Knight TM, Pullin AS. Urban greening to cool towns and cities: a systematic review of the empirical evidence. Landsc Urban Plan. 2010;97(3):147–55.

Braham RR Jr. Cloud physics of urban weather modification—a preliminary report. Bull Am Meteorol Soc. 1974:100–6.

Changnon SA. Inadvertent weather modification in urban areas: lessons for global climate change. Bull Am Meteorol Soc. 1992;73(5):619–27.

Ching J, Brown M, Burian S, Chen F, Cionco R, Hanna A, Hultgren T, McPherson T, Sailor D, Taha H, Williams D. National urban database and access portal tool. Bull Am Meteorol Soc. 2009;90(8):1157–68.

Chung M, Dufour A, Pluche R, Thompson S. How much does dry-season fog matter? Quantifying fog contributions to water balance in a coastal California watershed. Hydrol Process. 2017;31(22):3948–61.

Cohen B. Urbanization in developing countries: current trends, future projections, and key challenges for sustainability. Technol Soc. 2006;28(1–2):63–80.

Collier CG. The impact of urban areas on weather. Q J R Meteorol Soci : A journal of the atmospheric sciences, applied meteorology and physical oceanography. 2006;132(614):1–25.

Czerwińska J, Wielgosiński G, Szymańska O. Is the polish smog a new type of smog? Ecol Chem Eng S. 2019;26(3):465–74.

Davis K. The origin and growth of urbanization in the world. Am J Sociol. 1955;60(5):429–37.

Demuzere M, Harshan S, Järvi L, Roth M, Grimmond CSB, Masson V, Oleson KW, Velasco E, Wouters H. Impact of urban canopy models and external parameters on the modelled urban energy balance in a tropical city. Q J R Meteorol Soc. 2017;143(704):1581–96.

Ding AJ, Huang X, Nie W, Sun JN, Kerminen VM, Petäjä T, Su H, Cheng YF, Yang XQ, Wang MH, Chi XG. Enhanced haze pollution by black carbon in megacities in China. Geophys Res Lett. 2016;43(6):2873–9.

Duynkerke PG. Turbulence, radiation and fog in Dutch stable boundary layers. Bound-Layer Meteorol. 1999;90:447–77.

Earl N, Simmonds I, Tapper N. Weekly cycles in peak time temperatures and urban heat Island intensity. Environ Res Lett. 2016;11(7):074003.

Fernando HJS. Fluid dynamics of urban atmospheres in complex terrain. Annu Rev Fluid Mech. 2010;42:365–89.

Filonchyk M, Peterson MP, Sun D. Deterioration of air quality associated with the 2020 US wildfires. Sci Total Environ. 2022;826:154103.

Fitzjarrald DR, Lala GG. Hudson Valley fog environments. J Appl Meteorol Climatol. 1989;28(12):1303–28.

Forthun GM, Johnson MB, Schmitz WG, Blume J, Caldwell RJ. Trends in fog frequency and duration in the Southeast United States. Phys Geogr. 2006;27(3):206–22.

Gautam R, Singh MK. Urban heat Island over Delhi punches holes in widespread fog in the indo-Gangetic plains. Geophys Res Lett. 2018;45(2):1114–21.

Ghude SD, Bhat GS, Prabhakaran T, Jenamani RK, Chate DM, Safai PD, Karipot AK, Konwar M, Pithani P, Sinha V, Rao PSP. Winter fog experiment over the indo-Gangetic plains of India. Curr Sci. 2017;112:767–84.

Gonçalves FLT, da Rocha RP, Fernandes GP, Petto S Jr. Drizzle and fog analysis in the São Paulo metropolitan area: changes 1933-2005 and correlations with other climate factors. Erde. 2008;139(1–2):61–76.

Grimmond CSB, Oke TR. Aerodynamic properties of urban areas derived from analysis of surface form. J Appl Meteorol. 1999;38(9):1262–92.

Gu Y, Kusaka H, Tan J. Impacts of urban expansion on fog types in Shanghai, China: numerical experiments by WRF model. Atmos Res. 2019;220:57–74.

Gultepe I, Müller MD, Boybeyi Z. A new visibility parameterization for warm-fog applications in numerical weather prediction models. J Appl Meteorol Climatol. 2006;45(11):1469–80.

Gultepe I, Tardif R, Michaelides SC, Cermak J, Bott A, Bendix J, Müller MD, Pagowski M, Hansen B, Ellrod G, Jacobs W. Fog research: a review of past achievements and future perspectives. Pure Appl Geophys. 2007;164:1121–59.

Gultepe I, Pearson G, Milbrandt JA, Hansen B, Platnick S, Taylor P, Gordon M, Oakley JP, Cober SG. The fog remote sensing and modeling field project. BAMS. 2009;90(3):341–60.

Haberlie AM, Ashley WS, Pingel TJ. The effect of urbanisation on the climatology of thunderstorm initiation. Q J R Meteorol Soc. 2015;141(688):663–75.

Han X, Cao T. Urbanization level, industrial structure adjustment and spatial effect of urban haze pollution: evidence from China's Yangtze River Delta urban agglomeration. Atmos Pollut Res. 2022;13(6):101427.

Hingmire D, Vellore RK, Krishnan R, Ashtikar NV, Singh BB, Sabade S, Madhura RK. Widespread fog over the indo-Gangetic Plains and possible links to boreal winter teleconnections. Clim Dyn. 2019;52:5477–506.

Huff FA, Changnon SA Jr. Precipitation modification by major urban areas. Bull Am Meteorol Soc. 1973;54(12):1220–33.

Jacobson MZ. Atmospheric pollution: history, science, and regulation. Cambridge University Press; 2002.

Jain M. Urbanization and its manifestations on localized weather. [Ph.D. Thesis, Jawaharlal Nehru University]. 2019.

Jain M. Mitigation of urbanization ill-effects through urban agriculture inclusion in cities. In: New forms of urban agriculture: an urban ecology perspective. Springer Nature Singapore: Singapore; 2022a. p. 39–56.

Jain M. Increasing atmospheric extreme events and role of disaster risk management: dimensions and approaches. In: Extremes in atmospheric processes and phenomenon: assessment, impacts and mitigation. Springer Nature Singapore: Singapore; 2022b. p. 303–28.

Jain M. Two decades of nighttime surface urban heat Island intensity analysis over nine major populated cities of India and implications for heat stress. Front Sustain Cities. 2023;5:1084573.

Jain M. Future land use and land cover simulations with cellular automata-based artificial neural network: a case study over Delhi megacity (India). Heliyon. 2024;10(14):e34662.

Jain M, Dimri AP, Niyogi D. Urban sprawl patterns and processes in Delhi from 1977 to 2014 based on remote sensing and spatial metrics approaches. Earth Interact. 2016;20(14):1–29.

Jain M, Dimri AP, Niyogi D. Land-air interactions over urban-rural transects using satellite observations: analysis over Delhi, India from 1991–2016. Remote Sens. 2017;9(12):1283.

Jain M, Saxena P, Sonwani S. Characteristics, dynamics, and impact of the Thar Desert dust storms on air quality over northern India. Natural Hazards (Accepted); 2024.

Jia X, Quan J, Zheng Z, Liu X, Liu Q, He H, Liu Y. Impacts of anthropogenic aerosols on fog in North China plain. J Geophys Res Atmos. 2019;124(1):252–65.

Kim HH. Urban heat Island. Int J Remote Sens. 1992;13(12):2319–36.

Kim HS, Kim DS, Kim H, Yi SM. Relationship between mortality and fine particles during Asian dust, smog–Asian dust, and smog days in Korea. Int J Environ Health Res. 2012;22(6):518–30.

Kleinman PJ, Pimentel D, Bryant RB. The ecological sustainability of slash-and-burn agriculture. Agric Ecosyst Environ. 1995;52(2–3):235–49.
Koschmieder H. Theorie der horizontalen sichtweite, beiträge zur physik der freien atmosphäre. Meteorol Z. 1924;12:33–55.
Kulkarni R, Jenamani RK, Pithani P, Konwar M, Nigam N, Ghude SD. Loss to aviation economy due to winter fog in New Delhi during the winter of 2011–2016. Atmos. 2019;10(4):198.
Kumar A, Singh R, Singh S. Characteristic of fog, foggy weather and its impact on agriculture. Int J Appl Environ Sci Technol. 2015;3:1.
Kunkel BA. Parameterization of droplet terminal velocity and extinction coefficient in fog models. J Appl Meteorol Climatol. 1984;23(1):34–41.
Kutty SG, Agnihotri G, Dimri AP, Gultepe I. Fog occurrence and associated meteorological factors over Kempegowda International Airport, India. Pure Appl Geophys. 2019;176:2179–90.
Kutty SG, Dimri AP, Gultepe I. Climatic trends in fog occurrence over the indo-Gangetic plains. Int J Climatol. 2020;40(4):2048–61.
Kutty SG, Dimri AP, Gultepe I. Physical processes affecting radiation fog based on WRF simulations and validation. Pure Appl Geophys. 2021;178:4265–88.
LaDochy S. The disappearance of dense fog in Los Angeles: another urban impact? Phys Geogr. 2005;26(3):177–91.
Lankao PR, Qin H. Conceptualizing urban vulnerability to global climate and environmental change. Curr Opin Environ Sustain. 2011;3(3):142–9.
Li ZH, Yang J, Shi CE, Pu MJ. Urbanization effects on fog in China: field research and modeling. Pure Appl Geophys. 2012;169:927–39.
Li H, Meier F, Lee X, Chakraborty T, Liu J, Schaap M, Sodoudi S. Interaction between urban heat Island and urban pollution Island during summer in Berlin. Sci Total Environ. 2018;636:818–28.
Li J, Sun R, Liu T, Xie W, Chen L. Prediction models of urban heat Island based on landscape patterns and anthropogenic heat dynamics. Landsc Ecol. 2021;36:1801–15.
Lietzke B, Vogt R, Young DT, Grimmond CSB. Physical fluxes in the urban environment. In: Understanding urban metabolism. Routledge; 2014. p. 29–44.
Lindberg F, Grimmond CSB, Yogeswaran N, Kotthaus S, Allen L. Impact of city changes and weather on anthropogenic heat flux in Europe 1995–2015. Urban Clim. 2013;4:1–15.
Liu D, Yang J, Niu S, Li Z. On the evolution and structure of a radiation fog event in Nanjing. Adv Atmos Sci. 2011;28:223–37.
Liu D, Li Z, Yan W, Li Y. Advances in fog microphysics research in China. Asia-Pac J Atmos Sci. 2017;53:131–48.
Liu Q, Wang ZY, Wu BG, Nie HH, Chen DH, Gultepe I. Microphysics of fog bursting in polluted urban air. Atmos Environ. 2021;253:118357.
Mahrt L. Weak-wind mesoscale meandering in the nocturnal boundary layer. Environ Fluid Mech. 2007;7:331–47.
Mansouri ST, Zarghami E. Investigating the effect of the physical layout of the architecture of high-rise buildings, residential complexes, and urban heat islands. Energy Built Environ. 2025;6(1):1–17.
McLeod J, Shepherd M, Appelbaum M. Evidence of cloud and rainfall modification in a mid-sized urban area–a climatological analysis of Augusta, Georgia. City Environ Interact. 2024;21:100141.
Miller JD, Hutchins M. The impacts of urbanisation and climate change on urban flooding and urban water quality: a review of the evidence concerning the United Kingdom. J Hydrol Region Stud. 2017;12:345–62.
Mohajerani A, Bakaric J, Jeffrey-Bailey T. The urban heat Island effect, its causes, and mitigation, with reference to the thermal properties of asphalt concrete. J Environ Manag. 2017;197:522–38.
Musselman RC, Sterrett JL. Sensitivity of plants to acidic fog (Vol. 17, No. 2, pp. 329–333). American Society of Agronomy, Crop Science Society of America, and Soil Science Society of America; 1988.
Myers JN. Fog. Sci Am. 1968;219(6):74–83.

Niu F, Li Z, Li C, Lee KH, Wang M. Increase of wintertime fog in China: potential impacts of weakening of the eastern Asian monsoon circulation and increasing aerosol loading. J Geophys Res Atmos. 2010;115:D7.
Niyogi D, Bhalachandran S, Brousse O, Jain M, Shreevastava A, Dimri AP. Investigation of the impact of urbanization under the 2015 Delhi heat wave scenario. In *13th Symp. Of the urban environment* 2017, January.
Oke TR. City size and the urban heat Island. Atmos Environ (1967). 1973;7(8):769–79.
Oke TR. The energetic basis of the urban heat Island. Q J R Meteorol Soc. 1982;108(455):1–24.
Oke TR, Mills G, Christen A, Voogt JA. Urban Climates. Cambridge University Press; 2017.
Paschalis A, Chakraborty TC, Fatichi S, Meili N, Manoli G. Urban forests as main regulator of the evaporative cooling effect in cities. AGU Advances. 2021;2(2):e2020AV000303.
Pérez-Díaz JL, Ivanov O, Peshev Z, Álvarez-Valenzuela MA, Valiente-Blanco I, Evgenieva T, Dreischuh T, Gueorguiev O, Todorov PV, Vaseashta A. Fogs: physical basis, characteristic properties, and impacts on the environment and human health. Water. 2017;9(10):807.
Price JD. On the formation and development of radiation fog: an observational study. Bound-Layer Meteorol. 2019;172(2):167–97.
Quan J, Zhang Q, He H, Liu J, Huang M, Jin H. Analysis of the formation of fog and haze in North China plain (NCP). Atmos Chem Phys. 2011;11(15):8205–14.
Quan J, Liu Y, Jia X, Liu L, Dou Y, Xin J, Seinfeld JH. Anthropogenic aerosols prolong fog lifetime in China. Environ Res Lett. 2021;16(4):044048.
Raj S, Paul SK, Chakraborty A, Kuttippurath J. Anthropogenic forcing exacerbating the urban heat islands in India. J Environ Manag. 2020;257:110006.
Rizwan AM, Dennis LY, Chunho LIU. A review on the generation, determination and mitigation of Urban Heat Island. J Environ Sci. 2008;20(1):120–8.
Román-Cascón C, Yagüe C, Steeneveld GJ, Sastre M, Arrillaga JA, Maqueda G. Estimating fog-top height through near-surface micrometeorological measurements. Atmos Res. 2016;170:76–86.
Romanov P. Urban influence on cloud cover estimated from satellite data. Atmos Environ. 1999;33(24–25):4163–72.
Sachweh M, Koepke P. Radiation fog and urban climate. Geophys Res Lett. 1995;22(9):1073–6.
Salamanca F, Martilli A, Tewari M, Chen F. A study of the urban boundary layer using different urban parameterizations and high-resolution urban canopy parameters with WRF. J Appl Meteorol Climatol. 2011;50(5):1107–28.
Sawaske SR, Freyberg DL. Fog, fog drip, and streamflow in the Santa Cruz Mountains of the California Coast Range. Ecohydrology. 2015;8(4):695–713.
Saxena P, Sonwani S, Srivastava A, Jain M, Srivastava A, Bharti A, Rangra D, Mongia N, Tejan S, Bhardwaj S. Impact of crop residue burning in Haryana on the air quality of Delhi, India. Heliyon. 2021;7(5):e06973.
Schmid PE, Niyogi D. Modeling urban precipitation modification by spatially heterogeneous aerosols. J Appl Meteorol Climatol. 2017;56(8):2141–53.
Seinfeld JH, Pandis SN. Atmospheric chemistry and physics: from air pollution to climate change. 3rd ed. John Wiley & Sons; 2016.
Shahmohamadi P, Che-Ani AI, Maulud KNA, Tawil NM, Abdullah NAG. The impact of anthropogenic heat on formation of urban heat Island and energy consumption balance. Urban Stud Res. 2011;2011(1):497524.
Shao N, Lu C, Jia X, Wang Y, Li Y, Yin Y, Zhu B, Zhao T, Liu D, Niu S, Fan S. Radiation fog properties in two consecutive events under polluted and clean conditions in the Yangtze River Delta, China: a simulation study. Atmos Chem Phys. 2023;23(17):9873–90.
Shem W, Shepherd M. On the impact of urbanization on summertime thunderstorms in Atlanta: two numerical model case studies. Atmos Res. 2009;92(2):172–89.
Shepherd JM. A review of current investigations of urban-induced rainfall and recommendations for the future. Earth Interact. 2005;9(12):1–27.
Shepherd JM, Stallins JA, Jin ML, Mote TL. Urbanization: impacts on clouds, precipitation, and lightning. Urban Ecosyst Ecol. 2010;55:1–28.

Shi C, Roth M, Zhang H, Li Z. Impacts of urbanization on long-term fog variation in Anhui Province, China. Atmos Environ. 2008;42(36):8484–92.
Shi T, Lu G, Wen X, Liu L, Qi P. A systematic review of the potential influence of urbanization on the regional thunderstorm process and lightning activity. Atmos. 2024;15(3):374.
Singh RP, Chauhan A. Sources of atmospheric pollution in India. In: Asian atmospheric pollution. Elsevier; 2022. p. 1–37.
Stallins JA, Rose LS. Urban lightning: current research, methods, and the geographical perspective. Geogr Compass. 2008;2(3):620–39.
Stathopoulou M, Synnefa A, Cartalis C, Santamouris M, Karlessi T, Akbari H. A surface heat Island study of Athens using high-resolution satellite imagery and measurements of the optical and thermal properties of commonly used building and paving materials. Int J Sustain Energy. 2009;28(1–3):59–76.
Stewart ID, Oke TR. Local climate zones for urban temperature studies. Bull Am Meteorol Soc. 2012;93(12):1879–900.
Szidat S. Sources of Asian haze. Science. 2009;323(5913):470–1.
Taha H. Urban climates and heat islands: albedo, evapotranspiration, and anthropogenic heat. Energ Buildings. 1997;25(2):99–103.
Tardif R, Rasmussen RM. Event-based climatology and typology of fog in the new York City region. J Appl Meteorol Climatol. 2007;46(8):1141–68.
Tjernström M, Shupe MD, Brooks IM, Persson POG, Prytherch J, Salisbury DJ, Sedlar J, Achtert P, Brooks BJ, Johnston PE, Sotiropoulou G. Warm-air advection, air mass transformation and fog causes rapid ice melt. Geophys Res Lett. 2015;42(13):5594–602.
Umar M, Atif S, Hildebrandt ML, Tahir A, Azmat M, Zeeshan M. Trends of aerosol optical thickness using VIIRS S-NPP during fog episodes in Pakistan and India. Atmos. 2021;12(2):242.
United Nations. World population prospects: the 2012 revision: volume II: demographic profiles. United Nations Department of Economic and Social Affairs Rep. ST/ESA/SRE.A/345, 2013 844 pp. [Available online at http://esa.un.org/unpd/wpp/Publications/Files/WPP2012_Volume-II-Demographic-Profiles.pdf.]
United Nations. World urbanization prospects: the 2014 revision, highlights. United Nations Department of Economic and Social Affairs Rep. ST/ESA/SER.A/352, 2014 32 pp. [Available online at http://esa.un.org/unpd/wup/Highlights/WUP2014-Highlights.pdf.]
United Nations. World urbanization prospects: the 2018 revision Department of Economic and Social Affairs, population division (2019) (ST/ESA/SER.A/420). New York: United Nations; 2019.
van Oldenborgh GJ, Yiou P, Vautard R. On the roles of circulation and aerosols in the decline of mist and dense fog in Europe over the last 30 years. Atmos Chem Phys. 2010;10(10):4597–609.
Vo TT, Hu L, Xue L, Li Q, Chen S. Urban effects on local cloud patterns. Proc Natl Acad Sci. 2023;120(21):e2216765120.
Voogt JA, Oke TR. Thermal remote sensing of urban climates. Remote Sens Environ. 2003;86(3):370–84.
Wang Q, Zhang X, Wang J. Analysis of the influence of aerosol pollution on fog formation in an urban environment. Atmos Environ. 2015;119:122–9.
Wang Y, Niu S, Lu C, Lv J, Zhang J, Zhang H, Zhang S, Shao N, Sun W, Jin Y, Song Q. Observational study of the physical and chemical characteristics of the winter radiation fog in the tropical rainforest in Xishuangbanna, China. Sci China Earth Sci. 2021;64:1982–95.
Wessels HRA. Growth and disappearance of evaporation fog during the transformation of a cold air mass. Q J R Meteorol Soc. 1979;105(446):963–77.
Williams AP, Schwartz RE, Iacobellis S, Seager R, Cook BI, Still CJ, Husak G, Michaelsen J. Urbanization causes increased cloud base height and decreased fog in coastal Southern California. Geophys Res Lett. 2015;42(5):1527–36.
Witiw MR, LaDochy S. Trends in fog frequencies in the Los Angeles Basin. Atmos Res. 2008;87(3–4):293–300.
Yadav PK. Slash-and-burn agriculture in north-East India. J Expert Opin Environ Biol. 2013;2:2–5.

Yan S, Zhu B, Kang H. Long-term fog variation and its impact factors over polluted regions of East China. J Geophys Res Atmos. 2019;124(3):1741–54.
Yan S, Zhu B, Huang Y, Zhu J, Kang H, Lu C, Zhu T. To what extents do urbanization and air pollution affect fog? Atmos Chem Phys. 2020;20(9):5559–72.
Yang J, Xie YJ, Shi CE, Liu DY, Niu SJ, Li ZH. Ion composition of fog water and its relation to air pollutants during winter fog events in Nanjing, China. Pure Appl Geophys. 2012;169:1037–52.
Yang L, Niyogi D, Tewari M, Aliaga D, Chen F, Tian F, Ni G. Contrasting impacts of urban forms on the future thermal environment: example of Beijing metropolitan area. Environ Res Lett. 2016;11(3):034018.
Yang Y, Tang D, Yang X. Investigating the spatio-temporal variations of the impact of urbanization on haze pollution using multiple indicators. Stoch Env Res Risk A. 2021;35:703–17.
Ye X, Wu B, Zhang H. The turbulent structure and transport in fog layers observed over the Tianjin area. Atmos Res. 2015;153:217–34.
Yu X, Zhu B, Yin Y, Yang J, Li Y, Bu X. A comparative analysis of aerosol properties in dust and haze-fog days in a Chinese urban region. Atmos Res. 2011;99(2):241–7.
Zhang Y, Sun K, Gao Z, Pan Z, Shook MA, Li D. Diurnal climatology of planetary boundary layer height over the contiguous United States derived from AMDAR and reanalysis data. J Geophys Res Atmos. 2020;125(20):e2020JD032803.
Zhao L, Lee X, Smith RB, Oleson K. Strong contributions of local background climate to urban heat islands. Nature. 2014;511(7508):216–9.
Zhao-Ming L, Shou-Ting G, Ji-Song S. A numerical study of the urban intensity effect on fog evolution in the Beijing-Tianjin-Hebei region. Atmos Ocean Sci Lett. 2012;5(3):240–5.
Zhong S, Qian Y, Sarangi C, Zhao C, Leung R, Wang H, Yan H, Yang T, Yang B. Urbanization effect on winter haze in the Yangtze River Delta region of China. Geophys Res Lett. 2018;45(13):6710–8.
Zhou B, Ferrier BS. Asymptotic analysis of equilibrium in radiation fog. J Appl Meteorol Climatol. 2008;47(6):1704–22.
Zhou X, Bai Z, Yang Y. Linking trends in urban extreme rainfall to urban flooding in China. Int J Climatol. 2017;37(13):4586–93.

# Chapter 6
# Fog Dynamics in Affecting Climate and Air Quality

**Kalpana Rajouriya, Atar Singh Pipal, and Ajay Taneja**

## 6.1 Introduction

Fog is composed of very small water droplets (and sometimes minute ice particles) suspended in the atmosphere, reducing horizontal visibility at the Earth's surface to less than 1000 m (Hang et al. 2016). Water makes up 0.001% of the atmosphere (Gleick and Schneider 1996). The water droplets mainly found in the lowest layer in the atmosphere which is known as Troposphere. One of the main factors influencing climatic phenomena is water, which is abundant in the atmosphere and plays a crucial role in defining the Earth's heat budget (William 2014). The final visible representations of water in the atmosphere include fog near the surface of both land and water, as well as clouds at different altitudes. Since this is the only distinction between fog and clouds, fog is frequently described as a stratus cloud that is located slightly above the ground. When defining fog, the main criterion is lower visibility range <1 km (Lakra and Avishek 2022). When heat is removed or moisture is added (by evaporation or combining two air parcels of different temperatures), fog is created when the air gets closer to saturation. Factors including synoptic conditions, local climatology, season, air temperature, boundary layer stability, wind direction, wind speed, dew point depression, topography, terrain, and condensation nuclei properties play a significant role in predicting fog. Because of this, a wide range of physical processes, from the microscopic level of aerosol to the synoptic scale, are involved in fog prediction (Pipal et al. 2014, 2019). The study of Gultepe et al. (2007) provides a highly vivid summary of previous studies on fog and its future directions. Despite the extensive and varied history of fog foresting, precise

K. Rajouriya · A. Taneja (✉)
Department of Chemistry, Dr. Bhimrao Ambedkar University, Agra, India
e-mail: ataneja5@hotmail.com

A. S. Pipal
Indian Institute of Tropical Meteorology, Ministry of Earth Sciences, Pune, India

 
P. Saxena et al. (eds.), *Fog: Formation, Chemistry, Dynamics, Impact and Dissipation*,
https://doi.org/10.1007/978-981-96-4422-3_6

forecasting remains challenging due to the vast scope of physical and chemical processes associated with it (Koračin and Dorman 2017).

Fog affects people and related systems both directly and indirectly wherever it happens in the world. In addition to decreasing visibility, naturally occurring fog can have an impact on the vegetation, air quality, and radiation budget, all of which have a direct impact on temperature and surface–air interaction. Fog events occur in several places on the Earth; some are rare, while others are common. Globally, fog events affect several climate zones, such as tropical, polar, temperature, and moist to dry zones. Due to the presence of more aerosols, fog forms in cities where there is a higher concentration of haze particles as a result of elevated air pollution levels (Zheng et al. 2015; Pipal et al. 2016; Poku et al. 2019; Cheng et al. 2022; Pipal and Taneja 2023). The direct and indirect effects on the human health, animals (directly impacting respiration and dermatological symptoms), and plants are determined by the chemical, physical, and composition of fog condensation nuclei. Air quality scenarios demonstrate how fog combined with haze directly has a negative influence on humans, since haze particles provide a condensation surface for fog droplets to develop in metropolitan locations (Polivka 2018; Hutchings et al. 2010). Deaths are being caused by visual obstructions during periods of intense fog, whether on roads, trains, airplanes, or water transportation (Leung et al. 2020; Peng et al. 2018; Lakra and Avishek 2022).

Fog plays a significant role in both climate dynamics and air quality, though it is often overlooked compared to other weather phenomena like rain or storms. Understanding fog's impact requires a closer look at how it forms, how it interacts with atmospheric conditions, and its effects on local and regional air quality, as well as on the broader climate system. Fog can have a significant impact on both air quality and climate, with effects that vary depending on its nature and composition. Smog, a hazardous combination of smoke and fog, has arisen as a major public health concern, affecting both urban and rural air quality (Razzaq et al. 2024). When comparing the hourly average values of $SO_2$, PM10, and CO air pollutants for monthly, seasonal, and annual bases, Özdemir et al. (2018) discovered that the average concentrations of these pollutants were greater on foggy days than on other days. High aerosol concentrations and low supersaturation are common characteristics of radiation fog in contaminated locations (Mazoyer et al. 2019). Thus, in highly contaminated areas (as determined by Ghude et al. 2023), the concentration of droplet numbers in fogs is frequently significantly higher than in less polluted situations (as determined by Price 2011). At the same time, research shows that fog processing may increase particle toxicity after fog dissipation (Decesari et al. 2017; Patel and Rastogi 2018). Different types of fog can influence climate and air quality in various ways. For example, radiation fog often occurs in calm conditions and can collect pollutants, making the air feel heavier. Understanding how fog interacts with air quality is crucial for anyone concerned about health or safety during foggy weather. As fog rolls in, it is important to be aware that it does not create pollution itself, but rather worsens existing air quality issues. Fog can influence air quality in various ways. It often interacts with air pollutants, potentially leading to health risks and visibility issues. Understanding how fog affects pollutants and their impact on health is crucial for addressing air quality concerns. Fog is made up of tiny water droplets that condense in the air. When humidity is high, this moisture can trap air pollutants like nitrogen dioxide and sulfur dioxide. In urban areas, fog can hold on to fine particle pollution, worsening conditions. This trapped pollution can create a blanket effect, reducing visibility and causing unsafe driving conditions. Pollution can mix with fog during incidents like wildfires, further deteriorating air quality. The presence of smoke and other pollutants makes the situation worse, affecting both people and the environment. Fog, therefore, acts like a container for these harmful substances, trapping them closer to the ground. Fatigue and general discomfort are also common among those exposed to foggy, polluted environments. The combination of high humidity and air pollution makes it important for individuals to monitor their air quality, especially when fog is present.

Here's an overview of how fog dynamics affect both climate and air quality. This chapter provides a quick discussion of all the topics associated to fog dynamics in affecting air quality and climate. Fog formation; fog forecasting and detection; the impact of fog on human health, transportation, the economy, ecology, and the environment are also among the topics covered in this chapter.

## 6.2 Formation of Fog

Fog is formed by the interaction of several influencing elements, which can be synoptic [such as winter teleconnections and Pacific Decadal Oscillation (PDO)], mesoscale (such as the region's topography and orography), and microscopic [such as the chemical and physical characteristics of cloud condensation nuclei (CCN)]. The most important element in the formation and maintenance of fog is the amount of water vapor in the atmosphere. The creation, dissipation, and frequency of fog formation are influenced by the boundary layer condition, thermodynamic features of the atmosphere, and the microphysical characteristics of CCN. Fog droplets require aerosol, humidity, and chilling to form. Atmospheric aerosols function as CCN; their size and chemical makeup affect how quickly they grow and how easily they can produce secondary aerosols. Aerosol also controls fog's microphysical characteristics; particle size and concentration have a greater impact on CCN activation than particles' chemistry. Aerosol number concentration can reduce the saturation value required for fog formation. An increase in aerosol concentration causes a rise in small-sized cloud droplets, which in turn causes a decrease in the effective droplet radius. This increases the optical thickness of fog and, thus, its albedo. High aerosol concentrations and latent heat released during condensation exacerbate the Brownian collision, hastening the production of fog. According to Ahrens and Henson (2021), aerosol is an essential element for the formation of fog droplets; without it, relative humidity must be significantly higher than 100%. According to Ahrens and Henson (2021), fog is created in two fundamental ways: (A) when air temperatures fall below the dew point, radiation, advection, and upslope fog are created; and (B) when more water vapor is introduced to the air, frontal fog and

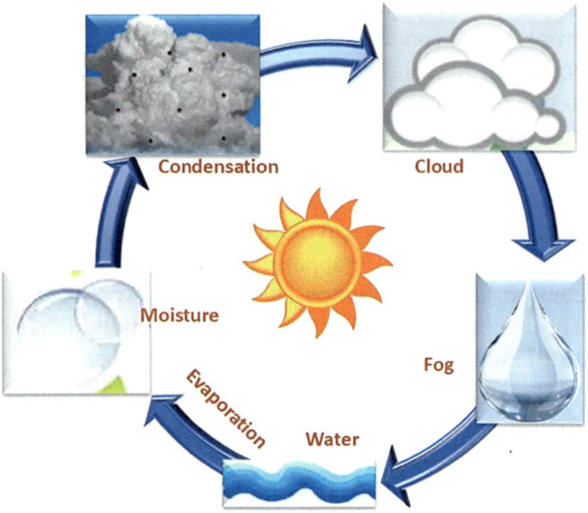


**Fig. 6.1** Formation of fog by cycling procedure. What is fog made of? (types, causes, formation)—JournalHow (Visited on 30/10/2024)

stem fog are created. The phases of the fog life cycle start with (a) the initial stage, when the air is saturated, temperatures are very low, and a strong inversion layer forms. This is followed by (b) the fog formation stage, when the winds are calm, the temperature drops more, and the activation of more aerosol particles makes room for the fog to form. (c) The fog grows and expands, keeping the saturation level constant either by adding more water vapor that requires some mixing or by continuously cooling the air. Finally, there is the dissipation phase. Gaining a comprehensive understanding of fog formation is essential, but so is knowing how and when fog dissipates. Or when does a fog get thick from being shallow? There have been attempts in this area, including the use of infrared photography to monitor the progression of fog at night (Price and Stokkereit 2020) and the function of turbulence. The process of fog formation is depicted in Fig. 6.1. The three primary processes that cause fog to form are now well understood: radiation-induced cooling, turbulence, and interactions between the land and atmosphere. Emerging technologies with significant potential for regional and global spatial and temporal features of fog include satellite detection of clouds and fog and the analysis and interpretation of the resulting data.

### *6.2.1 Factors Affecting Fog Formation*

Rather of focusing on each tiny component, this chapter has merely covered the influencing factors that taken together could affect the fundamental meteorological variables. Based on the extensive study that has been required and conducted over the previous few decades, these factors have been selected. Places such as the Northern Plains of India have extensive fog formation due to the combination of low temperatures, calm weather, heavy aerosol, and moisture brought in by Western disturbances from the Mediterranean Sea (Ghude et al. 2017). An episode of fog, particularly marine fog, can vary in strength and size depending on synoptic scale circulation. Similar to this, fog forms on the Korean Western Peninsula when a high-pressure zone forms in the Yellow Sea. At night, moisture-laden southwesterly winds flow toward the land where they meet with offshore land breezes and katabatic easterly winds. The air becomes saturated due to nocturnal cooling, which causes advection fog to form near the coast. The Iberian Peninsula's fog water collection rate is influenced by teleconnection patterns, specifically the North Atlantic, Mediterranean, and Western Mediterranean oscillations. A study by Corell et al. (2019) found that while the Mediterranean winds and cyclonic circumstances bring in more precipitation, the Atlantic winds and surface level anticyclonic conditions increase the number of foggy days. El Nino and La Nina (climatic patterns) have a big impact on how often winter fog occurs in Eastern China (Hu et al. 2020). The main factors that affect the fog formation are given as follows:

#### 6.2.1.1 Soil Properties

When combined with other favorable climatic circumstances, evaporation from damp soil surfaces delivers water vapor to the atmosphere, which aids in the creation and maintenance of fog. In addition to the local weather, the indo gangetic plain (IGP) region's wet soil—a consequence of irrigation—provides the atmosphere with enough moisture (Izhar et al. 2020). The region is covered in a dense layer of fog. Smith et al.'s research from 2021 revealed that the fog's beginning time might change by 5 h due to the soil's high thermal conductivity. The first attempt to investigate the relationship between soil moisture and fog was conducted by Adhikari and Wang (2020). Smith et al. (2021) have examined the effects of bare soil and grass on fog, but they have also indicated that more research is necessary to determine the degree to which grass can function as an insulator between the soil and the atmosphere.

#### 6.2.1.2 Climate Change

The nature and life cycle of fog are influenced by changes in the terrestrial, atmospheric, and hydrological systems, since each component interacts to determine a single fog episode. The most likely explanation for these patterns is thought to be a synergistic effect. An observational study conducted in a few key Taiwanese locations found that enhanced visibility, which was interpreted as a stand-in for fog and haze occurrences, was caused by decreased concentrations of PM10 and low relative humidity because of rising urban temperatures (Maurer et al. 2019). Due to anticyclonic conditions prevailing in the pacific northwest (PNW), Eastern China has been observed to have numerous winter fog events during El Nino years; in contrast, the opposite occurs during a La Nina event (Hu et al. 2020). According to

Bokwa et al. (2018), urbanization affects fog through changing local temperatures, which shows up as UHI (Urban Heat Island), moisture content, wind speed, and aerosol loading. A common characteristic of nearly all large cities and metropolitan settings is high temperatures, which lower relative humidity and prevent fog from forming. The most obvious illustration of how urbanization affects fog is a fog hole that can be observed in satellite images over cities.

#### 6.2.1.3 Aerosol and Other Air Pollutants Produced by Human Activities

According to Gautam and Singh's (2018) theory, the intense heat coming from cities either makes it possible for fog to form or makes it evaporate faster than it would otherwise. Gautam and Singh (2018) initially observed the fog holes in the Indian Subcontinent on June 30, 2014, MODIS photos across several cities in the Northern Plain, with Delhi showing the largest hole. Low-level fog is suppressed and delayed by UHI (Urban Heat Island) and UDI (Urban Dry Island), although vertical updraft encourages the creation of upper-level fog (Yan et al. 2020). In Shanghai, China, a study discovered that these urban features weakened the inversion layer and prevented ground-level moisture from reaching dew point, which is caused by a rise in nighttime temperature and made it harder for fog to form (Gu et al. 2019).

## 6.3 Detection of Fog Forecasting

With the development of modern technologies, monitoring and fog prediction are no longer restricted to in situ observations. Fog and its related elements are currently being thoroughly studied using many newer tools and approaches. The two main approaches in fog prediction and forecasting are statistical and dynamical forecasting.

### *6.3.1 Numerical Modeling, Satellite Remote Sensing, and Artificial Intelligence*

Numerical Weather Prediction (NWP) models have been used to forecast future weather conditions utilizing the most recent observational weather data through data assimilation since the 1950s, when computer simulation first became popular. The type of fog that forms determines whether a numerical model fog forecast is successful. Advection fog is driven by different processes, while radiative fog is primarily influenced by radiative processes, land–atmosphere interaction, and turbulence. Satellite data are extremely important for predicting in areas where in situ observations are lacking or were setting up observation sites is challenging. Infrared channels are used in most studies on fog identification in satellite photography. The heat from clouds and air particles that reemit solar energy is detected by the satellite through the infrared channel. One popular technique for identifying fog in satellite photos is the Brightness Temperature Difference (BTD) thresholding method. This approach is based on the interplay between the emissive and reflectance properties of the band and the fog droplet. The fog droplet's height, optical thickness, and size are all correlated with the brightness temperature. According to Amani et al. (2020), when it comes to fog detection at night, the temperature obtained at 3.8 μm is somewhat lower than at 11 μm for fog droplets, while it remains the same at both wavelengths for larger sized cloud droplets. Fog's brightness temperature is negated by sunlight reflectance, making it challenging to detect fog during the day using infrared bands (Weston and Temimi 2020). Because high-level clouds can hide ground-level fog, it can be challenging for remote sensing satellites to identify fog. In these situations, microwave telecommunication networks can be used as a supplementary method in addition to the currently available instruments and methodologies. Since atmospheric circumstances affect Received Signal Level (RSL), it makes sense to monitor meteorological events in addition to communication signals for fog identification. Microwave communications have been investigated for their potential in monitoring air pollution, rainfall, and the detection of near-surface moisture. By estimating LWC (Liquid Water Content), microwave links (ML) can be used to identify fog.

## 6.4 Fog Dynamics in Affecting Climate and Air Quality

Some specific effects of fog are given as follows

### *6.4.1 Fog's Impact on Climate*

#### 6.4.1.1 Fog's Influence on Climate

Although it is frequently overlooked, fog plays a crucial role in controlling regional temperatures. Long-term consequences on local weather patterns and ecosystems may result from its influence on temperature, cloud formation, and energy balance.

Regional Cooling and Temperature Regulation

**Albedo Effect** The *albedo effect* refers to the fraction of solar energy reflected by a surface back into space (Stephens et al. 2015). Different surfaces have varying albedo values, with lighter surfaces reflecting more sunlight and darker surfaces absorbing more (Taha et al. 1992). Fog droplets can increase the reflectivity of surfaces, especially in urban areas where buildings and roads are typically dark. This can lead to a higher albedo, reflecting more sunlight and potentially cooling the local environment. By reflecting more sunlight, fog can help cool the surface below it (Lakra and Avishek 2022). This is similar to how snow and ice have high albedo and keep temperatures lower by reflecting sunlight. The presence of fog can create localized cooling effects, which can influence microclimates and potentially affect weather patterns in the area. Since fog has a high albedo (reflectivity), less sunlight reaches the Earth's surface because it reflects sunlight. Cooling may result from this, particularly in temperate and coastal areas (Ching 2025). In coastal regions, for example, marine fog can considerably lower the surface temperature, particularly in the summer. This cooling impact is a crucial component of regional microclimates since it can assist reduce temperature swings (Hale et al. 2024).

**Nighttime Warming** The water droplets in the fog absorb and reradiate infrared radiation from the surface back down, preventing heat from escaping into the atmosphere. As a result, regions experiencing fog often have less extreme temperature drops during the night. This phenomenon can contribute to the creation of microclimates, where specific areas maintain distinct climatic conditions due to fog presence (Peck 1996). It's interesting to note that fog may warm things at night as well. Fog can slow down nighttime cooling by trapping heat close to the surface when it forms and settles over the ground. In places where fog is frequent, this may lead to milder nights (Zhang et al. 2024a).

**Moisture Regulation** Fog plays an important role in regulating moisture in the environment. Fog can lead to light precipitation, such as drizzle, which directly adds moisture to the ground and vegetation. When fog settles on plants and the ground, it can lead to the formation of dew, which provides a crucial source of moisture for many plants, especially in arid and semiarid regions (Vogel and Müller-Doblies 2011). By maintaining high humidity levels, fog can slow down the evaporation of water from the soil and vegetation, helping to conserve moisture in the ecosystem. Fog contributes to the local water cycle by returning moisture to the ground that might have otherwise been lost to the atmosphere. Local humidity levels can rise due to fog, which has an impact on regional precipitation patterns and evaporation rates (Alkandari 2024). High moisture levels in fog-prone places can affect local weather, plant development, and even agricultural cycles. In some places, persistent fog can limit the quantity of sunlight that reaches the surface, which can have an impact on how plants perform photosynthesis.

### Cloud Formation and Feedback Mechanisms

**Fog as a Precursor to Clouds** Fog can indeed act as a precursor to cloud formation under certain conditions. If the air containing fog is lifted, it can rise above the ground. This lifting can be caused by factors such as wind or changes in atmospheric pressure. As the fog rises, it can transition into low-lying clouds. This is because the process of lifting can cause further cooling and condensation, leading to the formation of cloud droplets. Once the fog has risen and transformed into clouds, these clouds can continue to develop and grow, influenced by various atmospheric conditions such as temperature, humidity, and wind patterns. Fog and low-level clouds share similar dynamics, and fog formation can lead to the creation of low clouds. These clouds also contribute to cooling by reflecting solar radiation back into space (Zhang et al. 2024b).

**Climate Feedback Loops** In fog-prone coastal areas, the combination of fog, low clouds, and the ocean can generate climate feedback loops. Fog and low clouds, for instance, can lower the quantity of solar radiation that the Earth's surface absorbs, resulting in lower surface temperatures. Fog development may then be strengthened by cooler temperatures. Regional climates can be impacted by this feedback process, particularly in coastal areas where fog is prevalent (Lakra and Avishek 2022).

#### 6.4.1.2 Fog and Climate Change

The interaction between fog and *climate change* is a topic of increasing interest, as changing temperature patterns, humidity levels, and atmospheric circulation could alter fog dynamics in various regions.

Impact of Warming on Fog Formation

**Warmer Air** As global temperatures rise, the capacity of air to hold moisture increases, which may lead to more fog formation in some areas. However, warmer air may also reduce the frequency of fog in regions where it is typically formed through cooling (e.g., radiational fog in inland areas) (Maurer et al. 2019).

**Changing Fog Patterns** In coastal areas where advection fog is common, warming ocean temperatures might enhance fog formation due to the increased moisture available in the atmosphere. Conversely, in more arid regions, warmer temperatures may reduce fog frequency as the air remains too warm to cool sufficiently for condensation.

**Urban Heat Island Effect** Cities are likely to see altered fog patterns due to the urban heat island effect, where cities are warmer than surrounding rural areas due to human activity, heat absorption by buildings, and decreased vegetation. This could prevent fog formation or alter its characteristics, affecting both local temperature regulation and air quality (Gautam and Singh 2018).

Fog as a Climate Indicator

**Environmental Shifts** Changes in fog frequency and intensity could serve as indicators of broader climate shifts. For example, an increase in fog may signal changes in ocean temperatures or atmospheric moisture levels, offering early warnings for shifts in regional weather patterns or climate systems (Syed et al. 2012).

## 6.4.2 Fog's Affects the Air Quality

### 6.4.2.1 Reduce Quality of Air Due to the Fog

Fog, especially in urban or industrial areas, can *trap pollutants* near the ground, particularly in areas with high levels of particulate matter (PM2.5), nitrogen oxides ($NO_x$), sulfur dioxide ($SO_2$), and other industrial emissions. This leads to *increased air pollution* and can contribute to the formation of *smog*. Urban air is typically full of contaminants that act as cloud condensation nuclei (CCN), which increases the likelihood of fog development when combined with other favorable climatic circumstances. What effect can fog have from contaminated air carrying an aerosol? The surface tension of CCN is impacted by pollution, which makes them activate even at lower relative humidity levels than in cleaner air. Smog is the most notorious result of the combination of pollution and fog. Dr. Harold Antoine Des Voeux first described the classical smog as a combination of smoke or soot and fog following the Glasgow, Scotland, smog incident in 1909. Subsequently, after the horrifying London smog incident in 1952, when sulfur dioxide emissions from excessive coal burning were found to be the culprit gas, smog containing relatively high concentrations of $SO_2$ formed in winter months with high humidity, a phenomenon also known as "London Smog".

Another type of smog is known as "photochemical smog," because it is created in sunlight and contains ozone and nitrogen oxide as its principal constituents. It is sometimes referred to as ground-level ozone or "Los Angeles Smog" after an incident that occurred in Los Angeles. Ozone is also produced by volatile organic compounds, or VOCs (Osibanjo et al. 2021). Recently, a new class of smog was discovered in Poland, earning the moniker "Polish Smog." This class of smog was differentiated from the two previously described classes based on important contaminants and climatic conditions. Additionally, pollutants like nitrogen oxides ($NO_x$), a precursor to ozone, can still accumulate during foggy conditions and contribute to secondary air pollution once the fog lifts. Due to the high amounts of dust (PM2.5 and PM10) and benzopyrene, they are also known as dusty smog. It has been determined that the dust particles contain a high percentage of ammonium sulfate and ammonium nitrate, which formed under anticyclonic conditions brought on by Easterly winds and temperatures below 0 °C (Wielgosiński and Czerwińska 2020). As CCN, aerosols are essential to the development of fog droplets. Even if they are not functioning as CCN, absorbing aerosols can still have an impact on the fog by increasing air temperature, lowering relative humidity, preventing fog from forming, and accelerating the evaporation of the fog droplets. In addition to the effects of UHI, this may be one of the causes of the declining frequency of fog and the formation of fog holes over urban areas. Elevated levels of air pollution have the direct effect of causing fog to form, while their indirect effect of raising air temperature and causing UHI can also result in the formation of urban fog holes. Thus, it is questionable if aerosols have a clear influence. According to Poku et al. (2019), the interaction between fog and aerosol particles can have an impact on the chemical composition, size, and distribution of the aerosol as well as cause secondary aerosol formation. Additionally, the concentration of aerosols in the atmosphere can control the optical thickness of fog.

Human Health Issues by Environmental Fog

Long-term impacts on the respiratory and cardiovascular systems, resulting in ischemic heart disease and cerebrovascular illness (Polivka 2018). A few studies (Bharadwaj et al. 2016) describe the effects of the 1952 fog on participants who were either still in the womb of their mothers at the time of the incident or had just turned one year old. According to Javed et al. (2021), the synergistic effect of COVID-19 and pollution is explained in depth in this chapter.

#### 6.4.2.2 Visibility Reduction

A significant meteorological factor that has a negative impact on aviation operations is visibility. In comparison to other weather-related mishaps, visibility-related fatalities occur most frequently (30–35%), according to the NTSB what is this data, and its consequences on aircraft-related delays are the highest (Gultepe et al. 2019). As a result, its estimation and measurements are crucial for algorithm development and aviation planning. Ceiling and visibility were reported as contributory variables in 24% of all general aviation incidents between 1989 and early 1997, according to a review of National Transportation Safety Board (NTSB) statistics (Kulesa 2002). According to the National Transportation Safety Board (NTSB), weather has a major role (about 23%) in all aviation accidents (Kulesa 2002). According to FAA (2010) research based on the NTSB data repository, low visibility and ceiling circumstances were contributing factors to at least 18% of weather-related mishaps. Aerial vehicle takeoff and landing delays can also be greatly impacted by visibility and ceiling conditions. The categories of fog in respect to visibility are given in Table 6.1.

From the previous literature, it was inferred that fog formation increases with rising pollution levels (Table 6.2). In the air, different types of pollutants like PM, nitrates, organic pollutants, and inorganic elements are present. From the literature,

**Table 6.1** Categories of fog in respect to visibility

| Visibility (m) | Fog density class | LWC (mg $m^{-3}$) |
|---|---|---|
| <100 | Very dense fog | 81.9 |
| 100–250 | Dense fog | 85.4 |
| 251–500 | Medium fog | 22.6 |
| 501–1000 | Light fog | 12.6 |
| 1000–2000 | Haze | 4.1 |
| >2000 | No fog | 10.3 |

**Table 6.2** Global data of visibility reduction due to the fog

| S.No. | References | Location | Findings |
|---|---|---|---|
| 1. | Nelli et al. (2024) | United Arab Emirates | Although the fine mode is associated with a high number of smaller fog droplets, medium to big droplets between 10 and 35 μm fundamentally contribute the most to the liquid water content (LWC) |
| 2. | Faryal et al. (2024) | Canada | As could be expected, changes in density and velocity are greater in dense fog because visibility is poor, whereas changes are smaller in thin fog because visibility is better |
| 3. | Tiwari et al. (2011) | Delhi, India | The outcome shows that lower depression temperatures and higher pollution levels (aerosol mass concentration) are conducive to fog formation. Thus, one of the key elements in visibility predictions is the mass concentration of aerosols |
| 4. | Quan et al. (2011) | China | A high concentration of aerosols encourages the development of many tiny fog drops, which more effectively decrease visibility |
| 5. | Shen et al. (2015) | China | It was discovered that nitrate was more crucial to the development of haze |
| 6. | Xue et al. (2015) | China | The findings demonstrated that the high quantities of air pollutants linked to low wind speed were the primary cause of the poor visibility above Shanghai, accounting for 44.99% of the variance |
| 7. | Boutle et al. (2018) | Cardington, UK | The quantity of buildup and coarse-mode aerosol present for activation, since sensitivity testing using the large eddy simulation (LES) showed that a faster transition to well-mixed fog occurs when initial concentrations of bigger aerosol are significantly increased |
| 8. | Calsavara et al. (2021) | Brazil | This study is among the first to use driving simulators to assess the environment of the driving infrastructure in Brazil when there is fog |
| 9. | Xu et al. (2015) | China | As the communication range is increased, the path loss will decrease initially before increasing in tandem with the aerosol density. This phenomenon also happens when fog drop sizes grow |
| 10. | Lin et al. (2012) | China | The findings indicate that the lower PM10 concentration was mostly responsible for the increase in vision in both Beijing and Guangzhou |
| 11. | Deng et al. (2016) | China | Even if precipitation causes a drop in PM2.5 mass, high relative humidity combined with precipitation would intensify the aerosol hygroscopic growth effect, resulting in additional deterioration of visibility |
| 12. | Peng et al. (2020) | China | Aerosol particles don't contribute as much to extremely low visibility as relative humidity does |
| 13. | Dey et al. (2018) | India | Satellites use a positive brightness temperature difference (BTD) to identify fog at night |
| 14. | Tyagi et al. (2017) | India | Ten hours a day on average, the National Capital Region (NCR) of Delhi sees fog with visibility below 1000 m, with daily average visibility fluctuating between 1300 and 550 m during the fog and extended fog (>14 h) phases |

it was observed that low visibility due to fog occurs, because air quality deteriorates and various types of pollutants are present in the air (Nelli et al. 2024; Tiwari et al. 2011; Quan et al. 2011; Xue et al. 2015; Boutle et al. 2018; Lin et al. 2012; Deng et al. 2016). In the literature, it was also shown that some meteorological parameters (temperature, humidity, and wind) are also responsible for the fog formation and the visibility reduction (Tiwari et al. 2011; Xue et al. 2015; Deng et al. 2016). According to Peng et al. (2020), humidity is more responsible for the fog formation compared to the aerosol.

## 6.5 Conclusions

Fog is a widespread and complex atmospheric phenomenon that forms when air near the Earth's surface becomes saturated with moisture, causing water vapor to condense into tiny droplets. There are three main processes in understanding fog formation: turbulence, radiation-induced cooling, and land–atmosphere interactions. Distinct meteorological conditions cause distinct types of fog, such as advection fog, valley fog, and radiation fog. Fog, which originates at the surface as opposed to higher clouds, can significantly affect daily life, transportation, and agriculture. While fog is often associated with reduced visibility, its dynamics also have important implications for both climate and air quality. Fog dynamics are integral to both climate regulation and air quality. On one hand, fog can moderate local temperatures, alter cloud formation, and influence regional weather patterns. On the other hand, it can trap pollutants, exacerbating air quality issues, especially in urban and industrial areas. Understanding fog's dual role in improving and worsening air quality is critical for managing pollution, especially in fog-prone regions, and for predicting its long-term impacts on climate. With ongoing climate change, the frequency, duration, and intensity of fog events may shift, further altering the balance between its beneficial and detrimental effects on air quality and climate.

## References

Adhikari B, Wang L. The potential contribution of soil moisture to fog formation in the Namib Desert. J Hydrol. 2020;591:125326.

Ahrens CD, Henson R. Meteorology today: an introduction to weather, climate, and the environment. Boston: Cengage Learning; 2021.

Ali F, Khan ZH, Khattak KS, Gulliver TA. The effect of visibility on road traffic during foggy weather conditions. IET Intell Transp Syst. 2024;18(1):47–57. https://doi.org/10.1049/itr2.v18.1; https://doi.org/10.1049/itr2.12432.

Alkandari AJ. Climate and climate change aspects of Kuwait. In: Terrestrial environment and ecosystems of Kuwait: assessment and restoration. Cham: Springer; 2024. p. 57–91.

Amani M, Mahdavi S, Bullock T, Beale S. Automatic nighttime sea fog detection using GOES-16 imagery. Atmos Res. 2020;238:104712.

Bharadwaj P, Zivin JG, Mullins JT, Neidell M. Early-life exposure to the great smog of 1952 and the development of asthma. Am J Respir Crit Care Med. 2016;194(12):1475–82.

Bokwa A, Wypych A, Hajto MJ. Impact of natural and anthropogenic factors on fog frequency and variability in Kraków, Poland in the years 1966–2015. Aerosol Air Qual Res. 2018;18(1):165–77.

Boutle I, Price J, Kudzotsa I, Kokkola H, Romakkaniemi S. Aerosol–fog interaction and the transition to well-mixed radiation fog. Atmos Chem Phys. 2018;18(11):7827–40.

Calsavara F, Kabbach Junior FI, Larocca APC. Effects of fog in a Brazilian road segment analyzed by a driving simulator for sustainable transport: drivers' visual profile. Sustainability. 2021;13(16):9448.

Cheng YH, Huang YC, Pipal AS, Jian MY, Liu ZS. Source apportionment of black carbon using light absorption measurement and impact of biomass burning smoke on air quality over rural central Taiwan: a yearlong study. Atmos Pollut Res. 2022;13(1):101264.

Ching J. Climate change and air quality: role of atmospheric aerosols and challenges to society. In: Routledge handbook of climate change and society. London: Routledge; 2025. p. 45–59.

Corell D, Estrela MJ, Valiente JA, Azorin-Molina C, Chen D. Influences of synoptic situation and teleconnections on fog-water collection in the Mediterranean Iberian Peninsula, 2003–2012. Int J Climatol. 2019;40:3297–317.

Decesari S, Sowlat MH, Hasheminassab S, Sandrini S, Gilardoni S, Facchini MC, Sioutas C. Enhanced toxicity of aerosol in fog conditions in the Po Valley, Italy. Atmos Chem Phys. 2017;17(12):7721–31.

Deng H, Tan H, Li F, Cai M, Chan PW, Xu H, Wu D. Impact of relative humidity on visibility degradation during a haze event: a case study. Sci Total Environ. 2016;569:1149–58.

Dey S. On the theoretical aspects of improved fog detection and prediction in India. Atmos Res. 2018;202:77–80.

FAA. Weather-related aviation accident study 2003–2007. Washington: Federal Aviation Administration; 2010. p. 51.

Gautam R, Singh MK. Urban heat island over Delhi punches holes in widespread fog in the Indo-Gangetic plains. Geophys Res Lett. 2018;45(2):1114–21.

Ghude SD, Bhat GS, Prabhakaran T, Jenamani RK, Chate DM, Safai PD, Rajeevan M. Winter fog experiment over the Indo-Gangetic plains of India. Curr Sci. 2017;112:767–84.

Ghude SD, Jenamani RK, Kulkarni R, Wagh S, Dhangar NG, Parde AN, Rajeevan M. WiFEX: walk into the warm fog over Indo-Gangetic Plain region. Bull Am Meteorol Soc. 2023;104(5):E980–E1005.

Gleick PH, Schneider S. Encyclopedia of climate and weather. Water Resour. 1996;2:817–23.

Gu Y, Kusaka H, Tan J. Impacts of urban expansion on fog types in Shanghai, China: numerical experiments by WRF model. Atmos Res. 2019;220:57–74.

Gultepe I, Tardif R, Michaelides SC, Cermak J, Bott A, Bendix J, Cober SG. Fog research: a review of past achievements and future perspectives. Pure Appl Geophys. 2007;164:1121–59.

Gultepe I, Sharman R, Williams PD, Zhou B, Ellrod G, Minnis P, Neto FA. A review of high impact weather for aviation meteorology. Pure Appl Geophys. 2019;176:1869–921.

Hale M, Merrifield M, Clemesha R, Gershunov A, Guirguis K, Benmarhnia T, Iacobellis S. Mean summer land temperatures in the Southern California coastal zone: connections with ocean processes. J Geophys Res Atmos. 2024;129(14):e040188.

Hang C, Nadeau DF, Gultepe I, Hoch SW, Román-Cascón C, Pryor K, Pardyjak ER. A case study of the mechanisms modulating the evolution of valley fog. Pure Appl Geophys. 2016;173:3011–30.

Hay WW. The accelerating rate of global change. Rend Fis Acc Lincei. 2014;25(1):29–48. https://doi.org/10.1007/s12210-014-0287-z

Hu S, Zhang W, Turner AG, Sun J. How does El Niño-Southern oscillation affect winter fog frequency over eastern China? Clim Dyn. 2020;54:1043–56.

Hutchings JW, Ervens B, Straub D, Herckes P. N-nitrosodimethylamine occurrence, formation and cycling in clouds and fogs. Environ Sci Technol. 2010;44(21):8128–33.

Izhar S, Gupta T, Panday AK. Scavenging efficiency of water soluble inorganic and organic aerosols by fog droplets in the Indo Gangetic Plain. Atmos Res. 2020;235:104767.

Javed A, Aamir F, Gohar UF, Mukhtar H, Zia-Ui-Haq M, Alotaibi MO, Pop OL. The potential impact of smog spell on humans' health amid COVID-19 rages. Int J Environ Res Public Health. 2021;18(21):11408.

Koračin D, Dorman CE. Introduction. In: Koračin D, Dorman C, editors. Marine fog: challenges and advancements in observations, modeling, and forecasting. Cham: Springer; 2017. p. 1–6.

Kulesa G. Weather and aviation: how does weather affect the safety and operations of airports and aviation, and how does FAA work to manage weather-related effects? J Potential Impacts Climate Change Transp. 2002;5:10.

Lakra K, Avishek K. A review on factors influencing fog formation, classification, forecasting, detection and impacts. Rendiconti Lincei. 2022;33(2):319–53.

Leung AC, Gough WA, Butler KA. Changes in fog, ice fog, and low visibility in the Hudson Bay region: impacts on aviation. Atmos. 2020;11(2):186.

Lin M, Tao J, Chan CY, Cao JJ, Zhang ZS, Zhu LH, Zhang RJ. Regression analyses between recent air quality and visibility changes in megacities at four haze regions in China. Aerosol Air Qual Res. 2012;12(6):1049–61.

Maurer M, Klemm O, Lokys HL, Lin NH. Trends of fog and visibility in Taiwan: climate change or air quality improvement? Aerosol Air Qual Res. 2019;19:896–910. https://doi.org/10.4209/aaqr.2018.04.0152.

Mazoyer M, Burnet F, Denjean C, Roberts GC, Haeffelin M, Dupont JC, Elias T. Experimental study of the aerosol impact on fog microphysics. Atmos Chem Phys. 2019;19(7):4323–44.

Nelli N, Francis D, Abida R, Fonseca R, Masson O, Bosc E. In-situ measurements of fog microphysics: Visibility parameterization and estimation of fog droplet sedimentation velocity. Atmos Re. 2024;309:107570.

Osibanjo OO, Rappenglück B, Retama A. Anatomy of the March 2016 severe ozone smog episode in Mexico-City. Atmos Environ. 2021;244:117945.

Özdemir ET, Deniz A, Yavuz V, Doğan N, Akbayır İ. Investigation of fog-air quality relationship in Istanbul. Fresenius Environ Bull. 2018;27(1):30–6.

Patel A, Rastogi N. Oxidative potential of ambient fine aerosol over a semi-urban site in the Indo-Gangetic Plain. Atmos Environ. 2018;175:127–34.

Peck MS. Transforming the globe. In: Designing the molecular world: chemistry at the frontier, vol. 19. Princeton: Princeton University Press; 1996. p. 323.

Peng Y, Abdel-Aty M, Lee J, Zou Y. Analysis of the impact of fog-related reduced visibility on traffic parameters. J Transp Eng Part A Syst. 2018;144(2):04017077.

Peng Y, Wang H, Hou M, Jiang T, Zhang M, Zhao T, Che H. Improved method of visibility parameterization focusing on high humidity and aerosol concentrations during fog–haze events: application in the GRAPES_CAUCE model in Jing-Jin-Ji, China. Atmos Environ. 2020;222:117139.

Pipal AS, Taneja A. Measurements of indoor air quality: science and applications. In: Handbook of metrology and applications. Singapore: Springer; 2023. p. 1621–55.

Pipal AS, Jan R, Bisht DS, Srivastava AK, Tiwari S, Taneja A. Day and night variability of atmospheric organic and elemental carbon during winter of 2011-12 in Agra, India. Sustain Environ Res. 2014;24(2):107–16.

Pipal AS, Tiwari S, Satsangi PG. Seasonal chemical characteristics of atmospheric aerosol particles and its light extinction coefficients over Pune, India. Aerosol Air Qual Res. 2016;16(8):1805–19.

Pipal AS, Singh S, Satsangi GP. Study on bulk to single particle analysis of atmospheric aerosols at urban region. Urban Clim. 2019;27:243–58.

Poku C, Ross AN, Blyth AM, Hill AA, Price JD. How important are aerosol–fog interactions for the successful modelling of nocturnal radiation fog? Weather. 2019;74(7):237–43.

Polivka BJ. The great London smog of 1952. Am J Nurs. 2018;118(4):57–61.

Price J. Radiation fog. Part I: observations of stability and drop size distributions. Bound-Layer Meteorol. 2011;139:167–91.

Price J, Stokkereit K. The use of thermal infra-red imagery to elucidate the dynamics and processes occurring in fog. Atmos. 2020;11(3):240.
Quan J, Zhang Q, He H, Liu J, Huang M, Jin H. Analysis of the formation of fog and haze in North China Plain (NCP). Atmos Chem Phys. 2011;11(15):8205–14.
Razzaq A, Zafar MM, Zahra LT, Qadir F, Qiao F, Jiang X. Smog: Lahore needs global attention to fix it. Environ Challenges. 2024;16:100999.
Shen XJ, Sun JY, Zhang XY, Zhang YM, Zhang L, Che HC, Zhang YW. Characterization of submicron aerosols and effect on visibility during a severe haze-fog episode in Yangtze River Delta, China. Atmos Environ. 2015;120:307–16.
Smith DK, Renfrew IA, Dorling SR, Price JD, Boutle IA. Sub-km scale numerical weather prediction model simulations of radiation fog. Q J R Meteorol Soc. 2021;147(735):746–63.
Stephens GL, O'Brien D, Webster PJ, Pilewski P, Kato S, Li JL. The albedo of Earth. Rev Geophys. 2015;53(1):141–63.
Syed FS, Körnich H, Tjernström M. On the fog variability over south Asia. Clim Dyn. 2012;39:2993–3005.
Taha H, Sailor DJ, Akbari H. High-albedo materials for reducing building cooling energy use. Berkeley: Lawrence Berkeley National Laboratory; 1992.
Tiwari S, Payra S, Mohan M, Verma S, Bisht DS. Visibility degradation during foggy period due to anthropogenic urban aerosol at Delhi, India. Atmos Pollut Res. 2011;2(1):116–20.
Tyagi S, Tiwari S, Mishra A, Singh S, Hopke PK, Singh S, Attri SD. Characteristics of absorbing aerosols during winter foggy period over the National Capital Region of Delhi: impact of planetary boundary layer dynamics and solar radiation flux. Atmos Res. 2017;188:1–10.
Vogel S, Müller-Doblies U. Desert geophytes under dew and fog: The "curly-whirlies" of Namaqualand (South Africa). Flora-Morphol Distrib Funct Ecol Plants. 2011;206(1):3–31.
Weston M, Temimi M. Application of a nighttime fog detection method using SEVIRI over an arid environment. Remote Sens. 2020;12(14):2281.
Wielgosiński G, Czerwińska J. Smog episodes in Poland. Atmos. 2020;11(3):277.
Xu C, Zhang H, Cheng J. Effects of haze particles and fog droplets on NLOS ultraviolet communication channels. Opt Express. 2015;23(18):23259–69.
Xue D, Li C, Liu Q. Visibility characteristics and the impacts of air pollutants and meteorological conditions over Shanghai, China. Environ Monit Assess. 2015;187:1–10.
Yan S, Zhu B, Huang Y, Zhu J, Kang H, Lu C, Zhu T. To what extents do urbanization and air pollution affect fog? Atmos Chem Phys. 2020;20(9):5559–72.
Zhang L, Shi H, Gao S, Li S. Overestimated fog-top entrainment in WRF simulation leading to unrealistic dissipation of sea fog: a case study. Remote Sens. 2024a;16(10):1656.
Zhang G, Peng X, Sun W, Fu Y, Yang Y, Liu D, Bi X. Fog/cloud processing of atmospheric aerosols from a single particle perspective: a review of field observations. Atmos Environ. 2024b;329:120536.
Zheng W, Li X, Xie J, Yin L, Wang Y. Impact of human activities on haze in Beijing based on grey relational analysis. Rendiconti Lincei. 2015;26:187–92.

# Chapter 7
# Impact of Fog on Physiology and Growth of Plant Species

**Diksha Singh and Bhanu Pandey**

## 7.1 Introduction

Fog forms when atmospheric water vapor condenses around particles such as dust or salt, known as condensation nuclei, producing minute droplets that provide a cloud-like appearance at ground level (Herndon and Whiteside 2018). Fog can be classified by its formation mechanisms, including radiation, advection, and upslope fog (Hang et al. 2016; Price 2019). Radiative cooling at night cools the Earth's surface and nearby air, commonly forming radiation fog (Lin et al. 2021). Advection cooling occurs when warm, moist air passes over cooler surfaces, like snow or cold water, cooling to its dew point and forming advection fog (Yang et al. 2018). Upslope cooling happens as air rises along a mountain slope, cooling adiabatically and forming upslope fog (Chen et al. 2021). The homogenous and heterogenous nucleation and nuclei diversity during condensation was due to activation, growth, and deposition of fog droplets (Pérez-Díaz et al. 2017). This fog droplet formation occurs by condensation of water vapor in the absence of foreign condensation nuclei (i.e., homogenous condensation) with supersaturation higher than several 100%, which usually does not occur in the atmosphere. Fog droplets condense onto tiny particles (like dust, salt, or pollutants) in the atmosphere, which act as condensation nuclei. Without these, it would be difficult for water vapor to condense (Gill et al. 1983). When air cools to its dew point, it reaches saturation and excess water vapor

D. Singh
CSIR-Central Institute of Mining and Fuel Research, Barwa Road Campus, Dhanbad, Jharkhand, India

B. Pandey (✉)
CSIR-Central Institute of Mining and Fuel Research, Barwa Road Campus, Dhanbad, Jharkhand, India

Academy of Scientific and Innovative Research (AcSIR), Ghaziabad, Uttar Pradesh, India
e-mail: bhanu@cimfr.nic.in; bhanubot@gmail.com

 
P. Saxena et al. (eds.), *Fog: Formation, Chemistry, Dynamics, Impact and Dissipation*,
https://doi.org/10.1007/978-981-96-4422-3_7

condenses, often forming fog near the surface (Gultepe et al. 2007). For fog to form, relative humidity must be close to 100% (Cáceres et al. 2007). As air cools to the dew point, water vapor condenses into droplets that scatter light, reducing visibility. As condensation continues, the fog layer can deepen. The fog will persist as long as cooling, high relative humidity, and adequate moisture are maintained. The height of the fog layer is determined by the extent of cooling and the amount of moisture available. Fog often forms near the ground and can vary in thickness depending on local conditions like wind pattern, topography temperature, and humidity. Fog formation is also influenced by the presence of water bodies or distinct landscapes, where warm, moist air comes into contact with a colder surface, causing condensation (Gultepe et al. 2007). High relative humidity in the lowest atmospheric layers is crucial; during radiative cooling, fog is more likely to form than dew in stagnant air (Hůnová et al. 2018). Fog varies by region due to climate, topography, and land use and can be studied with satellite and drone data (Akimoto and Kusaka 2015).

Population growth and urbanization contribute to the urban heat island effect by increasing heat absorption due to lower albedo in urban areas, while the reduction of cropland, water bodies, and forests decreases water vapor. As a result, cities experience higher temperatures and lower humidity, reducing fog formation (Yan et al. 2020). Over time, urban fog days have declined (Sachweh and Koepke 1995; LaDochy 2005; Shi et al. 2008; Guo et al. 2017; Yan et al. 2019). While urban heat islands inhibit near-surface fog, upward air movements can transfer water vapor at higher levels, creating a "wet island" effect in the upper atmosphere (Niu et al. 2010; Li et al. 2012; Kang et al. 2022). Aerosols, particularly from pollution, have a complex impact on fog, affecting it directly through radiation and indirectly by altering droplet formation (Khain and Pinsky 2018). Aerosols that scatter sunlight modify how the Earth absorbs and reflects energy (Tie et al. 2019). During the day, they block sunlight, cooling the surface and allowing fog to linger longer (Shi et al. 2008; Maalick et al. 2016). At night, aerosols have less impact, since surface cooling drives fog formation (Maalick et al. 2016). Aerosols above the fog layer create a "dome effect," blocking sunlight and preventing fog evaporation (Ding et al. 2016), while aerosols within the fog layer absorb heat and cause quicker evaporation (Maalick et al. 2016). Higher aerosol concentrations lead to more numerous, smaller fog droplets, making fog denser and more resistant to evaporation (Fan et al. 2018; Rosenfeld et al. 2008). However, excessive aerosol levels can thin or prevent fog by limiting available moisture (Quan et al. 2011; Wang et al. 2015; Guo et al. 2017). Aerosols can also alter weather patterns, creating land–sea temperature differences and shifting wind systems, which can promote fog in new areas (Niu et al. 2010; Ding et al. 2019).

Fog plays a crucial role in regulating the water balance and photosynthetic function of vegetation and contributing to biogeochemical cycles (Carmichael et al. 2020; Bagukas et al. 2021; Berrones et al. 2022). Fog is indispensable for moisture supply to plants in arid and semiarid regions during dry spells, thus lowering water stress, hydrating, and helping retain turgor pressure (Negret et al. 2013; Merrium et al. 2022a, b). In the Atacama and Namib Deserts, fog lowers temperatures, increasing humidity and reducing evapotranspiration and conserving soil moisture, which generally promotes a stable microclimate conducive to growth (Soderberg 2010; Azúa-Bustos et al. 2011; Roth-Nebelsick et al. 2012; Glaser et al. 2022; Sun et al.

2024). Fog also provides nitrogen, sulfur, and phosphorus to leaves, enriching nutrient uptake, a process that favors nutrient-poor soils (Nyaga 2013; Gottlieb et al. 2019). Trichomes and waxy cuticles of the coastal redwoods ensure coherent assimilation of moisture in foggy conditions (Dawson and Goldsmith 2018; Chin 2021). The moisture present in the soil enhances, assisting the shallow-rooted plants and maintaining the symbiotic association with the mycorrhizal fungi. It thus prevents extreme water deficit conditions in the drought-stressed plants and maintains photosynthesis and respiration (Dawson and Goldsmith 2018; Gerber 2018). The process supports biodiversity by using absorbed moisture in leaves, specializing in cloud forest epiphytes. However, fog can inhibit photosynthesis by reducing sunlight and promote fungal infections like mildew, which harm plant tissues (Mieslerová et al. 2022). Fog accumulation on leaves can impede stomatal gas exchange, reducing $CO_2$ uptake and oxygen release, affecting plant growth (Pariyar et al. 2017). In polluted areas, acidic fog from SOx and NOx damages plant tissues, disrupts cell membranes, alters soil chemistry, and increases toxic metals like aluminum and cadmium, harming plant growth and vitality. Acid fog strips away critical nutrients, degrading the fertility of the soil and further elevating metal toxicity, which inhibits plant growth and enhances oxidative stress (Bytnerowicz and Grulke 1992; Grodzińska-Jurczak 1998). Soil acidification by polluted fog mobilizes hazardous metals further, and these modifications also inhibit the action of plant growth promoting rhizospheric microorganisms (PGPR), which are the key organisms involved in the uptake of nutrients and pathogen resistance of the plants (Dar et al. 2019). These interactions must be understood to improve climate-change-related agriculture and ecosystem management. Wheat, for example, has adapted to catch and use fog water, making it drought-resistant (Hakeem et al. 2021; Ali et al. 2022). These patterns have ecological ramifications, especially in climate-change-induced fog reductions. Comprehensive research is needed, since plant communities are interdependent and can adapt to fog and acidity. Research on sustainable agriculture and resource management can strengthen vulnerable ecosystems. The complicated relationship between plant physiology and factors like fog and acidic fog needs additional study. Therefore, understanding the physiological and biochemical mechanisms by which plants respond to these influences is essential for effective agriculture.

## 7.2 Impact of Fog on Different Categories of Plants

Seasonal variation and soil patterns significantly influence the plant community structure (Pandey et al. 2014). Plant communities are classified as tree, shrubs and herbs/grasses based on size. The climatic and topographical pattern distributes these plant diversity and clustering of same community leads to landscape formation. Fog regime affects plant biodiversity providing water as sustainable resource as well as toxic aerosolic compounds, such as sulfate ($SO_4^{2-}$), nitrate ($NO_3^-$), and heavy metals like lead (Pb) and cadmium (Cd), which can be harmful to plant health and ecosystem balance. Dominant species of vegetation can alter microclimatic conditions along gradients of fog (Sotomayor and Drezner 2019). However, the effect of fog varies with different plant communities and affects them differently.

### 7.2.1 Effects of Fog on Tree Species

Fog plays a crucial role for the ecology of tree species, especially in arid and semiarid regions. This serves as a source of moisture, potentially improving growth and survival rates in times of drought. Tree species that thrive in foggy environments have evolved unique adaptations, including the ability to intercept fog and store water in their leaves, allowing them to make the most of this moisture. The processes of fog water deposition and interception are crucial, underscoring the ecological significance of fog in sustaining vegetation, especially in regions that have low rainfall. The process of fog water deposition involves the collection of moisture from fog onto various surfaces, including soil, vegetation, and structure. This phenomenon takes place, as water droplets that are suspended in fog gradually settle and merge on various surfaces, primarily influenced by gravitational forces or adhesive properties. Fog precipitation is highly site specific based on vegetation and climatic factors. Vegetation factors encompass the height, size, and structure of the canopy, tree location and population density, leaf arrangement and shape, as well as the presence of epiphytes. Regional vegetation pattern defines fog density more accurately than altitude and terrains (Ball and Tzanopoulos 2020). The Landsat fog detection technique with satellite derived vegetation pattern shows that more fog density results in higher greenness (Ball Tzanopoulos 2020). The climatic factors that majorly have an impact on trees are fog frequency, fog properties, and wind speed and direction (Lakra and Avishek 2022). Besides that, terrain orientation and elevation also affect fog precipitation. *Aextoxicon punctatum*, a rainforest tree, is profoundly abundant in Chile's semiarid zone along with xerophytic vegetation nearby (del-Val et al. 2006). It has been suggested that this forest (*A. punctatum*) persists due to fog water deposition determining trees regeneration pattern and forest structure showing edge effect. That means the decrease in tree population structure at the edge of xerophytic vegetation.

Fog water interception is another phenomenon of fog water harvesting through trees and shrubs. It refers to the process by which vegetation, particularly trees and shrubs, captures moisture from fog. This phenomenon takes place, when tiny water droplets present in fog attach themselves to the surfaces of leaves, branches, and stems. The captured water can subsequently percolate into the soil, enhancing moisture levels and fostering plant development, particularly in arid or semiarid regions where conventional rainfall is limited. *Anogeissus dhofarica*, a dominant endemic tree of Dhofar cloud forest shows highest fog water interception (250% of observed rainfall) among tree species (Friesen et al. 2018). Therefore, it collects filtered fog water forming larger droplets dripping on the forest floor. Monsoon precipitation is another fog water interception impacting cloud forest of semiarid Dhofar mountains, recharging its groundwater up to 24% and groundwater of Salalah coastal plain up to 7%. Both depict reasonably high groundwater conservation through fog water interception, despite being semiarid region regardless of land cover, transpiration, etc. Canopy wetting and dripping for more than half of the time was caused by fog water interception. The harvested fog water was utilized for plant or saplings growth in water-deficient condition. A study on coastal endemic bishop pine (*Pinus muricata* D. Don) shows that shallow soil moisture due to fog deposition positively affects sapling trees growth (Baguskas et al. 2016a, b). Also, water status for sapling increased more than adult tree in foggy condition than in non-foggy condition. The fog event greatly lessens water loss in plant due to transpiration maintaining temperature and vapor pressure. Regalado and Ritter (2021) reported ten-fold lower sap flow and enhanced foliar water uptake associated with downward xylemic sap transport activating nocturnal transpiration.

Fog also do promote ecophysiological benefit with enhanced foliar water uptake as studied in four tree species (*Taxodium distichum* (L.) Rich., *Nyssa aquatica* L., *Nyssa biflora* Walter, and *Liquidambar styraciflua* L.) (Carmichael et al. 2020). All four tree species are treated with stable isotopes of water for foliar and bark water uptake. Up to three hours of submerged treatment of excised leaf from saplings with deionized water shows that foliar water uptake was enhance with approximately 5–10% across leaf surface giving total foliar water capacity. After 24 hour of fogging event, there is a positive uptake of bark water in *T. distichum* sapling enriching xylem water potential and leaf conductance. Besides being a part of auxiliary water pool system, fog also contributes to preventing excess solar radiation fluxes, as it appears to reflect less and absorb more due to the high canopy of cloud forests. This way fog blocks the transmission of solar fluxes of incident radiation providing radiative cooling and reduces surface heating impacts (Yan et al. 2021). Foggy environment decreased 45% global radiation mitigating water loss in Macaronesian cloud forest (Regalado and Ritter 2021). Further foggy environment leads to 71.4% reduction in net radiation of tree species with carbon assimilation and better water use strategy (León-González et al. 2023). Likewise different condition favors different physiological changes. For instance, fog situation under drought condition decreases photosynthetic activity due to decreased photon flux density and increases maximal stomatal conductance as studied in *Drimys brasiliensis* (Eller et al. 2013). The hydrological role of fog depicts lower radiation, decreased evapotranspiration rate, and lesser $CO_2$ intake due to wetness of tree canopy in the Andean páramo (Berrones et al. 2022).

Fog can provide a source of essential nutrients for plants, including nitrogen, sulfur, and phosphorus, though this contribution may vary depending on environmental conditions and fog composition. The horizontal and vertical inputs of nitrogen during different climate (summer fog, winter rain, and precipitation throughout the year) was recorded (Templer et al. 2015). It was found that summer fog is the direct and indirect source of nitrogen altering physiologic functions of redwood trees. Acidic fog also results in nitrogen and sulfur deposition in the stems of Jack pine showing fertilization effect (Bartels et al. 2019). The Douglas fir (*Pseudotsuga menziesii*) shows very low amount of the root uptake of Ca, Mg, and K collected through fog drip during acid fog (Turner and Tingey 1990). Acidic fog alternatively promotes visible injury to experimental trees saplings with decrease in dry weight and photosynthesis activity of tree species (Yoshida and Takenaka 2004). It degrades the epicuticular wax and also alters nutrient status. The 9-day fog exposure to saplings of *Abies fraseri* (Pursh) Poir. and *Picea rubens* Sarg. records positive relation between leaf conductance and foliar uptake with working stomata and carbon gain (Berry et al. 2014). Table 7.1 shows the negative impact of acidic fog on different plants including tree and shrubs with respect to their parts.

### *7.2.2 Effects of Fog on Shrubs*

Fog and dew have lately proven to play a potent role in agriculture and economic sector. Fog occurrence greatly influences shrub plants too, as it enhances soil water potential (Macek et al. 2018). In semiarid ecosystems that depend on fog, shrubs help support the growth of tree species by interacting with them and providing water through traditional means (Macek et al. 2018). However, there are a lot of shrub's physiological traits which is affected by coastal fog. The coastal fog during summer and spring seasons play a major role in shrubland vegetation and distribution, especially larger plant distribution, with some sensitivity to drought stress. *Prunus ilicifolia*, *Rhus integrifolia*, and *Lyonothamnus floribundus* are mostly distributed in San Clemente Island, California, with lower elevation and higher cloud within ravines (Clemesha et al. 2021). One of the cactus species, *Cylindropuntia prolifera*, is distributed with higher elevation and lowest annual fog. Coastal fog positively influences plant growth and physiology. A coastal plant, *Artemisia californica*, uses fog water opportunistically, as it enhances plant water content during summer drought condition (Emery and Lesage 2015). Coastal fog in California results in downwelling long wave radiations on foggy day than clear-sky day increasing water use and light use efficiency in plants (Baguskas et al. 2021). This suggests that agriculture system was benefitted with increasing irrigation and photosynthetic output. Advancing our basic knowledge about coastal fog, the seaside daisy (*Erigeron glaucus*) is a fog-influenced nectar plant (Gomes & Baguskas 2022). Monitoring soil moisture and microclimatic conditions in each chamber under simulated fog event, the photosynthesis and stomatal conductance increased more in non-irrigated seaside daisy plants than irrigated plants. Therefore, the foliar uptake of fog water surges the possibility of the survival of this nectar plant.

Fog results in natural irrigation, adding 10% of crop watering (Moratiel et al. 2016). The raise in soil water content due to fog also helps overcome accumulated crop evapotranspiration between irrigations. Assessment reports for the Jizan region indicate that fog contributes to 55 mm of water per year, which can help in irrigation of barley, millet, and wheat crops. However, these crops require a minimum of 200 mm of water per year to thrive (Valjarević et al. 2023). The water collected in Jizan region by fog ranges from 20 mm to 110 mm. The maximum collection was from hilly areas and shorelines. The vegetation influenced by fog water includes apple, banana, grapes, papaya, plum, and citrus varieties, which dwindle from coast to mainland. *Phoenix dactylifera* (palm), *Olea europaea* (olive), and *Triticum durum* (wheat) also utilize water from fog in their growing period. Hence, Jizan region is

**Table 7.1** Effect of acid fog on plants

| S. No. | Plant | Plant part | Acidity | Symptoms | Reference |
|---|---|---|---|---|---|
| 1. | Oak (*Quercus acutissima*) and Chinese fir (*Cunninghamia lanceolata*) | Root | 4.5 & 2.5 | Injured roots with reduced root length and area | Liu et al. 2022 |
| 2. | Pine (*Pinus massoniana*) | Roots of seedling | 4.6 | Well-developed primary lateral roots | Zhou et al. 2022 |
| 3. | Rice (*Oryza sativa* L.) | Root | 4.5–3.5 | Reduced root growth | Liu et al. 2018 |
| 4. | Sweet gum (*Liquidambar styraciflua* L.—Altingiaceae) | Leaf | 2.5–3.8 | Change in cuticular morphology | Rodríguez-Sánchez et al. 2020 |
| 5. | Pak choi (*Brassica rapa* subsp. *chinensis*) | Leaf | 3.5 | Reduced growth and yellowing of leaves | Zha et al. 2022 |
| 6. | Mustard varieties (*Brassica integrifolia Brassica rapa Brassica juncea*) in Hanoi | Leaf | 3.0 | Discoloring of leaves turning yellow; white spots; curling of leaf margins showing necrosis | Pham et al. 2022 |
| 7. | Bean plant (*Phaseolus vulgaris* L.) | Pollen | 2.0–4.5 | Decreased pollen germination; change from tetrad shape to polygonal form | Chehregani et al. 2006 |
| 8. | Tea (*Camellia sinensis*) | Photosynthesis | 2.5–3.5 | Inhibition of photosynthesis, antioxidant regulation, and metabolism | Zhang et al. 2020 - China |
| 9. | Four o' clock plant (*Mirabilis Jalapa* L.) | Photosynthesis | 2.0–4.0 | Decreased chlorophyll content | Chen et al. 2023 |
| 10. | Maple tree (*Acer rubrum* L.) White oak (*Quercus alba* L.) | Plant physiology | 4.5 | Decreased leaf nutrient content, water relations, leaf-level and canopy-level gas exchange, total biomass, and allocation | Medeiros et al. 2016 |
| 11. | Tea (*Camellia sinensis*) | Plant physiology | 3.5, 2.5 | Proline, malondialdehyde, antioxidant enzyme activity, total nitrogen, caffeine, catechins, and free amino acids increase catechin and decreased caffeine contents | Zhang et al. 2020 |

one of the most important regions of Saudi Arabia, providing better agricultural practices with the utilization of fog water for crop watering and preserving plants in arid region.

However, the pH of fog is influenced by the surrounding air quality, industrial emissions, agricultural activities, and natural environmental conditions. Acidic fog is more frequently studied due to its greater environmental impact. Plants experience acidic fog from pH range 5.6 to pH 2.5 and lower. Acidic fog changes morphological character of crops, altering leaf anatomy as well as stomata structure and affecting seed as well as pollen germination (Shafiq et al. 2023). Acid fog declines biodiversity and yield, causing leaf injury to certain extent. The growth parameters of two crop plants, brinjal (*Solanum melongena* Linn.) and cowpea (*Vigna unguiculata* sp., *cylindrica* L.), were adversely affected by acidic fog ranging from pH 2.5 to 5.6 (Arora et al. 2024). Acid fog leads to higher cuticular permeability and $Mg^{2+}$ and $Ca^{2+}$ leaching in holly plant (*Ilex aquifolium*) (Barker and Ashenden 1992).

The application of artificial fogging system also affects the growth of shrub plants along with managing environmental cooling during heat stress. For instances, Laur et al. (2021) installed a fogging system at shaded and non-shaded tunnel without disturbing air temperature, soil temperature and relative humidity to check the yield of *Ocimum basilicum* (basil), *Eruca vesicaria* subsp. *Sativa* L. (arugula) and *Lactuca sativa* L.(lettuce). The observation shows enhanced soil water status and canopy leaf wetness. The result also depicts that canopy-level fogging system is limited in affecting air temperatures, and shaded plants are highly potent in managing heat in high tunnel. The effect of fogging treatment on different plants is discussed in Table 7.2.

The studies till now prove that cloud-immersed environment (fog) is better for plants to thrive due to intermittent leaf wetting, balanced transpiration, and limited growth. Laurel forest fog water interception leads to a decrease in global solar radiation, which decreases mass sap flow in *Myrica faya* (Regalado and Ritter 2021). Under the same conditions, an isotypic behavior of *M. faya*, balancing carbon assimilation rate and growth against higher risk of hydraulic degradation, was reimbursed by foggy environment. Hence it was decisive that under limited supply of water, persistent fog at the site was a good source of nonconservative water.

### 7.2.3 Effects of Fog on Grasslands

The studies on grasslands are very scarce, but fog in grasslands also influences growth of biodiversity like *Baccharis vernalis* which nurses the sapling of mosses through fog dripping, leading to the generation of moss below the plants (Chung et al. 2017; Hernandez and Vita 2004). Cuevas et al. (2023) reported that xerophytic vegetation such as *Eulychnia breviflora, Adesmia microphylla, Escallonia pulverulenta,* and *Echinopsis chiloensis* harvests fog. Some plant species like *Tillandsia*

**Table 7.2** Fogging impact on plant physiology

| S. No. | Plant | Fogging period | Fog effect | Reference |
|---|---|---|---|---|
| 1. | Rosebay (*Nerium oleander*) | 15 min/h for 28–30 days | Taller plants with high flower yield, high percentage of unopened flower buds | Prabakaran et al. 2021 |
| 2. | Strawberry cultivars: 'Maehyang' and 'Seolhyang' | 9–12 days | Enhanced rooting, early growth of daughter plants, new leaves, long roots with increased fresh and dry weight | Wei et al. 2020 |
| 3. | Basil (*Ocimum basilicum* L.), Arugula (*Eruca vesicaria* subsp. *Sativa* L.), and Lettuce (*Lactuca sativa* L.) | 6 min/h for 6 h daily for approx. 15 days | Increased canopy leaf wetness, soil moisture, and foliar macronutrients like N, P, Mg, and Ca, and no impact on yield | Laur et al. 2021 |
| 4. | Tomato (*Solanum lycopersicon esculentum*) | 15 min/h for approx. 8 h for 3 months (during summer) | Increased pollen grain viability and improved fruit set with high yield | Harel et al. 2013 |
| 5. | Tomato (*solanum lycopersicon* L.) | 10 min/h for 6 h daily for approx. 60 days | Increased plant leaf area, cell membrane efficiency, maximized the contents of chlorophyll, relative water, and proline, and high fruit yield | Sharaf-Eldin et al. 2023 |
| 6. | Strawberry cultivars: 'Seolhyang' and 'Kuemsil' | 30 days | Older leaves with high transpiration rate, better growth and physiology at the initial phase, and 'Seolhyang' with high energy fluxes and chlorophyll *a* | Choi 2022 |
| 7. | Banda nutmeg (*Myristica fragrans*) | 5 min/h for 3 h daily for 2.5 months | Successful grafting, increased average shoot length and number of leaves | Ruhnayat et al. 2023 |
| 8. | Wheat plant (200 genotypes) | Approx. 8 h naturally for 3 months | Leaf rolling inward to twisted, leaf angle changing from erect to semi-erect, and high moisture capturing and yield | Merrium et al. 2022a, b |
| 9. | Cucumber plants (*Cucumis sativus* L. 'Natsu Suzumi') | 4 h daily for 3 sunny days | Improved greenhouse humidity and increased leaf conductivity and gaseous exchange | Yasutake et al. 2014 |

and *Helichrysum* are found to be covered with fruticose lichens which are also able to harvest fog (Cuevas et al. 2023; Kolita and Gogoi 2024). Fog intercepted by tussock grasses in New Zealand considerably benefits water yield (Ingraham and Mark 2000). Due to lack of canopy, tussock grasses have very low water storage capacity with no drip. Fog event contributes about 24.4 mm to soil moisture of the Andean páramos (Monte grasslands), which are dominated by tussock grasses and cushion plants (Berrones et al. 2022). Increasing global warming leads to dry spell in temperate grasslands, where fog plays very crucial role for plants such as maintaining foliar water balance, leaf gaseous exchange, and radiative cooling. The fog as a local moisture offers 9–42% of total foliage other than rainfall water and provides water cycle complexity with fog water inputs (Li et al. 2021). Leaf wetting during air–leaf and leaf–air water exchange, and radiative cooling also retains the survival of many species in temperate grasslands (Li et al. 2023). Non-rainwater inputs can affect leaf energy balance and water budget. Reduced precipitation or fog enhanced the microbial colonization in the roots of *Holcus lanatus,* a perennial pasture grass increasing its relationship with fog stress (Ayling et al. 2021). Hence, apart from impacting grasslands, fog also enhances soil condition with microbiomes, especially mycorrhiza fungi, which are suitable for agricultural grasslands.

Very low annual precipitation at Namib Desert is suitable for growth of *Stipagrostis sabulicola, an endemic grass* (Ebner et al. 2011). With the help of fog, it can harvest 4–5 L fog water per fog event, increasing leaf water potential as well as soil water content. Ebner et al. (2011) report shows that fog harvest of S. sabulicola happens by stem flow with absorption by root system. Also, its wider leaf structure helps in fog catchment and further in irrigation. California coastal prairie grasses can utilize up to 28–66% of water via root uptake during summer from fog only (Corbin et al. 2005). Therefore, summer fog can ameliorate summer drought utilizing fog water.

## 7.3 Plant Growth Controlling Factors

There are three critical plant growth-promoting factors: soil, leaf, and air, which are majorly influenced by fog. Soil is directly linked to plant providing interface and substrate for plant growth and nutrition. Thereby, leaf is the part of plant which produces food, and air is a basic requirement for the survival of all living beings including plants. Focusing on the modulation on these three criteria with respect to fog, plant growth and survival physiology can be better understood (Fig. 7.1).

### *7.3.1 Role of Fog in Soil Moisture, Ecosystem Productivity, and Nutrient Cycling*

Fog plays a critical role in soil physiology, as soil water is a vital component of Earth's ecosystems. By increasing soil moisture content, fog helps recharge groundwater systems to some extent, significantly contributing to land–atmosphere interactions. It contains hydrogen and oxygen isotopes, which are linked to the formation of radiation fog, such as in the Namib Desert (Kaseke et al. 2017). Soil can act as both a sink and a source for fog. While fog provides water to the soil surface,

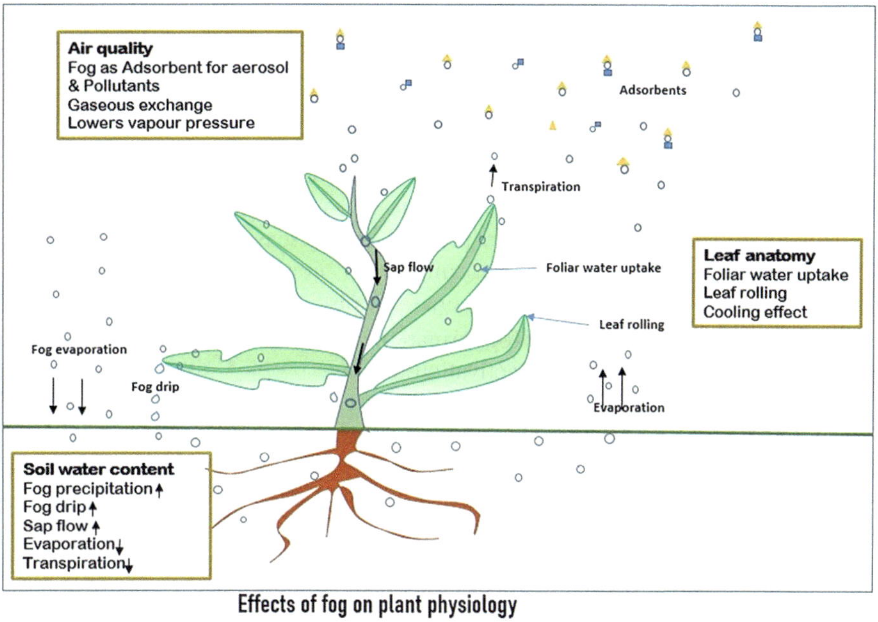


**Fig. 7.1** Effects of fog on plant physiology

evaporation from the soil also contributes water vapor to the atmosphere, aiding in fog formation (Qiao et al. 2020). Understanding the interactions between climate, soil, and vegetation in drylands is key to maintaining ecosystem productivity. Fog serves as a crucial water source for ecological processes, especially in areas with sparse rainfall. In drylands, factors such as precipitation, sap flow, and liquid water availability are all vital to soil water content, with fog also being an integral part of the hydrological cycle (Runyan et al. 2019). Fog facilitates groundwater recycling through evapotranspiration and redistributes water vapor to the soil surface. For instance, more than 50% of cloud formation events in the Namib Desert are derived from non-oceanic sources (Kaseke et al. 2017). Given the impacts of global warming and climate change, it is essential to evaluate the quantity and origin of fog, as declining rainfall may intensify reliance on fog for maintaining ecological functions in the future. Additionally, fog acts as a vector for nutrients and pollutants, influencing the hydrological inputs of ecosystems.

One of the key factors influencing fog formation is the coupling of soil moisture and heat. Due to water's high latent heat of vaporization, simple conduction drives heat flow within the soil, maintaining the water and energy balance in terrestrial ecosystems. Field observations and the HYDRUS-1D model indicate that surface flux and cumulative evaporation contribute to fog formation in fog-dominated drylands (Adhikari and Wang 2020). The interaction between soil moisture and heat not only controls energy and hydrological processes but also plays a pivotal role in regulating ecosystem functions, enhancing water balance, and benefiting organisms that depend on these inputs.

Fog impacts terrestrial ecosystems by modulating the coupling between soil processes, vegetation, and atmospheric conditions. It affects both the canopy (forest or tree crowns) and the soil surface through the advection of water droplets by wind, influencing ecosystem diversity (Weathers et al. 2020). Key processes such as crown wetting, elemental deposition, and radiation scattering and absorption are affected, which in turn regulate productivity, water fluxes, and nutrient acquisition in the soil. Fog drip, in particular, exerts a greater influence on soil physiology than foliar flow, contributing 30–40% of annual precipitation during seasonal droughts on Santa Cruz Island, CA. This fog-induced shallow soil moisture supports tree sapling growth (Carbone et al. 2013; Baguskas et al. 2016b).

Fog-dominated areas often support a disproportionate number of rare and endemic plant species, particularly in regions like coastal California, where fog ameliorates drought stress. Increased fog drip during summer enhances soil respiration, promoting microbial biomass and improving plant and microbial metabolism through consistent water inputs. This highlights fog's crucial role in maintaining ecosystem resilience in regions subject to seasonal water scarcity.

Fog contributes to the deposition of higher concentrations of chemicals in soil compared to precipitation (Pérez-Díaz et al. 2017). Studies have shown that acidic precipitation leaches more potassium ($K^+$) and magnesium ($Mg^{2+}$) ions from unhealthy trees than from healthy ones. Biomass burning in agricultural lands further adds to the accumulation of potassium, calcium ($Ca^{2+}$), and magnesium in fog (Baumgardner et al. 2003). The chemical composition of fog is more concentrated than that of rain due to deposition processes in the ecosystem, with over 80% of ions such as $SO4^{2-}$, $H^+$, and $NH4^+$ being deposited in areas like the Appalachian Mountains during summer. These ions supply essential nutrients to biological systems from the canopy to the soil, but they can also contribute to ecosystem pollution.

As coastal fog rises from the ocean and envelops forests, it acts as a vector for sea salts, nutrients, and microbes, transporting them from the ocean to forests and then to the land. Fog also influences hypolithic ecology, promoting microbial communities by increasing water availability in regions where rainfall is scarce. Studies show that smaller rocks are colonized more frequently in these areas due to the moisture provided by fog (Warren-Rhodes et al. 2013). Fog serves as a dynamic connector between aquatic and terrestrial ecosystems, nourishing forests while also transporting pollutants. Understanding these cross-system dynamics is key to uncovering how beneficial and pathogenic microbes are transported and accumulated in the environment.

Bacteria, being ubiquitous and extremophilic, are the most abundant organisms found in fog water and are capable of surviving the harsh conditions of aerosolization (Amato et al. 2017). Fog also impacts rhizospheric microbial communities and influences soil factors and fine root functional traits, especially in tropical cloud forests (Long et al. 2020). In the coastal Atacama Desert, fog plays a significant role in sustaining soil biology by regulating nutrient cycling, particularly phosphorus. The concentration of phosphorus is the highest near the coast, ranging from 76% to 100%, and decreases exponentially inland. Satellite-based fog detection has confirmed this trend, showing a marked reduction in fog intensity as it moves inland.

### 7.3.2 Leaf Anatomy

Current research emphasizes that leaf physiology is intrinsically linked to climate change adaptation strategies. Global warming has altered precipitation patterns, influencing various aspects of plant function, such as water relations, energy balance, carbon metabolism, nutrient leaching, and reproductive processes. Leaf wetting from precipitation can have both positive and negative effects, depending on the timing, frequency, duration, intensity, and variability of rainfall. Fog and dew, key sources of leaf wetness, play an essential role in plant water uptake, which is a fundamental characteristic of plants. In recent years, fog-induced leaf wetness has been found to enhance water movement through cuticles, hydathodes, and stomata, either in liquid or vapor forms. The genetic response of aquaporins (AQPs) to high humidity alters molecular pathways, and fog-induced wetting of leaves positively impacts forest canopy trees by balancing water levels within the ecosystem. Coastal fog enhances plant water status through mechanisms such as foliar application and fog drip, which directly supply moisture and cascade benefits that improve flower development, pollen viability, seed formation, and mycorrhizal associations, all of which are crucial for plant reproduction and growth. Therefore, leaves serve as a primary physiological tool for harvesting fog and balancing ecosystem fluxes.
Leaf wetness confers numerous benefits to plants, including alterations in cuticular membrane thickness and needle wettability (Pérez-Díaz et al. 2017). However, these cascading effects can disrupt soil moisture dynamics, leading to root hypoxia and altering groundwater dynamics, which can be detrimental to the ecosystem. Prolonged wetness can lead to morphological changes in leaves. Oversaturation modifies hydraulic conductance at the cellular level, while excess water accumulation may cause cell lysis or disrupt solute concentrations, thereby affecting both primary and secondary metabolic pathways (Nguyen et al. 2017). Reduced transpiration, caused by leaf wetting, can negatively impact photosynthesis and gas exchange by slowing the diffusion of gases through water, limiting $CO_2$ uptake, and reducing photosynthetic efficiency. Changes in transpiration also influence nutrient transport from the soil and alter stomatal conductance. The leaf surface structure, including waxy cuticles, leaf angles, and leaf rolling, demonstrates how fog water can affect plants. For instance, leaf rolling and inclination toward the root zone in wheat enhances fog capture under water-deficient stress (Merrium et al. 2022a, b). A significant correlation has been observed between leaf rolling, moisture capture, plant physiology, and grain yield.

Technological advancements such as triboelectric nanogenerators (TENGs) have been used to analyze leaf surface morphology and precipitation patterns, providing insights into improving fog collection to mitigate water shortages in arid regions. By mimicking natural leaf surface features with leaf films, TENG technology enhances fog capture, helping alleviate water scarcity and promote environmental sustainability (Begum and Chandrasekhar 2024). Xerophytic plants, native to semiarid regions, are natural fog harvesters with specialized adaptations for efficient water capture. Leaf inclination, surface traits, and hydrophilic properties enable these plants to retain water more effectively, with species such as *E. breviflora*, *A. microphylla*, *E. pulverulenta*, and *E. chiloensis* demonstrating high water retention through fog drip (Cuevas et al. 2023). These species retain water that would otherwise be lost through evaporation from leaf surfaces. Xerophytic plants often have tortuous, dry, and cracked leaves that accumulate water, which subsequently contributes to stem flow. Features like erect leaves, grooves, and prickle hairs further enhance water harvesting and direct the collected water to the plant base, facilitating self-irrigation (Ali et al. 2022). These adaptations highlight the role of xerophytes in sustaining ecosystems by efficiently capturing and utilizing fog water.

Fog significantly influences energy balance, affecting leaf temperature, light reflection due to water droplet accumulation, and reducing vapor pressure deficit (Gerlein-Safdi et al. 2018). Temperature increases as a result of condensation, while conduction of heat away from the leaf surface leads to cooling. The presence of fog droplets creates an albedo effect and refracts light, ultimately impacting both the quantity and quality of light available for photosynthesis, which is critical for plant function. Changes in leaf temperature also affect the vapor pressure gradient and carbon balance, thereby influencing the available energy for metabolic processes (O'Sullivan et al. 2017). Additionally, fog moisture can promote the growth of plant pathogens and endophytes on leaf surfaces and contribute to leaf leaching and salt deposition (Dawson and Goldsmith 2018).

### 7.3.3 Air Quality

Air is essential for the survival of all living beings, including plants, which rely on it for respiration and gas exchange through their leaf surfaces. Air plays a critical role in the physical, chemical, and biological processes of plants by contributing various parameters essential for soil health. While air is vital for plant life, plants also significantly improve air quality by purifying it and removing pollutants. Through stomatal uptake and non-stomatal deposition (adsorption), plants can effectively produce secondary gas-phase pollutants, while their associated microbiomes help eliminate contaminants without adversely affecting human health (Brilli et al. 2018). Consequently, plants are increasingly utilized indoors in urban environments to enhance air quality. Additionally, fog in forested areas improves air quality as tree leaves—particularly those of species such as *Butea monosperma*, *Tectona grandis*, and *Diospyros melanoxylon*—have the capacity to collect particulates, thereby acting as barriers to pollutants (Sahu et al. 2021). These trees also contribute to refining air quality in urban settings.

The gaseous exchange occurring at the plant surface also affects temperature regulation, with fog providing a notable cooling effect. Numerous studies have highlighted and patented artificial fogging systems designed to create controlled fog cooling environments within greenhouses, yielding significant benefits for crop growth (Park et al. 2020; Al-Ismaili et al. 2021; Soussi et al. 2022). However, fog can also contribute to air quality issues by introducing minute pollutants that increase density and decrease visibility. Major air pollutants include sulfur dioxide ($SO_2$), carbon monoxide (CO), particulate matter (PM), nitrogen dioxide ($NO_2$), ozone ($O_3$), and hydrocarbons, collectively referred to as smog (Gera and Bhasin 2023). Fog can be characterized as a visible aerosol containing tiny water droplets, gases, and a variety of small particles. A report on air quality in Istanbul indicates that concentrations of $SO_2$, $PM_{10}$, and CO are notably higher on foggy days compared to clear days (Özdemir et al. 2018). The formation of coastal fog positively influences the deposition of $SO_2$ and ammonium ($NH_4^+$) while negatively impacting nitrate deposition in the atmosphere (Jung et al. 2021). Fog scavenges air pollutants released from industrial activities, vehicular emissions, and coal combustion from thermal power plants. The atmospheric boundary layer retains fog and air pollutants, including the wet scavenging of $PM_{2.5}$ and other aerosol particles (Qian et al. 2023). The scavenging effect is most pronounced during fog formation and diminishes during fog dissipation.

Strong urbanization alters the structure of fog through condensation and evaporation processes, delaying its formation and dissipation due to reduced saturation levels. Research indicates that fog formation originating from the South China Sea carries polluted air masses and $PM_{2.5}$ concentrations, which further lead to reduced visibility through turbulent mixing (Ju et al. 2020). In addition to contributing to air pollution, fogging systems serve as an effective method for dust suppression (Saurabh et al. 2022). For example, the implementation of fogging systems has been shown to reduce dust concentration in mining and related industries from 0.62–1.73 mg $m^{-3}$ to 0.10–0.17 mg $m^{-3}$ in active work areas. Dust containing $PM_{2.5}$, $PM_{10}$, and silica is minimized, creating a smog-controlled environment that supports plant health.

In greenhouses equipped with ventilation and fogging systems, the rate of plant transpiration contributes approximately 55.14–84.53% of atmospheric water vapor (Villarreal-Guerrero et al. 2020). Fogging systems effectively lower vapor pressure deficits in the air, enhance humidity, and reduce temperatures. For instance, fogging systems or evaporative cooling techniques mitigate heat stress, thereby improving physiological and biochemical parameters and preventing diseases in tomato plants (Sharaf-Eldin et al. 2023). The collection of air humidity serves to enhance crop irrigation efforts. Additionally, reduced solar radiation and temperature during the high-heat summer months facilitate the propagation of melons and other fruits, resulting in optimal growth conditions (Lim et al. 2021).

## 7.4 Alteration in Plant Growth Pattern

### 7.4.1 Metabolism

Fog is a frequent phenomenon during both dry and rainy seasons, which can significantly alter plant metabolism. During the dry season, drought conditions lead to increased evaporative demand, which in turn elevates oxidative stress in plants (Boanares et al. 2020). This oxidative stress response is closely linked to foliar water uptake, particularly from fog. Plants that experience slower, more sustained absorption of water are better able to mitigate oxidative damage, such as lipid peroxidation, compared to plants with faster, less efficient water absorption mechanisms. In this context, *Tillandsia landbeckii*, a fog-dependent species in the arid Atacama Desert, exhibits metabolic adaptations modulated by fog availability. The evaporation of leaf water influences cellular oxidation and hydration processes, which impacts carbon and energy metabolism (Jaeschke et al. 2024). Enhanced cellulose levels in this species suggest its involvement in various biosynthetic pathways and enzymatic reactions, including hydrogen transfer. Fog, therefore, plays a key role in driving biosynthesis by cycling organic precursor pools. Changes in cellulose content in *T. landbeckii* are associated with $CO_2$ fixation and sucrose synthesis before export to leaf tissues (Baan et al. 2023). In addition to its effects on plant metabolism, *T. landbeckii* enhances the soil microbiome compared to bare soil. In the Atacama Desert, bacterial community structures vary with elevation and across different marine fog zones, creating microbial hotspots (Alfaro et al. 2021). Many microbiomes thrive under low temperatures, displaying adaptations in lifestyle and metabolism (Singh et al. 2022). Similarly, *Myrica faya* exhibits reduced transpiration rates and less efficient water use under high evaporative demand, highlighting the impact of fog on plant water management. The combined effects of radiation and wetness lead to improved stomatal conductance by increasing leaf surface area (Garcia-Tejera et al. 2024). Simulations suggest that a wetted canopy enhances transpiration efficiency irrespective of radiation levels, increasing photosynthetic activity in *M. faya*. These studies also highlight the importance of foliar water uptake in maintaining hydration in adult trees under natural conditions (Dawson and Goldsmith 2018). This uptake influences transpiration, $CO_2$ diffusion, and photosynthesis, further affecting leaf hydrophobicity, stomatal structure, and positioning in tropical forest and savanna plants (Aparecido et al. 2017). Coastal pines subjected to greater cloud cover exhibited larger basal area growth and increased sap flow, allowing them to better withstand summer stress (Carbone et al. 2013). Similarly, overnight fog in *Salvia leucophylla* reduces xylem tension, benefiting drought-deciduous shrubs in shrublands (Emery 2016). This overnight fog is particularly advantageous for species such as *Artemisia californica* and *S. leucophylla*, compared to evergreen species like *Adenostoma fasciculatum* and *Ceanothus megacarpus*. By mitigating xylem tension, fog water helps maintain vascular epiphytes, which are crucial for controlling water budgets during the dry season (Wu et al. 2018).

Studies examining the physiological effects of fog on rubber plantations revealed a decrease in carbon assimilation and evapotranspiration during the cool dry season, which ultimately enhanced crop water productivity (Gnanamoorthy et al. 2022). The combination of foggy days and low temperatures results in reduced plant physiological activity, characterized by low vapor pressure deficits, atmospheric water potential, high relative humidity, and moist canopies. While many studies focus on natural fogging events, controlled fogging systems under low relative humidity have been employed in greenhouses to study crops such as strawberries (Choi 2022). For instance, increased transpiration was observed in older leaves with lower photosynthetic rates, suggesting that fog was most beneficial during the initial growth phases of the strawberry plants.

Continuous moisture provided by fog fosters microbial growth on leaf surfaces, including fungi, bacteria, and arbuscular mycorrhizal fungi. Scanning electron micrographs reveal fungal hyphae spreading across leaf surfaces and penetrating stomatal pores, sometimes covering epicuticular wax plugs (Wu et al. 2018). These fungal endophytes play a role in plant water management and exhibit mutualistic relationships with their host plants. The study of phyllosphere microbiology is critical, as it demonstrates how these microorganisms help plants manage foliar water and survive under different temperature and fog conditions. While leaf veins facilitate water transport, the presence of bacterial strains on leaf surfaces enhances water retention and shapes the phyllosphere environment. The microbial colonization of the phyllosphere supports plant survival by improving nutrient uptake and providing growth-promoting factors, particularly in diverse and challenging climates.

Fog water can sometimes exhibit a lower pH, resulting in what is known as acidic fog. This acidic fog increases stomatal resistance and causes damage to both the upper and lower epidermis of leaves, with minimal damage to the xylem. Additionally, the rhizosphere pH plays a crucial role in shaping the below-ground plant microbiome. Similarly, the interaction on the phylloplane—often referred to as the phyllosphere—above ground, particularly in the plant canopy, influences microbial communities. Acidic fog has also been shown to stimulate foliar cation leaching in conifer species within rainforest ecosystems (Gilbert and Renner 2021). The pH of the bark can influence the composition of slime mold communities, demonstrating how plant–microbe interactions can impact a plant's adaptability to fog.

In addition to natural fog events, artificial fogging systems have been employed to enhance plant physiological responses. For instance, in cherry tomato plants, artificial fogging improves antioxidant responses while serving as a cooling mechanism, mitigating evaporation and heat stress during the hot summer conditions typical of the Mediterranean basin (Leyva et al. 2013). This highlights the importance of fog, both natural and artificial, in influencing plant–environment interactions and promoting plant resilience.

### 7.4.2 Genetic Modulations

Genes in plants govern a wide range of biological functions, from major physiological activities to minor biochemical processes. Both conserved genes, which are common across species, and unique, species-specific genes are of interest when exploring plant proteomes to uncover specific mechanisms. To understand evolutionary changes and genetic regulation, studying plant genomes (including genetic markers) is key. Abiotic stresses, such as fog, also influence genetic regulation. While there is limited research specifically addressing genetic modulation in response to fog, it is clear that fog impacts various molecular mechanisms.

One such mechanism involves the regulation of proton flux, which balances plant pH. Acidic fog has been shown to upregulate $H^+$-ATPase gene expression in the plasma membrane, helping stabilize $H^+$ ion concentration (Ren et al. 2018). This gene activity facilitates the outward flow of excess $H^+$ ions, maintaining pH homeostasis on the leaf surface. Fog exposure also promotes greater stomatal opening, where guard cells excrete protons, creating an ionic flux (Gilbert and Renner 2021). As $CO_2$ assimilation and photosynthesis produce carbonic acid, further aids in neutralizing acidic pH on the leaf surface. Thus, the regulation of ATPase plays a vital role in maintaining phylloplane pH during acidic fog events.

Aquaporins (AQPs), which are membrane proteins regulating water channels, also play a key role in foliar water uptake. Specific members of the aquaporin family, such as PIP-2 (plasma membrane intrinsic proteins) and TIP (tonoplast intrinsic proteins), have been shown to influence water channel activity under fog conditions (Vignesh and Palanisamy 2021). These aquaporins contribute to the regulation of stomatal movement and mechanisms associated with water transport. Research has found that high humidity conditions, akin to those in foggy environments, alter ethylene signaling pathways, including the expression of *EIN2* and *EIN3/EIL1* genes, which are involved in plant stress responses and stomatal regulation (Jiang et al. 2024). This modulation enhances leaf movements, such as hyponasty, and reactive oxygen species (ROS) responses. Furthermore, high humidity triggers ethylene biosynthesis, suggesting that fog may contribute to plant evolutionary processes. Heat shock proteins, particularly Hsp70, are another example of genes that are upregulated during cold and drought stress, conditions often associated with fog (Howlader et al. 2017; Singh et al. 2022).

The interaction between genotype and environment (G × E) is critical for evaluating the genetic variation in plants as influenced by environmental factors like fog. Quantifying genetic responses to fog requires targeting its impact on genotypes and understanding how allele frequencies shift over time, with fog acting as a key environmental driver (Czech et al. 2022). Microarray analyses have shown that fog treatments induce the expression of defense and stress-related genes in plants such as rice. For instance, genes encoding NADPH oxidase and germin proteins were upregulated, which helped balance oxidative stress (Satoh et al. 2014).

### 7.4.3 Plant Vigor and Yield

Foggy conditions have been shown to significantly enhance the radial growth of both evergreen and deciduous forests. Fog also mitigates frost stress, which in turn improves fruit quality and yield, as observed in the 'Naomi' mango cultivar (Abdel-Aziz et al. 2023). Fog exposure promotes leaf area expansion, thereby enhancing photosynthesis, increasing the membrane stability index, and improving both the yield and the physical and chemical characteristics of crops.

In a study of winter wheat, supplementary irrigation with 30 mm and 60 mm of collected fog water addressed 26% and 34% of water deficiencies, respectively, resulting in yield increases of 0.6 and 1.7 tons per hectare (Kanooni and Kohan 2023). Similarly, seasonal fog improved water productivity in tropical rubber plantations, contributing to better crop performance under water-limited conditions (Gnanamoorthy et al. 2022). Fog enhances plant yield by increasing resilience against wilting, drought stress, and pathogenic attacks. In red spruce, exposure to acid fog (pH 4.2) initially stimulated wax production; however, long-term exposure eventually reduced wax synthesis due to a decline in the rate of de novo wax production (Pérez-Díaz et al. 2017). Additionally, prolonged exposure to fog led to reduced production of secondary alcohols in the plants.

Hot fog treatment combined with pretreatments like potassium humate and Azotovit was found to improve crop productivity in potato seed tubers, leading to better product quality (Kostenko et al. 2020). Fog capture through leaf rolling and leaf angle alteration also enhanced root development and increased grain yield in wheat, especially during critical growth stages (Merrium et al. 2022a, b). In the context of tomato cultivation, evaporative cooling systems increased the yield of sellable fruits by 31.5%, while fogging systems and shaded tunnels led to yield increases of 28.8% and 17%, respectively, by mitigating heat stress in plants (Sharaf-Eldin et al. 2023). Moreover, fog water has been utilized as a sustainable water source for the soilless cultivation of lettuce (*Lactuca sativa*) in the Atacama Desert, with the experiment monitoring both fog water collection and its quality for agricultural use (Albornoz et al. 2024).

Fog events have been instrumental in optimizing irrigation practices, such as enabling double cropping in regions like Delhi. By incorporating fog forecasting and weather prediction models, more efficient irrigation habits have been developed (Smith et al. 2024). In Southern Bolivia, fog moisture harvesting has proven to be a viable strategy for enhancing food security (Castelli et al. 2023). In conclusion, fog plays a critical role in enhancing agricultural productivity, improving crop resilience, and serving as a sustainable water resource, especially in arid and semiarid regions.

## 7.5 Conclusions and Future Prospects

This chapter emphasizes the significant importance of fog within arid and semiarid ecosystems, drawing attention to its various contributions to plant physiology, soil moisture dynamics, and the overall health of the ecosystem. Fog serves as an essential source of moisture, especially in areas characterized by limited conventional rainfall, thus fostering a variety of plant communities and boosting agricultural productivity. However, the dual role of fog, which can also carry harmful pollutants and cause acidic deposition, poses considerable challenges that need further study. This chapter suggests that fog plays a crucial role in increasing water availability, fostering plant growth, and promoting nutrient cycling. Various species of trees, shrubs, and grasses demonstrate differing responses to fog, which illustrates their adaptations that allow them to succeed in water-limited environments. Additionally, the ability of fog to affect leaf structure and improve air quality underscores its significance in maintaining plant health and supporting agricultural systems. The complex relationships between fog and these environmental elements present a valuable domain for further investigation. As climate change continues to alter precipitation and atmospheric conditions, the time is getting ever more urgent for understanding the complexity of fog formation and its ecological implications. Future research should focus on the long-term effects of both neutral and acid fog on plant metabolism and gene expression, as well as on community processes. This includes the exploration of potential options available in new fog capture systems and artificial fog systems designed to maximize efficient use of water and enhance crop resistance to drought and pest stress. Additionally, the relationship between urbanization and fog dynamics warrants further exploration, particularly regarding its implications for air quality and urban agriculture. As cities expand, understanding how anthropogenic factors influence fog formation could inform sustainable urban planning and agricultural practices.

**Acknowledgments** The authors sincerely acknowledge the support of the Director, CSIR-Central Institute of Mining and Fuel Research, Dhanbad, for providing all the necessary laboratory facilities and financial assistance.

## References

Abdel-Aziz HF, Hamdy AE, Sharaf A, Abd El-wahed AEWN, Elnaggar IA, Seleiman MF, Omar M, Al-Saif AM, Shahid MA, Sharaf M. Effects of fogging system and nitric oxide on growth and yield of 'Naomi' mango trees exposed to frost stress. Life. 2023;13(6):1359.

Adhikari B, Wang L. The potential contribution of soil moisture to fog formation in the Namib Desert. J Hydrol. 2020;591:125326.

Akimoto Y, Kusaka H. A climatological study of fog in Japan based on event data. Atmos Res. 2015;151:200–11.

Albornoz F, Vilches I, del Río C, Carter V. Soilless production of lettuce (*Lactuca sativa*) in the Atacama Desert using fog water: water quality and produce mineral composition. Horticulturae. 2024;10(9):957.

Alfaro FD, Manzano M, Almiray C, García JL, Osses P, del Rio C, Vargas C, Latorre C, Koch MA, Siegmund A, Abades S. Soil bacterial community structure of fog-dependent *Tillandsia landbeckii* dunes in the Atacama Desert. Plant Syst Evol. 2021;307(5):56.

Ali Z, Merrium S, Habib-ur-Rahman M, Hakeem S, Saddique MAB, Sher MA. Wetting mechanism and morphological adaptation; leaf rolling enhancing atmospheric water acquisition in wheat crop—a review. Environ Sci Pollut Res. 2022;29(21):30967–85.

Al-Ismaili AM, Al-Maimani AM, Al-Saadi AK, Al-Kindi FS, Al-Manthri IA, Al-Jabri KM, Al-Dairi MK, Al-Mahroqi SH. Recent patents on greenhouse cooling techniques: a review. Recent Patent Eng. 2021;15(3):338–55.

Amato P, Joly M, Besaury L, Oudart A, Taib N, Moné AI, Deguillaume L, Delort AM, Debroas D. Active microorganisms thrive among extremely diverse communities in cloud water. PLoS One. 2017;12(8):e0182869.

Aparecido LM, Miller GR, Cahill AT, Moore GW. Leaf surface traits and water storage retention affect photosynthetic responses to leaf surface wetness among wet tropical forest and semiarid savanna plants. Tree Physiol. 2017;37(10):1285–300.

Arora V, Singh BJ, Bithel N, Malik N, Upadhyay SK, Singh R. Effect of simulated acid rain on plant growth behaviour of *Solanum melongena* Linn. and *Vigna unguiculata* ssp *cylindrica* (L.) Walp. Environ Dev Sustain. 2024;26(1):627–55.

Ayling SM, George BH, Rogers JB. Mycorrhizal colonisation in roots of Holcus lanatus (Yorkshire fog) in a permanent pasture under conditions of reduced precipitation. Botany. 2021;99(4):199–208.

Azúa-Bustos A, González-Silva C, Mancilla RA, Salas L, Gómez-Silva B, McKay CP, Vicuña R. Hypolithic cyanobacteria supported mainly by fog in the coastal range of the Atacama Desert. Microb Ecol. 2011;61:568–81.

Baan J, Holloway-Phillips M, Nelson DB, Kahmen A. The metabolic sensitivity of hydrogen isotope fractionation differs between plant compounds. Phytochemistry. 2023;207:113563.

Baguskas SA, King JY, Fischer DT, D'Antonio CM, Still CJ. Impact of fog drip versus fog immersion on the physiology of Bishop pine saplings. Funct Plant Biol. 2016a;44(3):339–50.

Baguskas SA, Still CJ, Fischer DT, D'Antonio CM, King JY. Coastal fog during summer drought improves the water status of sapling trees more than adult trees in a California pine forest. Oecologia. 2016b;181:137–48.

Baguskas SA, Oliphant AJ, Clemesha RE, Loik ME. Water and light-use efficiency are enhanced under summer coastal fog in a California agricultural system. J Geophys Res Biogeosciences. 2021;126(5):e2020JG006193.

Ball L, Tzanopoulos J. Interplay between topography, fog and vegetation in the central south Arabian mountains revealed using a novel Landsat fog detection technique. Remote Sens Ecol Conserv. 2020;6(4):498–513.

Barker MG, Ashenden TW. Effects of acid fog on cuticular permeability and cation leaching in holly (Ilex aquifolium). Agric Ecosyst Environ. 1992;42(3–4):291–306.

Bartels SF, Gendreau-Berthiaume B, Macdonald SE. The impact of atmospheric acid deposition on tree growth and forest understory vegetation in the Athabasca Oil Sands region. Sci Total Environ. 2019;696:133877.

Baumgardner RE Jr, Isil SS, Lavery TF, Rogers CM, Mohnen VA. Estimates of cloud water deposition at mountain acid deposition program sites in the Appalachian Mountains. J Air Waste Manage Assoc. 2003;53(3):291–308.

Begum SR, Chandrasekhar A. Biomimicking hydrophobic leaf structure using soft lithography for fog harvesting, triboelectric nanogenerators as a self-powered rain sensor. Iscience. 2024;27(2):108878.

Berrones G, Crespo P, Ochoa-Sánchez A, Wilcox BP, Célleri R. Importance of fog and cloud water contributions to soil moisture in the Andean Páramo. Hydrology. 2022;9(4):54.

Berry ZC, White JC, Smith WK. Foliar uptake, carbon fluxes and water status are affected by the timing of daily fog in saplings from a threatened cloud forest. Tree Physiol. 2014;34(5):459–70.

Boanares D, da-Silva CJ, dos Santos Isaias RM, França MGC. Oxidative metabolism in plants from Brazilian rupestrian fields and its relation with foliar water uptake in dry and rainy seasons. Plant Physiol Biochem. 2020;146:457–62.

Brilli F, Fares S, Ghirardo A, de Visser P, Calatayud V, Muñoz A, Annesi-Maesano I, Sebastiani F, Alivernini A, Varriale V, Menghini F. Plants for sustainable improvement of indoor air quality. Trends in Plant Science, 2018;23(6):507–512.

Bytnerowicz A, Grulke NE. Physiological effects of air pollutants on western trees. In: The response of western forests to air pollution. New York: Springer; 1992. p. 183–233.

Cáceres L, Gómez-Silva B, Garró X, Rodríguez V, Monardes V, McKay CP. Relative humidity patterns and fog water precipitation in the Atacama Desert and biological implications. J Geophys Res Biogeo. 2007;112(G4)

Carbone MS, Park Williams A, Ambrose AR, Boot CM, Bradley ES, Dawson TE, Schaeffer SM, Schimel JP, Still CJ. Cloud shading and fog drip influence the metabolism of a coastal pine ecosystem. Glob Chang Biol. 2013;19(2):484–97.

Carmichael MJ, White JC, Cory ST, Berry ZC, Smith WK. Foliar water uptake of fog confers ecophysiological benefits to four common tree species of southeastern freshwater forested wetlands. Ecohydrology. 2020;13(7):e2240.

Castelli G, Sanchez AC, Mestrallet A, Montaño LC, de Armentia TL, Salbitano F, Bresci E. Fog as unconventional water resource: mapping fog occurrence and fog collection potential for food security in southern Bolivia. J Arid Environ. 2023;208:104884.

Chehregani A, Malayeri BE, Kavianpour F, Yazdi HL. Effect of acid rain on the development, structure and viability of pollen grains in bean plants (*Phaseolus vulgaris*). Pak J Biol Sci. 2006;9:1033–6.

Chen CL, Chen TY, Hung HM, Tsai PW, Chou CCK, Chen WN. The influence of upslope fog on hygroscopicity and chemical composition of aerosols at a forest site in Taiwan. Atmos Environ. 2021;246:118150.

Chen LY, Wang L, Wang HY, Zhu X. Characterising the influence of acid rain on the growth and physiological characteristics of Mirabilis Jalapa Linn. In southern China. Acta Physiol Plant. 2023;45(1):5.

Chin A. Absorption of water by the leaves of coast redwood. Davis: University of California; 2021.

Choi HG. Effect of fogging on survival rate and photophysiological response during the initial growth period of strawberries in a greenhouse. In: XXXI international horticultural congress (IHC2022): international symposium on adaptation of horticultural plants to abiotic 1372; 2022, August, p. 325–30.

Chung M, Dufour A, Pluche R, Thompson S. How much does dry-season fog matter? Quantifying fog contributions to water balance in a coastal California watershed. Hydrol Process. 2017;31(22):3948–61.

Clemesha RE, Gershunov A, Lawson DM, Longcore T, MacDonald B, Booker M, Munson B, O'Connor K. A high-resolution record of coastal clouds and fog and their role in plant distributions over San Clemente Island, California. Environ Res Commun. 2021;3(10):105003.

Corbin JD, Thomsen MA, Dawson TE, D'Antonio CM. Summer water use by California coastal prairie grasses: fog, drought, and community composition. Oecologia. 2005;145:511–21.

Cuevas JG, Bustamante C, Ostria-Gallardo E, Hernández P. The role of xerophytic vegetation on fog harvest. J Arid Environ. 2023;209:104887.

Czech L, Peng Y, Spence JP, Lang PL, Bellagio T, Hildebrandt J, Fritschi K, Schwab R, Rowan BA, Weigel D, Scheepen JF, Vasseur F, Exposito-Alonso M. Monitoring rapid evolution of plant populations at scale with pool-sequencing. BioRxiv. 2022:2022–02.

Dar ZA, Bhat RA, Bhat JI, Mir SA, Amin A, Rashid A, Rifat B, Lone R. Microbial diversity and their role in plant and soil health under stress conditions. In vitro plant breeding towards novel agronomic traits: biotic and abiotic stress tolerance; 2019. p. 149–66.

Dawson TE, Goldsmith GR. The value of wet leaves. New Phytol. 2018;219(4):1156–69.

del-Val EK, Armesto JJ, Barbosa O, Christie DA, Gutiérrez AG, Jones CG, Marquet PA, Weathers KC. Rain forest islands in the Chilean semiarid region: fog-dependency, ecosystem persistence and tree regeneration. Ecosystems. 2006;9:598–608.

Ding AJ, Huang X, Nie W, Sun JN, Kerminen VM, Petäjä T, Su H, Cheng YF, Yang XQ, Wang MH, Chi XG, Wang JP, Virkkula A, Guo WD, Yuan J, Wang SY, Zhang RJ, Wu YF, Song Y, Zhu T, Zilitinkevich S, Kulmala M, Fu CB. Enhanced haze pollution by black carbon in megacities in China. Geophys Res Lett. 2016;43(6):2873–9.

Ding Q, Sun J, Huang X, Ding A, Zou J, Yang X, Fu C. Impacts of black carbon on the formation of advection–radiation fog during a haze pollution episode in eastern China. Atmos Chem Phys. 2019;19(11):7759–74.

Ebner M, Miranda T, Roth-Nebelsick A. Efficient fog harvesting by Stipagrostis sabulicola (Namib dune bushman grass). J Arid Environ. 2011;75(6):524–31.

Eller CB, Lima AL, Oliveira RS. Foliar uptake of fog water and transport belowground alleviates drought effects in the cloud forest tree species, *Drimys brasiliensis* (W interaceae). New Phytol. 2013;199(1):151–62.

Emery NC. Foliar uptake of fog in coastal California shrub species. Oecologia. 2016;182(3):731–42.

Emery N, Lesage J. Late summer fog use in the drought deciduous shrub, Artemisia californica (Asteraceae). Madrono. 2015;62(3):150–7.

Fan J, Rosenfeld D, Zhang Y, Giangrande SE, Li Z, Machado LA, Martin ST, Yang Y, Wang J, Artaxo P, Barbosa HMJ, Braga RC, Comstock JM, Zhe Feng Z, Gomes GWHB, Mei F, Ulrich Pöschl CPMLP, de Souza RA. Substantial convection and precipitation enhancements by ultrafine aerosol particles. Science. 2018;359(6374):411–8.

Friesen J, Zink M, Bawain A, Müller T. Hydrometeorology of the Dhofar cloud forest and its implications for groundwater recharge. J Hydrol: Regional Studies. 2018;16:54–66.

Garcia-Tejera O, Ritter A, Regalado CM. The combined effect of diffuse radiation and leaf wetness on functional traits and transpiration efficiency on a cloud forest species. Tree Physiol. 2024;44(6):tpae050.

Gera H, Bhasin M. Photochemical smog-a review. J Sci Res Teach. 2023;2(7):210–28.

Gerber, M. (2018). Determining strategies of Acanthosicyos horridus (! nara) to exploit alternative atmospheric moisture sources in the hyper-arid Namib Desert. (Doctoral dissertation, North-West University).

Gerlein-Safdi C, Gauthier PP, Caylor KK. Dew-induced transpiration suppression impacts the water and isotope balances of Colocasia leaves. Oecologia. 2018;187:1041–51.

Gilbert KJ, Renner T. Acid or base? How do plants regulate the ecology of their phylloplane? AoB Plants. 2021;13(4):plab032.

Gill PS, Graedel TE, Weschler CJ. Organic films on atmospheric aerosol particles, fog droplets, cloud droplets, raindrops, and snowflakes. Rev Geophys. 1983;21(4):903–20.

Glaser DM, Hartnett HE, Finn DR, Perez-Montaño S, Cadillo-Quiroz H, Desch S. Water vapor adsorption provides daily, sustainable water to soils of the hyperarid Atacama Desert. Astrobiology. 2022;22(10):1222–38.

Gnanamoorthy P, Song Q, Zhao J, Zhang Y, Zhang J, Lin Y, Zhou L, Bibi S, Sun C, Yu H, Zhou W, Sha L, Wang S, Chakraborty S, Burman PKD. Seasonal fog enhances crop water productivity in a tropical rubber plantation. J Hydrol. 2022;611:128016.

Gomes SE, Baguskas SA. Coastal fog enhances physiological function of seaside daisies (*Erigeron glaucus*). Western North Am Natural. 2022;82(3):537–48.

Gottlieb TR, Eckardt FD, Venter ZS, Cramer MD. The contribution of fog to water and nutrient supply to *Arthraerua leubnitziae* in the central Namib Desert, Namibia. J Arid Environ. 2019;161:35–46.

Grodzińska-Jurczak M. Conifer epicuticular wax as a biomarker of air pollution: an overview. Acta Soc Bot Pol. 1998;67(3–4):291.

Gultepe I, Tardif R, Michaelides SC, Cermak J, Bott A, Bendix J, Muller MD, Pagowski M, Hansen B, Ellrod G, Jacobs W, Toth G, Cober SG. Fog research: a review of past achievements and future perspectives. Pure Appl Geophys. 2007;164:1121–59.

Guo J, Su T, Li Z, Miao Y, Li J, Liu H, Xu H, Cribb M, Zhai P. Declining frequency of summertime local-scale precipitation over eastern China from 1970 to 2010 and its potential link to aerosols. Geophys Res Lett. 2017;44(11):5700–8.

Hakeem S, Ali Z, Saddique MAB, Habib-ur-Rahman M, Trethowan R. Leaf prickle hairs and longitudinal grooves help wheat plants capture air moisture as a water-smart strategy for a changing climate. Planta. 2021;254(1):18.

Hang C, Nadeau DF, Gultepe I, Hoch SW, Román-Cascón C, Pryor K, Fernando HJS, Creegan ED, Leo LS, Silver Z, Pardyjak ER. A case study of the mechanisms modulating the evolution of valley fog. Pure Appl Geophys. 2016;173:3011–30.

Harel D, Fadida H, Gantz S, Shilo K, Yasuor H. Evaluation of low pressure fogging system for improving crop yield of tomato (*Lycopersicon esculentum* Mill.): grown under heat stress conditions. Agronomy. 2013;3(2):497–507.

Hern'andez I, Vita A. Capítulo 18. Reforestaci'on para la expansi´ on de los bosquetes de Olivillo. In: Squeo FA, Guti´errez JR, Hern'andez IR. (eds) Historia Natural del Parque Nacional Bosque Fray Jorge. Ediciones Universidad de La Serena, La Serena, Chile; 2004. p. 307–19.

Herndon JM, Whiteside M. California wildfires: role of undisclosed atmospheric manipulation and geoengineering. J Geogr Environ Earth Sci Int. 2018;17(3):1–18.

Howlader J, Park JI, Robin AHK, Sumi KR, Nou IS. Identification, characterization and expression profiling of stress-related genes in easter lily (*Lilium formolongi*). Genes. 2017;8(7):172.

Hůnová I, Brabec M, Malý M, Valeriánová A. Revisiting fog as an important constituent of the atmosphere. Sci Total Environ. 2018;636:1490–9.

Ingraham NL, Mark AF. Isotopic assessment of the hydrologic importance of fog deposition on tall snow tussock grass on southern New Zealand uplands. Austral Ecol. 2000;25(4):402–8.

Jaeschke A, Böhm C, Schween JH, Schefuß E, Koch MA, Latorre C, Contreras S, Rethemeyer J, Wissel H, Lücke A. Evaluating the isotopic composition of leaf organic compounds in fog-dependent Tillandsia landbeckii across the coastal Atacama Desert: implications for hydroclimate reconstructions at the dry limit. Glob Planet Chang. 2024;235:104393.

Jiang Z, Yao L, Zhu X, Hao G, Ding Y, Zhao H, Wang S, Wen CK, Xu X, Xin XF. Ethylene signaling modulates air humidity responses in plants. Plant J. 2024;117(3):653–68.

Ju T, Wu B, Wang Z, Liu J, Chen D, Zhang H. Relationships between low-level jet and low visibility associated with precipitation, air pollution, and fog in Tianjin. Atmos. 2020;11(11):1197.

Jung J, Choi Y, Wong DC, Nelson D, Lee S. Role of sea fog over the Yellow Sea on air quality with the direct effect of aerosols. J Geophys Res Atmos. 2021;126(5):e2020JD033498.

Kang H, Zhu B, de Leeuw G, Yu B, van der A, R. J, Lu W. Impact of urban heat Island on inorganic aerosol in the lower free troposphere: a case study in Hangzhou, China. Atmos Chem Phys. 2022;22(16):10623–34.

Kanooni A, Kohan MR. Fog water harvesting potential and its use in supplementary irrigation of rainfed crops (winter wheat) in Abi-beyglu, Ardabil (Iran). Water Supply. 2023;23(9):3675–93.

Kaseke KF, Wang L, Seely MK. Nonrainfall water origins and formation mechanisms. Sci Adv. 2017;3(3):e1603131.

Khain AP, Pinsky M. Physical processes in clouds and cloud modeling. Cambridge University Press; 2018.

Kolita B, Gogoi RK. Lichen as pioneer of natural ecosystem. Chemistry, Biology and Pharmacology of Lichen; 2024. p. 81–9.

Kostenko MY, Goryachkina IN, Rembalovich GK, Drozhzhin KN, Beznosyuk RV. Studying the influence of treating potato seed tubers with hot fog of protective-stimulating preparations. In: IOP conference series: earth and environmental Science, vol. 488, No. 1. IOP Publishing; 2020. p. 012024.

LaDochy S. The disappearance of dense fog in Los Angeles: another urban impact? Phys Geogr. 2005;26(3):177–91.

Lakra K, Avishek K. A review on factors influencing fog formation, classification, forecasting, detection and impacts. Rendiconti Lincei Scienze Fisiche e Naturali. 2022;33(2):319–53.

Laur S, da Silva ALBR, Díaz-Pérez JC, Coolong T. Impact of shade and fogging on high tunnel production and mineral content of organically grown lettuce, basil, and arugula in Georgia. Agriculture. 2021;11(7):625.

León-González M, Regalado CM, Ritter A. Carbon fluxes associated with fog in an elfin cloud forest in Anaga (Tenerife, Canary Islands). J Hydrol. 2023;621:129534.

Leyva R, Constán-Aguilar C, Blasco B, Sánchez-Rodríguez E, Soriano T, Ruíz JM. A fogging system improves antioxidative defense responses and productivity in tomato. J Am Soc Hortic Sci. 2013;138(4):267–76.

Li Y, Aemisegger F, Riedl A, Buchmann N, Eugster W. The role of dew and radiation fog inputs in the local water cycling of a temperate grassland during dry spells in Central Europe. Hydrol Earth Syst Sci. 2021;25(5):2617–48.

Li Y, Riedl A, Eugster W, Buchmann N, Cernusak LA, Lehmann MM, Werner RA, Aemisegger F. The role of radiative cooling and leaf wetting in air–leaf water exchange during dew and radiation fog events in a temperate grassland. Agric For Meteorol. 2023;328:109256.

Li ZH, Yang J, Shi CE, Pu MJ. Urbanization effects on fog in China: field research and modeling. Pure Appl Geophys. 2012;169:927–39.

Lim MY, Jeong HJ, Roh MY, Choi GL, Kim SH, Choi SH. Changes in greenhouse temperature and solar radiation by fogging and shading during hydroponics in summer season. J Bio-Environ Control. 2021;30(3):230–6.

Lin KT, Han J, Li K, Guo C, Lin H, Jia B. Radiative cooling: fundamental physics, atmospheric influences, materials and structural engineering, applications and beyond. Nano Energy. 2021;80:105517.

Liu H, Ren X, Zhu J, Wu X, Liang C. Effect of exogenous abscisic acid on morphology, growth and nutrient uptake of rice (*Oryza sativa*) roots under simulated acid rain stress. Planta. 2018;248:647–59.

Liu X, Ma S, Jia Z, Ramzan M, Meng M, Wang J, Li C, Zhang Y, Zhang J. Complex effects of different types of acid rain on root growth of *Quercus acutissima* and *Cunninghamia lanceolata* saplings. Ecol Process. 2022;11(1):8.

Long W, Zhou Y, Schamp BS, Zang R, Yang X, Poorter L, Xiao C, Xiong M. Scaling relationships among functional traits are similar across individuals, species, and communities. J Veg Sci. 2020;31(4):571–80.

Maalick Z, Kühn T, Korhonen H, Kokkola H, Laaksonen A, Romakkaniemi S. Effect of aerosol concentration and absorbing aerosol on the radiation fog life cycle. Atmos Environ. 2016;133:26–33.

Macek P, Schöb C, Núñez-Ávila M, Hernández Gentina IR, Pugnaire FI, Armesto JJ. Shrub facilitation drives tree establishment in a semiarid fog-dependent ecosystem. Appl Veg Sci. 2018;21(1):113–20.

Medeiros JS, Tomeo NJ, Hewins CR, Rosenthal DM. Fast-growing *Acer rubrum* differs from slow-growing *Quercus alba* in leaf, xylem and hydraulic trait coordination responses to simulated acid rain. Tree Physiol. 2016;36(8):1032–44.

Merrium S, Ali Z, Habib-ur-Rahman M, Hakeem S, Khalid MA. Leaf rolling and leaf angle improve fog capturing and transport in wheat; adaptation for drought stress in an arid climate. Bot Stud. 2022a;63(1):13.

Merrium S, Ali Z, Tahir MHN, Habib-ur-Rahman M, Hakeem S. Leaf rolling dynamics for atmospheric moisture harvesting in wheat plant as an adaptation to arid environments. Environ Sci Pollut Res. 2022b;29(32):48995–9006.

Mieslerová B, Cook RT, Wheater CP, Lebeda A. Ecology of powdery mildews–influence of abiotic factors on their development and epidemiology. Crit Rev Plant Sci. 2022;41(6):365–90.

Moratiel Yungueros, R., Martínez-Cob, A., Tarquis, A. M., & Snyder, R. L. (2016). Soil water balance correction due to light rainfall, dew and fog in Ebro river basin (Spain).

Negret BS, Pérez F, Markesteijn L, Castillo MJ, Armesto JJ. Diverging drought-tolerance strategies explain tree species distribution along a fog-dependent moisture gradient in a temperate rain forest. Oecologia. 2013;173:625–35.

Nguyen HT, Meir P, Wolfe J, Mencuccini M, Ball MC. Plumbing the depths: extracellular water storage in specialized leaf structures and its functional expression in a three-domain pressure–volume relationship. Plant Cell Environ. 2017;40(7):1021–38.

Niu S, Lu C, Yu H, Zhao L, Lü J. Fog research in China: an overview. Adv Atmos Sci. 2010;27:639–62.

Nyaga, J. M. (2013). Nutritional contribution of atmospheric deposition to the Strandveld vegetation of west coast South Africa. (Doctoral dissertation, University of Cape Town).

O'sullivan OS, Heskel MA, Reich PB, Tjoelker MG, Weerasinghe LK, Penillard A, Zhu L, Egerton JJ, Bloomfield KJ, Creek D, Griffin KL, Hurry V, Meir P, Turnbull MH, Atkin OK. Thermal limits of leaf metabolism across biomes. Glob Chang Biol. 2017;23(1):209–23.

Özdemir ET, Deniz A, Yavuz V, Doğan N, Akbayır İ. Investigation of fog-air quality relationship in Istanbul. Fresenius Environ Bull. 2018;27(1):30–6.

Pandey B, Agrawal M, Singh S. Coal mining activities change plant community structure due to air pollution and soil degradation. Ecotoxicology. 2014;23:1474–83.

Pariyar S, Chang SC, Zinsmeister D, Zhou H, Grantz DA, Hunsche M, Burkhardt J. Xeromorphic traits help to maintain photosynthesis in the perhumid climate of a Taiwanese cloud forest. Oecologia. 2017;184:609–21.

Park SH, Moon JP, Kim JK, Kim SH. Development of fog cooling control system and cooling effect in greenhouse. J Bio-Environ Control. 2020;29(3):265–76.

Pérez-Díaz JL, Ivanov O, Peshev Z, Álvarez-Valenzuela MA, Valiente-Blanco I, Evgenieva T, Dreischuh T, Gueorguiev O, Todorov PV, Vaseashta A. Fogs: physical basis, characteristic properties, and impacts on the environment and human health. Water. 2017;9(10):807.

Pham HT, Nguyen LN, Le TT, Le MQ, Tran TK. Impact of simulated acid rain on the growth of three species Brassica integrifolia, Brassica rapa, Brassica juncea in Hanoi, Vietnam. Environ Sci Pollut Res. 2022;29(28):42090–101.

Prabakaran TS, Velmurugan M, Thamaraiselvi SP, Kalarani MK, Rajamani K, Pugalendhi L. Influence of different lighting and fogging system on growth, flower yield and bud opening in Nerium (*Nerium oleander*). Res Crop. 2021;22(4):913–9.

Price JD. On the formation and development of radiation fog: an observational study. Bound-Layer Meteorol. 2019;172(2):167–97.

Qian J, Liu D, Yan S, Cheng M, Liao R, Niu S, Yan W, Zha S, Wang L, Chen X. Fog scavenging of particulate matters in air pollution events: observation and simulation in the Yangtze River Delta, China. Sci Total Environ. 2023;876:162728.

Qiao N, Zhang L, Huang C, Jiao W, Maggs-Kölling G, Marais E, Wang L. Satellite observed positive impacts of fog on vegetation. Geophys Res Lett. 2020;47(12):e2020GL088428.

Quan J, Zhang Q, He H, Liu J, Huang M, Jin H. Analysis of the formation of fog and haze in North China Plain (NCP). Atmos Chem Phys. 2011;11(15):8205–14.

Regalado CM, Ritter A. Fog mitigates the consequences of a profligate water use strategy in a Macaronesian cloud forest tree species. J Hydrol. 2021;593:125871.

Ren X, Zhu J, Liu H, Xu X, Liang C. Response of antioxidative system in rice (*Oryza sativa*) leaves to simulated acid rain stress. Ecotoxicol Environ Saf. 2018;148:851–6.

Rodríguez-Sánchez VM, Rosas U, Calva-Vásquez G, Sandoval-Zapotitla E. Does acid rain alter the leaf anatomy and photosynthetic pigments in urban trees? Plan Theory. 2020;9(7):862.

Rosenfeld D, Lohmann U, Raga GB, O'Dowd CD, Kulmala M, Fuzzi S, Reissell A, Andreae MO. Flood or drought: how do aerosols affect precipitation? Science. 2008;321(5894):1309–13.

Roth-Nebelsick A, Ebner M, Miranda T, Gottschalk V, Voigt D, Gorb S, Stegmaier T, Sarsour J, Linke M, Konrad W. Leaf surface structures enable the endemic Namib desert grass *Stipagrostis sabulicola* to irrigate itself with fog water. J R Soc Interface. 2012;9(73):1965–74.

Ruhnayat A, Rostiana O, Syakir M. The improvement of grafting success of nutmeg (Myristica fragrans Hout.) by hooding and fogging. In: IOP conference series: earth and environmental Science, vol. 1172, No. 1. IOP Publishing; 2023, May. p. 012016.

Runyan C, Wang L, Lawrence D, D'Odorico P. Ecohydrological controls on the deposition of non-rainfall water, N, and P to dryland ecosystems. Dryland Ecohydrol. 2019:121–37.

Sachweh M, Koepke P. Radiation fog and urban climate. Geophys Res Lett. 1995;22(9):1073–6.
Sahu C, Basti S, Sahu SK. Particulate collection potential of trees as a means to improve the air quality in urban areas in India. Environ Process. 2021;8:377–95.
Satoh K, Saji S, Ito S, Shimizu H, Saji H, Kikuchi S. Gene response in rice plants treated with continuous fog influenced by pH, was similar to that treated with biotic stress. Rice. 2014;7:1–11.
Saurabh K, Chaulya SK, Singh RS, Kumar S, Mishra KK. Intelligent dry fog dust suppression system: an efficient technique for controlling air pollution in the mineral processing plant. Clean Techn Environ Policy. 2022;24:1–15.
Shafiq M, Iqbal MZ, Athar M. The effects of acid mist environment on plant growth: a review. J Plant Develop. 2023;30:149.
Sharaf-Eldin MA, Yaseen ZM, Elmetwalli AH, Elsayed S, Scholz M, Al-Khafaji Z, Omar GF. Modifying walk-in tunnels through solar energy, fogging, and evaporative cooling to mitigate heat stress on tomato. Horticulturae. 2023;9(1):77.
Shi C, Roth M, Zhang H, Li Z. Impacts of urbanization on long-term fog variation in Anhui Province, China. Atmos Environ. 2008;42(36):8484–92.
Singh D, Thapa S, Pandiyan K, Das MK, Pramanik A, Chakdar H. Microbial adaptations under low temperature. Survival Strategies in Cold-adapted Microorganisms. 2022;187–208
Smith DK, Reka S, Dorling SR, Ross AN, Renfrew IA, Jayakumar A, Anurose TJ, Parde AN, Ghude SD, Rumbold H. Forecasts of fog events in northern India dramatically improve when weather prediction models include irrigation effects. Commun Earth Environ. 2024;5(1):141.
Soderberg, K. S. (2010). The role of fog in the ecohydrology and biogeochemistry of the Namib Desert. (Doctoral dissertation, University of Virginia).
Sotomayor DA, Drezner TD. Dominant plants alter the microclimate along a fog gradient in the Atacama Desert. Plant Ecol. 2019;220(4):417–32.
Soussi M, Chaibi MT, Buchholz M, Saghrouni Z. Comprehensive review on climate control and cooling systems in greenhouses under hot and arid conditions. Agronomy. 2022;12(3):626.
Sun X, Amelung W, Klumpp E, Walk J, Mörchen R, Böhm C, Moradi G, May SM, Tamburini F, Wang Y, Bol R. Fog controls biological cycling of soil phosphorus in the Coastal Cordillera of the Atacama Desert. Glob Chang Biol. 2024;30(1):e17068.
Templer PH, Weathers KC, Ewing HA, Dawson TE, Mambelli S, Lindsey AM, Webb J, Boukili VK, Firestone MK. Fog as a source of nitrogen for redwood trees: evidence from fluxes and stable isotopes. J Ecol. 2015;103(6):1397–407.
Tie X, Long X, Li G, Zhao S, Cao J, Xu J. Ozone enhancement due to the photodissociation of nitrous acid in eastern China. Atmos Chem Phys. 2019;19(17):11267–78.
Turner DP, Tingey DT. Foliar leaching and root uptake of Ca, Mg and K in relation to acid fog effects on Douglas-fir. Water Air Soil Pollut. 1990;49:205–14.
Valjarević A, Algarni S, Morar C, Grama V, Stupariu M, Tiba A, Lukić T. The coastal fog and ecological balance for plants in the Jizan region, Saudi Arabia. Saudi J Biolog Sci. 2023;30(1):103494.
Vignesh MR, Palanisamy S. Aquaporin and its effect on foliar uptake to overcome drought stress in plants. Plant Physiol Rep. 2021;26(2):193–9.
Villarreal-Guerrero F, Pinedo-Alvarez A, Flores-Velázquez J. Control of greenhouse-air energy and vapor pressure deficit with heating, variable fogging rates and variable vent configurations: simulated effectiveness under varied outside climates. Comput Electron Agric. 2020;174:105515.
Wang F, Guo J, Zhang J, Huang J, Min M, Chen T, Liu H, Deng M, Li X. Multi-sensor quantification of aerosol-induced variability in warm clouds over eastern China. Atmos Environ. 2015;113:1–9.
Warren-Rhodes KA, McKay CP, Boyle LN, Wing MR, Kiekebusch EM, Cowan DA, Stomeo F, Pointing SB, Kaseke KF, Eckardt F, Henschel JR, Anisfeld A, Seely M, Rhodes KL. Physical ecology of hypolithic communities in the central Namib Desert: the role of fog, rain, rock habitat, and light. J Geophys Res Biogeo. 2013;118(4):1451–60.

Weathers KC, Ponette-González AG, Dawson TE. Medium, vector, and connector: fog and the maintenance of ecosystems. Ecosystems. 2020;23(1):217–29.

Wei H, Liu C, Ryong Jeong B. An optimal combination of the propagation medium and fogging duration enhances the survival, rooting and early growth of strawberry daughter plants. Agronomy. 2020;10(4):557.

Wu Y, Song L, Liu W, Liu W, Li S, Fu P, Shen Y, Wu J, Wang P, Chen Q, Lu H. Fog water is important in maintaining the water budgets of vascular epiphytes in an Asian tropical karst forests during the dry season. Forests. 2018;9(5):260.

Yan S, Zhu B, Kang H. Long-term fog variation and its impact factors over polluted regions of East China. J Geophys Res Atmos. 2019;124(3):1741–54.

Yan S, Zhu B, Huang Y, Zhu J, Kang H, Lu C, Zhu T. To what extents do urbanization and air pollution affect fog? Atmos Chem Phys. 2020;20(9):5559–72.

Yan Y, Liu Y, Liu X, Wang X. Effects of cloud microphysics on the vertical structures of cloud radiative effects over the Tibetan Plateau and the Arctic. Remote Sens. 2021;13(14):2651.

Yang L, Liu JW, Ren ZP, Xie SP, Zhang SP, Gao SH. Atmospheric conditions for advection-radiation fog over the western Yellow Sea. J Geophys Res Atmos. 2018;123(10):5455–68.

Yasutake D, Yu X, Asano T, Ishikawa M, Mori M, Kitano M, Ishikawa K. Control of greenhouse humidity and airflow with fogging and circulation systems and its effect on leaf conductance in cucumber plants. Environ Control Biol. 2014;52(2):101–5.

Yoshida K, Takenaka C. Effects of acid fog on tree physiology. J Japanese Forest Soc. 2004;86(1):54–60.

Zha Y, Tang J, Pan Y. The effects of simulated acid rain and cadmium-containing atmospheric fine particulate matter on the pakchoi (Brassica campestris. L) Seedlings growth and physiology. Soil Sci Plant Nutr. 2022;68(2):317–28.

Zhang X, Guo J, Vogt RD, Mulder J, Wang Y, Qian C, Wang J, Zhang X. Soil acidification as an additional driver to organic carbon accumulation in major Chinese croplands. Geoderma. 2020;366:114234.

Zhou S, Wang P, Ding Y, Xie L, Li A. Modification of plasma membrane H+-ATPase in Masson pine (*Pinus massoniana* Lamb.) seedling roots adapting to acid deposition. Tree Physiol. 2022;42(7):1432–49.

# Chapter 8
# Impact of Fog on Human Health

**Eliani Ezani**

## 8.1 Introduction

Fog is a prevailing atmospheric phenomenon, which suspended in a certain air volume near the ground level approximately 1 km. It consists of water droplets or vapors and ice particles. This chapter explores how fog affects human health, from the medieval miasma theory to the industrial revolution and modern era of technology, which brought notable episodes such as the Meuse Valley fog and London fog. These events transformed the landscape of air pollution epidemiology. This chapter will also focus on both the acute and chronic health effects of fog, examining topics such as the hidden air pollutants in fogs that affects health, mechanism of exposure that leads to acute and chronic health effects, and epidemiological evidence. Fog events not only impact health by trapping air pollutants at ground level but they also disrupt aviation and road transportation, increasing the risk of traffic accidents and compromising overall public safety.

### *8.1.1 From Miasmas to Modern Fog Air Pollution Disaster*

The fog phenomenon, previously referred to as "night air," was once associated with periods of poor air quality and potentially linked to outbreaks of infectious diseases. Before the germ theory emerged, the assumption of miasma as the cause of sickness during the Medieval period influenced public health practices in Europe, the Middle East, and China. It was hypothesized that the London cholera was due

---

E. Ezani (✉)
Department of Environmental and Occupational Health, Universiti Putra Malaysia, Serdang, Selangor, Malaysia
e-mail: elianiezani@upm.edu.my



P. Saxena et al. (eds.), *Fog: Formation, Chemistry, Dynamics, Impact and Dissipation*, https://doi.org/10.1007/978-981-96-4422-3_8

to the fact that the concentration of miasma, the airborne organic particles resulting from these processes, was the greatest at a lower elevation (Liu 2018). Interestingly, in 1717, a physician claimed that the "evil fogs" might induce disease and would appear during sunset over the riverbank. Thirty-seven years later, West Africa blamed a deadly pandemic for the poisonous and smelling fog (Chatfield 2020). In the nineteenth century, the appearance of so-called unhealthy fog was indicative of miasma.

In December 1930, following the Industrial Revolution, Belgium's Meuse Valley was blanketed in a thick fog containing high levels of particles and acidic gases. It has been reported that more than 60 people died due to severe respiratory health outcomes. The weather was characterized by anticyclonic conditions, with high atmospheric pressure and a light easterly wind blowing from the city of Liège into the narrow Meuse Valley. These conditions led to persistent fog, with temperatures within the fog ranging from 1 to 2 °C. A temperature inversion occurred at around 70–80 m above ground level, preventing the fumes from rising (Nemery et al. 2001). This atmospheric cap caused the industrial pollutants to accumulate in the air from steel mills, coke ovens, foundries, and smelters in Liège (Sokhi et al. 2022). The most common respiratory problems reported by physician who treated the patients during the fog episode including laryngeal irritation, retrosternal pain, coughing, and shortness of breath (Nemery et al. 2001).

Patients who were directly exposed to the outside air during the Meuse Valley fog event had signs and symptoms that correlated with local irritation on their respiratory mucosae. The postmortem specimens from Meuse valley's fog victim observed from histology of airways and alveolar region were shown for bronchial mucosa and alveolar damages. The specimen respiratory airway of alveolar filled with particles were observed in the histology of postmortem specimens' fog victim (Nemery et al. 2001). These earliest scientific proofs demonstrated the relationship between the fog phenomenon and atmospheric pollution, which resulted in respiratory diseases and death. A similar air pollution event occurred in Donora, Pennsylvania, USA, in October 1948, highlighting a link between short-term declines in air quality and increased mortality during acute episodes. Twenty people died, 1440 experienced serious illness, and another 4470 suffered mild to moderate symptoms. While a plethora of scientific literature reported on fogs in London, dense fog has been linked to higher mortality rates as far back as the seventeenth century (Brimblecombe 1999), with reports continuing up until the notorious London fog of 1952 (Brimblecombe 1982). Moreover, the well-documented presence of fog has portrayed in several poems and the iconic House of Parliament paintings of Monet (Novakov and Novakov 2006).

The dense fog of sulfur dioxide and particles accumulated as the lack of wind hindered the dispersion of the fog into the upper atmosphere during 1952 London fog event. The correlation between deaths and sulfur dioxide pollutant concentrations during the 5-day fog episode in London was reported by Brimblecombe (1987) in the famous book of the *Big Smoke*. Approximately 12,000 excess deaths were estimated to have occurred within three months as a result of prolonged exposure during that period. Bell et al. (2003) revisited the health record, suggesting that this excess mortality may be attributable to either through a delayed effect from the smog itself or due to elevated pollution levels that persisted through January 1953. Their work also points out that the combination of air pollution and influenza may have worsened the health outcomes, as those who survived the severe fog episode might have been more vulnerable to respiratory infections during that period. This interaction emphasizes the complex ways pollution can impact public health. Population-based studies assessing respiratory symptoms through questionnaires and lung function testing were initially influenced by the London fog events. These methods were later adopted by researchers in the USA and Canada. The landmark *Harvard Six Cities Study* (Dockery et al. 1993) expanded on this early research by incorporating detailed exposure assessments into its approach (Samet 2016). It is clear that air pollution events before the modern era of the twenty-first century differed in both intensity and the mix of pollutants, primarily due to the shift away from coal as a main fuel source.

Modern fog events, such as those in China and Delhi in 2022, are the result of ongoing climate change and fog formation that traps accumulated pollutants, further intensified by the industrialization in these countries. Recently, an unhealthy fog event was reported in Beijing in November 2024, during which $PM_{2.5}$ levels rose to 109.5 μg/m$^3$ (China Daily 2024). However, this is not new to China, as a large-scale, persistent fog enveloped multiple cities there back in 2013 (Zhang et al. 2013). The relationship between fog exposure and health impacts has been demonstrated through data on outpatient visits and hospitalizations for cardiovascular (Tan et al. 2019) and respiratory diseases, such as asthma and chronic obstructive pulmonary disease (COPD) (Park et al. 2015). This provides new evidence linking fog events to health outcomes in modern times, as hospital records serve as secondary data to associate high pollution levels with adverse health effects. Such analyses offer valuable insights into the broader health implications of severe fog events.

## 8.2 How Fog's Hidden Physicochemical and Biological Properties Affect Our Health

With the right balance of atmospheric conditions, fog will emerge. There are four crucial conditions for the formation of fogs; low temperature (between 3 °C and 13 °C), low wind speed (WS <2 m/s), high relative humidity (RH > 84%), and shallow boundary layer height (~500–800 m) (Agarwal et al. 2017). These factors will contribute to physical and chemical changes in particulate matter which led to higher concentrations of inorganic ions, polycyclic aromatic hydrocarbons (PAHs), and black carbon (BC) during fog events (Li et al. 2023). In addition, heavy metals in fine particulates are less likely to undergo chemical transformation due to their non-volatile nature. Sources such as coal and biomass combustion, vehicular emissions, crustal dust, and iron-producing industrial processes have been identified as the primary contributors during winter months (Agarwal et al. 2017).

Recent studies have shown a growing interest in estimating cancer risk by analyzing background PAH concentrations obtained from samples collected during both foggy and non-foggy days (Singh and Gupta 2016; Liu et al. 2017; Rajeev et al. 2021; Verma et al. 2022). Typical health risk assessments, such as Lifetime Average Daily Dose (LADD) and Incremental Lifetime Cancer Risk (ILCR) for children and adults, have been reported. But is that sufficient to establish causal effects? We will revisit this question in the upcoming subsection on the short- and long-term exposures to fog. During fog events, nucleation-mode particles shift within the aerosol size distribution, leading to increased concentrations of particles with a radius less than 0.1 μm, which form freshly nucleated aerosols. Top-of-atmosphere forcing was observed to increase on foggy days due to heightened backscattering of radiation into space. As the day progresses, both relative humidity and radiative forcing decrease, in contrast solar elevation increases the forcing. Hence, when fog events persist longer into the day, the diurnal average aerosol radiative forcing experiences a stronger effect compared to events confined to early morning hours (Das et al. 2008).

The oxidation of dissolved gases in fog droplets and their subsequent scavenging leads to the chemical transformation of gaseous species. In *Smog Formation*, Brimblecombe (2016) explains sulfur dioxide dissolved in fog droplets and underwent oxidation initiated by the hydroxyl radical during the London winter fog event, sulfur trioxide will react with water to produce sulfuric acid (Eq. (8.1)).

$$SO_2 + OH + M \rightarrow HOSO_2 \rightarrow SO_3 + OH \tag{8.1}$$

While, in modern winter smog, nitrogen dioxide predominates due to increased traffic emissions. Rapid oxidation appears to result from the direct oxidation of NO by oxygen at high $NO_x$ concentrations (primarily NO and $NO_2$). The secondary aerosol will form after fog evaporates following the condensation process of primary aerosols in the gaseous and aqueous phases. The acidity of fog water is a significant concern, as it poses potential risks, including respiratory irritation through the airways and direct effects from dermal contact and eye exposure. The lowest pH reported has been as low as 1.7, a finding first raised by Hackney et al. (1985) nearly four decades ago. In addition, Nieberding et al. (2018) indicated that fog acidification is primarily driven by emissions of $SO_2$ and $NO_x$. The high ion concentration levels are believed to result from the long-range transport of air pollutants from various sources in East Asia, as highlighted in their study. Their research also found a median pH of 4.05 in fog samples collected from rural mountainous areas of southwest China.

The mean pH of 4.78 for acid fog recorded over nearly half of 731 days in Kushiro, Japan, has been associated with an increase in hospital visits due to asthma, particularly in nonatopic asthmatic subjects. Tanaka et al. (1998) suggested that the weak bronchoconstrictive effect of acid fog may be more influential for these individuals. This finding is particularly relevant in the context of climate change, as increasing industrialization and fluctuating meteorological factors may intensify the health impacts of air pollutants. In the post-COVID-19 era, these dynamics might further alter exposure to such pollutants, making it essential to study the health effects of acid fog in today's evolving environmental landscape. Given the changing climate and increasing industrial activities, this area of research is becoming even more crucial, as acid fog's potential health effects, particularly on respiratory conditions like asthma, need to be further explored with updated experimental studies.

### *8.2.1 From Acid Fog to Bioaerosols Fog*

In addition to the complex physicochemical components of fog and their potential health impacts, there's another fascinating story to uncover which is the role of aerosols associated with living and dead organisms. These bioaerosols, originating from microorganisms such as bacteria, fungi, and viruses, can be carried within fog droplets or released into the environment by biological processes. Particulate-bound bioaerosols account for 15–25% of particulate matter by mass and are composed of particulate matter (PM) combined with pathogens (Shammi et al. 2021). Besides, the concentration of bacterial aerosols was observed to increase by up to 40% during foggy days compared to pre-fog conditions and decreased by 48% during post-fog periods in rural Arthauli, India (Saikh and Das 2023).

During foggy days, the droplets in the air can serve as a growth medium for microorganisms, leading to an increase in bacterial and fungal concentrations, particularly in the coarse size fraction (approximate less than 10 μm) when suspended in the air for a longer time. This growth is facilitated by high relative humidity (RH) and the presence of water-soluble ions during severe fog events, which enhance the survival of certain microbial species. High RH also prevents bacteria attached to dust particles from becoming airborne, as wet surfaces reduce their dispersal into the atmosphere (Qi et al. 2018). The increase of bioaerosols in the air during fog days is significant due to the stable atmospheric condition where no vertical mixing happens. A recent study by Vishwakarma et al. (2024) observed the presence of bacteria, such as *Bacillus*, *Enterobacter*, and *Cocci*, as well as fungi, including *Aspergillus*, *Cladosporium*, and *Penicillium*, during three episodes of foggy winter days between 2021 and 2024 in Varanasi, India. Table 8.1 shows example of species of bacteria and fungi known to cause health effects.

**Table 8.1** Example of pathogens and their health effects

| Pathogen | Species | Health effects |
|---|---|---|
| *Aspergillus* | Fungi | Aspergillosis |
| *Cladosporium* | Fungi | Respiratory allergy such as asthma |
| *Periconia* | Fungi | Eye infection |
| *Bacillus* | Bacteria | Ear infection |
| *Enterobacter* | Bacteria | Lower respiratory tract infections |

## 8.3 Exposure Pathway and Health Impacts

When fog is suspended in the air at or near the Earth's surface, it can have significant impacts on human health, particularly for those exposed outdoors, as individuals typically move through space and time with their daily activities. Factors such as age, nutritional status, and preexisting illnesses may also influence the severity and type of health effects experienced (Kampa and Castanas 2008). Males typically spend more time outdoors compared to females, and during fog episodes, hospital visits were observed to be more frequent for male patients than female patients. However, to better understand the total exposure profiles during foggy events, it is essential to determine how much time individuals actually spend outdoors on such days (Park et al. 2015).

The combined effects of smog and exposure to environmental tobacco smoke (ETS), such as secondhand smoke, significantly increase the risk of respiratory diseases, cerebrovascular diseases, and ischemic heart diseases. This becomes particularly problematic in countries with large smoking populations, such as China, India, the USA, and Russia. In these countries, secondhand smokers are exposed to both environmental pollution and tobacco smoke, which intensifies health risks. Cai and Wang (2017) highlighted the severity of this issue, noting that individuals exposed to both smog and secondhand smoke face greater health challenges and higher morbidity and mortality rates. Additionally, Javed et al. (2021) noted the synergistic effects of COVID-19 and smog, particularly among the elderly and individuals with comorbidities, who are at higher risk of severe health effects. Similar to COVID-19, smog poses a significant threat to those already suffering from cardiovascular or respiratory diseases, increasing their vulnerability to mortality risk.

Children are more likely to be exposed to higher levels of polluted air compared to adults, mainly because their immune systems are still developing. Additionally, due to their smaller body weight, children inhale a greater volume of air relative to their size, which increases the concentrations of pollutants they breathe in. Epidemiological studies have highlighted the short-term effects of air pollution on childhood asthma during foggy events. Kashiwabara et al. (2002, 2003) observed this particularly during the midnight-to-dawn period on foggy nights with elevated atmospheric temperatures, indicating a higher saturation of airborne water droplets that may exacerbate respiratory symptoms in children. Respiratory infections caused by pathogens can occur within a short time frame, sometimes just a few hours after exposure to air pollution, particularly during pollution episodes like fog. Cheng et al. (2021) examined sub-daily scales of acute upper respiratory infections (AURI) and acute lower respiratory infections (ALRI) in children. Their study found a higher risk of upper respiratory infections during cold seasons and nighttime. Their approach could prove valuable for future research on the health impacts of fog, whether linked to infectious or noninfectious respiratory diseases, especially in areas with high levels of air pollution.

The primary pathways of exposure are the inhalation of air pollutants and bioaerosols carried by the fog, as well as dermal contact. Ocular contact can also occur due to direct exposure to the fog, which may cause eye diseases. This occurrence has been observed during the 2016 Lahore smog. The highest concentration of pollutants analyzed was $NO_x$, which coincided with a 60% increase in the number of patients reporting ocular surface diseases compared to the same period in 2015. These conditions included dry eyes, irritation, lid erosion, corneal diseases, conjunctival disorders, uveitis, and excessive tearing (lacrimation) (Ashraf et al. 2019).

Figure 8.1 presents the formation of fog, from primary to secondary pollutants, and its vertical distribution within 1 km of the Earth's surface. It highlights how fog acts as a carrier for pollutants and bioaerosols, exposing individuals to health risks through various pathways, including inhalation, dermal contact, and ocular exposure. The multifaceted impact of fog on human health, such as respiratory, dermatological, and ocular effects, is demonstrated alongside its role in impairing visibility. This diagram, adapted from Pérez-Díaz et al. (2017), highlights various air pollutants, such as $NO_x$, $SO_2$, $O_3$, CO, VOCs, dioxins, heavy metals, and particulate

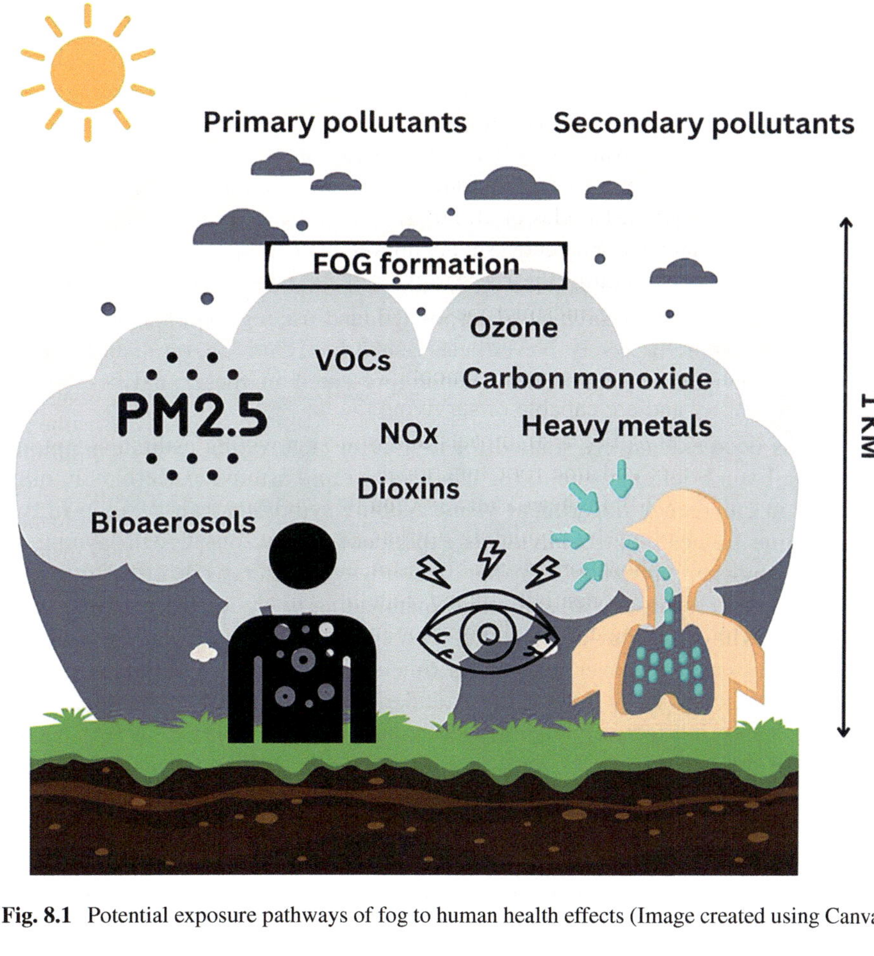


**Fig. 8.1** Potential exposure pathways of fog to human health effects (Image created using Canva)

matter, which can adversely affect the human body systems and contribute to associated diseases.

### 8.3.1 Acute and Long-Term Exposure

The dose and duration of exposure can lead to varying impacts on human health, ranging from acute to chronic effects, including cancer. Short-term exposure effects are typically assessed by analyzing the relationship between daily counts of morbidity or mortality in a given population and daily pollutant and meteorological variables recorded at central monitoring sites. For estimating long-term average pollution exposure, survival analysis is used to determine the contribution of geographical variations in exposure to the risk of adverse events over the follow-up period. This analysis adjusts for factors that may influence disease risk (Beverland et al. 2012).

In fog exposure and health assessments, studies of acute effects often report hospital visits for asthma symptoms within their study domains. During these episodes, pollutant concentrations and meteorological variables, such as temperature changes, relative humidity, cloud cover, and the presence of fog, are typically monitored. However, this approach introduces exposure error, likely resulting in downward bias in effect estimates and wider confidence intervals (CIs). For instance, the lack of spatially resolved daily $PM_{2.5}$ concentration data limits most studies to areas near monitoring sites, predominantly in urban settings. This constraint may not accurately represent the broader population. While such studies generally use city-level contrasts, more recent research has incorporated land-use regression models to estimate exposure more precisely. Nevertheless, suburban, exurban, rural, and semirural populations remain significantly underrepresented in these analyses (Kloog et al. 2013).

Fog has been extensively studied for its role in aggravating asthma symptoms (Fig. 8.2). Even brief exposure to fog can lead to rapid asthma exacerbation, often resulting in emergency department visits. Asthma symptoms can develop in two distinct time frames: some individuals experience a slow onset, with symptoms appearing anywhere from 6 to 30 h after exposure to a trigger, while others may face a much quicker response, with symptoms manifesting in less than 6 h, and in some cases, even within an hour of exposure. Two common study designs for transient outcomes are case-crossover design and time-series Poisson regression analyses. The case-crossover study design is particularly recommended for asthma event studies, because it effectively controls for confounding factors such as seasonality, underlying disease severity, the level of disease control, self-management behavior, personal asthma symptom triggers, and access to primary care. Although temperature change is a relatively consistent environmental exposure across small geographic areas, individuals may experience varying degrees of exposure. Additionally, the time between the onset of an asthma episode and the patient's registration at the emergency department remains unclear (Wasilevich et al. 2012). Table 8.2

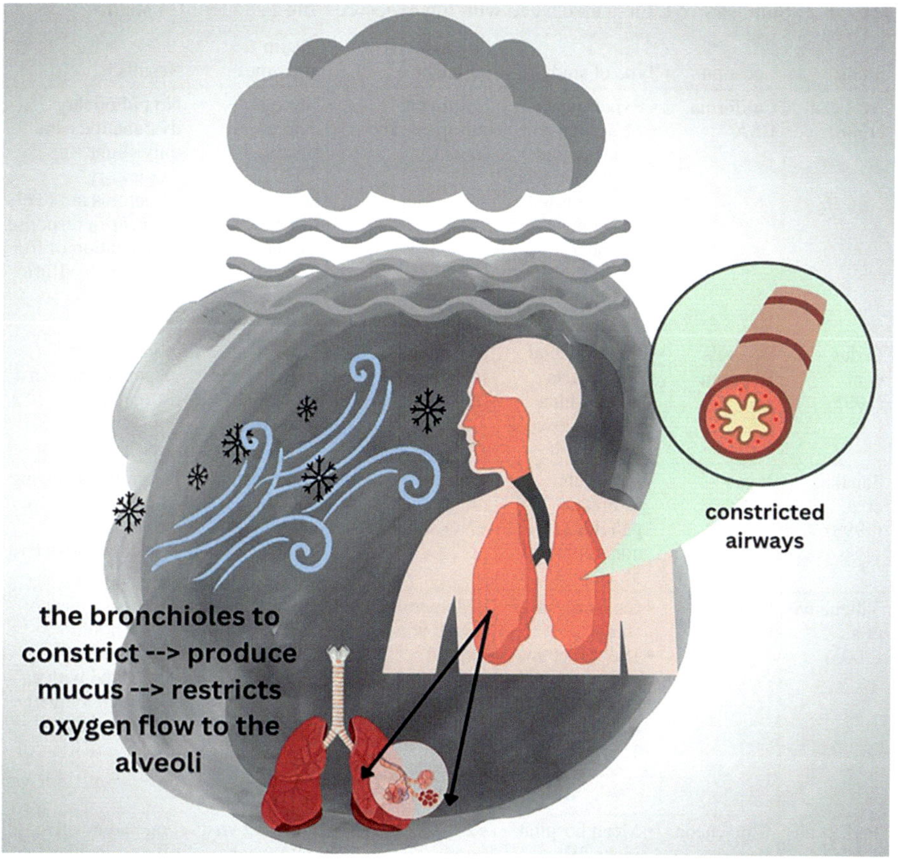


**Fig. 8.2** Asthma symptoms when inhaling polluted fogs (Image created using Canva)

summarizes the respiratory health risks, specifically asthma, associated with fog exposure over the past four decades (from 1984 to 2024).

The potential carcinogenic effects of fog-related pollutants have received increasing attention in recent years. Various studies have conducted human cancer risk assessments related to these pollutants, emphasizing the long-term health risks associated with prolonged exposure to foggy episodes (Agarwal et al. 2017; Chakraborty et al. 2023). Hazard Quotient (Hqi), Hazard Index (HI), and the Incremental Lifetime Cancer Risk (ILCR) were calculated to evaluate the level of toxicity of air pollutants. Another interesting approach to studying human respiratory deposition of bioaerosols during fog events was applied by Wei et al. (2022). They employed the multiple-path particle dosimetry model (version 3.04) to calculate the deposition of bioaerosols of varying sizes, estimating the health effects on different regions of the respiratory tract, such as the extrathoracic (ET) and tracheobronchial (TB) regions. It is found that inhalable bioaerosols are predominantly deposited in the extrathoracic region of the respiratory tract.

**Table 8.2** Summary of asthma associated with fog exposure from 1984 to 2024 studies

| Study | Location | Type of study | Fog-associated pollutants | Health outcomes | Results |
|---|---|---|---|---|---|
| Avol et al. (1988) | California, USA | • Experimental<br>• 22 normal and 22 asthmatic adult volunteers were exposed in a control chamber | Hydrogen sulfuric acid ($H_2SO_4$) | Forced expiratory function (FEV), airway resistance, irritant symptoms bronchial reactivity | No pulmonary dysfunctions and only slight respiratory symptoms are likely to occur in response to inhalation of fog containing sulfuric acid |
| Leduc et al. (1995) | Brussels, Belgium | • Experimental<br>• 14 subjects with asthma were exposed at rest for 1 h | Hydrogen sulfuric acid ($H_2SO_4$) | Pulmonary function and bronchial reactivity | No bronchial responsiveness in asthmatics |
| Tanaka et al. (1998) | Japan | • 102 adult asthmatic patients (44 nonatopic and 58 atopic) | • Acidity of the fog<br>• $NO_2$ and $SO_2$ | Hospital visit data | Naturally occurring acid fog may have a weak bronchoconstrictive effect |
| Villeneuve et al. (2005) | Ottawa, Canada | • Case-crossover study<br>• 18,970 asthma visits among children between 2 and 15 years of age | • Meteorological data: wind speed, temperature, atmospheric pressure, relative humidity, and visibility | Hospital visit data (year between 1992 and 2000) | The occurrence of fog or liquid precipitation was associated with an increased number of asthma visits |
| Park et al. (2015) | Chuncheon, Korea | • Mean hospital visits for asthma were 59.37 ± 34.01 per day | • Air pollution and Meteorological data | Hospital visit data (year between 2006 and 2012) | Fog had positive effects on hospital visits for airway diseases after adjusting for air pollutants |
| Kwon et al. (2016) | Chuncheon, Korea | • 583 patients with asthma exacerbations | • Air pollution and Meteorological data | Hospital visit data (year between 2006 and 2011) | Fog (lag 2) has shown protective effects against asthma exacerbations |

## 8.4 The Visible Fog

From visual perspective, fogs can influence speed-related behaviors related to potential car accident. In foggy conditions, the driver's brain perceives the reduced visibility as a sign that objects are farther away, which causes the driver to feel as though they are driving more slowly than usual. This happens because the brain processes the visual uncertainty of fog as a cue to adjust speed perception, even when the actual speed remains unchanged (Yan et al. 2014). Table 8.3 shows the

**Table 8.3** Summary of classification of haze, fog, and haze-fog days based on strength level and visibility

| Type | Strength level | Visibility |
|---|---|---|
| Haze | Slight haze (I) | ≥5 km and <10 km |
| | Mild haze (II) | ≥3 km and <5 km |
| | Moderate haze (III) | ≥2 km and <3 km |
| | Severe haze (IV) | <2 km |
| Fog | Light fog (I) | ≥1 km and <10 km |
| | Medium fog (II) | <1 km |
| Haze-fog | Light pollution (100 <AQI <150) | <10 km |
| | Heavy pollution (200 <AQI <300) | |

classification of haze, fog, and haze-fog days classified by Wei et al. (2022). Their classification is based on weather phenomena defined by the Meteorological Information Comprehensive Analysis and Process System (MICAPS) of the Meteorological Administration of China.

## 8.5 Vector-Borne Disease and Fog

Contrary to the common belief that fog primarily has negative effects on human health, a study revealed that fog day frequency could serve as a predictor of malaria incidence in rainforest areas. For example, in Mengla County, China, researchers found that fog day frequency exhibited a one-year delayed effect on annual malaria incidence (Tian et al. 2008). Malaria, caused by the *Plasmodium* parasite and transmitted by *Anopheles* mosquitoes, involves a complex interplay between human hosts and environmental factors. The transmission is significantly influenced by climatic conditions that promote the breeding and survival of *Anopheles* mosquitoes in water bodies. This environmental dependence classifies malaria as a climate-sensitive disease (Xiang et al. 2018), as variations in temperature, rainfall, and humidity can greatly affect mosquito life cycles, trend, and distribution of disease.

## 8.6 Conclusion

This chapter explores the impact of fog on human health, tracing its history from the medieval miasma theory to significant modern events, such as the Beijing fog in 2024. Fog-related air pollution worsens respiratory conditions, such as asthma, and increases the risk of acute respiratory infections, particularly among vulnerable groups like children and the elderly. Children, in particular, are more likely than adults to be exposed to higher levels of polluted air due to their activity patterns and physiological characteristics. Pollutants and bioaerosols deposited during fog events primarily accumulate in the extrathoracic region of the respiratory tract, contributing to both immediate and long-term health effects.

Emerging research underscores the potential carcinogenic risks associated with prolonged exposure to fog-related pollutants, highlighting the critical need for comprehensive health risk assessments. However, within the context of climate change, it is imperative to further investigate fog's effects on diseases such as asthma and COPD. Another promising area for expanding research on the health effects of fog is its connection to the One Health concept, particularly through the lens of vector-borne disease transmission influenced by climate change. Experimental studies and advanced modeling techniques are especially vital to understanding the mechanisms of fog-related health impacts, enabling more effective public health interventions and policies. These efforts must be integrated into broader climate action strategies to mitigate the compounded effects of pollution, fog, and global warming on human health.

## References

Agarwal A, Mangal A, Satsangi A, Lakhani A, Kumari KM. Characterization, sources and health risk analysis of $PM_{2.5}$ bound metals during foggy and non-foggy days in sub-urban atmosphere of Agra. Atmos Res. 2017;197:121–31. https://doi.org/10.1016/j.atmosres.2017.06.027.

Ashraf A, Butt A, Khalid I, Alam RU, Ahmad S. Smog analysis and its effect on reported ocular surface diseases: a case study of 2016 smog event of Lahore. Atmos Environ. 2019;198:257–64. https://doi.org/10.1016/j.atmosenv.2018.10.029.

Avol EL, Linn WS, Wightman LH, Whynot JD, Anderson KR, Hackney JD. Short-Term respiratory effects of sulfuric acid in fog: A laboratory study of healthy and asthmatic volunteers. JAPCA. 1988:38(3):258–63. https://doi.org/10.1080/08940630.1988.10466375

Bell ML, Davis DL, Fletcher T. A retrospective assessment of mortality from the London smog episode of 1952: The role of influenza and pollution. Environ Health Perspect. 2003;112(1):6–8. https://doi.org/10.1289/ehp.6539.

Beverland IJ, Cohen GR, Heal MR, Carder M, Yap C, Robertson C, Hart CL, Agius RM. A comparison of short-term and long-term air pollution exposure associations with mortality in two cohorts in Scotland. Environ Health Perspect. 2012;120(9):1280–5. https://doi.org/10.1289/ehp.1104509.

Brimblecombe P. Long term trends in London fog. Sci Total Environ. 1982;22:19–29.

Brimblecombe P. The big smoke: a history of air pollution in London since medieval times. 1st ed. London: Routledge; 1987.

Brimblecombe P. Air pollution and health history. New York: Academic Press; 1999. p. 5–18.

Brimblecombe P. Smog formation. In: Handbook of combustion. Hoboken: Wiley; 2016. p. 1–15. https://doi.org/10.1002/9783527628148.

Cai H, Wang C. Surviving with smog and smoke: precision interventions? Chest. 2017;152(5):925–9. https://doi.org/10.1016/j.chest.2017.06.030.

Chakraborty A, Gupta T, Mandaria A, Tripathi S. Trace elements in ambient aerosols and size-resolved fog droplets: trends, enrichment, and risk assessment. Heliyon. 2023;9(6):e16400. https://doi.org/10.1016/j.heliyon.2023.e16400.

Chatfield C. The gallery of natural phenomenon. 2020. http://www.phenomena.org.uk/index.html

Cheng J, Su H, Xu Z. Intraday effects of outdoor air pollution on acute upper and lower respiratory infections in Australian children. Environ Pollut. 2021;268:115698. https://doi.org/10.1016/j.envpol.2020.115698.

China Daily. Fog hits Beijing. 2024. https://www.chinadailyhk.com/hk/article/596778

Das SK, Jayaraman A, Misra A. Fog-induced variations in aerosol optical and physical properties over the Indo-Gangetic Basin and impact to aerosol radiative forcing. Ann Geophys. 2008;26:1345–54.

Dockery DW, Pope CA 3rd, Xu X, Spengler JD, Ware JH, Fay ME, Ferris BG Jr, Speizer FE. An association between air pollution and mortality in six U.S. cities. N Engl J Med. 1993;329(24):1753–9. https://doi.org/10.1056/NEJM199312093292401.

Hackney JD, Linn WS, Avol EL. Potential risks to human respiratory health from "acid fog": evidence from experimental studies of volunteers. Environ Health Perspect. 1985;63:57–61. https://doi.org/10.1289/ehp.856357.

Javed A, Aamir F, Gohar UF, Mukhtaramid H, Zia-UI-Haq M, Alotaibi MO, Bin-Jumah MN, Pop OL. The potential impact of smog spell on humans' health amid COVID-19 rages. Int J Environ Res Public Health. 2021;18(21):11408. https://doi.org/10.3390/ijerph182111408.

Kampa M, Castanas E. Human health effects of air pollution. Environ Pollut. 2008;151:362–7.

Kashiwabara K, Kohrogi H, Ota K, Moroi T. High frequency of emergency room visits of asthmatic children on misty or foggy nights. J Asthma. 2002;39(8):711–7. https://doi.org/10.1081/jas-120015794.

Kashiwabara K, Itonaga K, Moroi T. Airborne water droplets in mist or fog 05 affect nocturnal attacks in asthmatic children. J Asthma. 2003;40(4):405–11. https://doi.org/10.1081/JAS-120018710.

Kloog I, Ridgway B, Koutrakis P, Coull BA, Schwartz JD. Long- and short-term exposure to PM2.5 and mortality: using novel exposure models. Epidemiology. 2013;24(4):555–61. https://doi.org/10.1097/EDE.0b013e318294beaa.

Kwon J, Han Y, Oh MK, Lee CY, Kim JY, Kim EJ, Kim H, Kim WJ. Emergency department visits for asthma exacerbation due to weather conditions and air pollution in chuncheon, Korea: A case-crossover analysis. Allergy Asthma Immunol Res. 2016;8(6):512. https://doi.org/10.4168/aair.2016.8.6.512.

Leduc D, Fally S, De Vuyst P, Wollast R, Yernault JC. Acute exposure to realistic acid fog: effects on respiratory function and airway responsiveness in asthmatics. Environ Res. 1995;71(2):89–98. https://doi.org/10.1006/enrs.1995.1070.

Li R, Yan C, Meng Q, Yue Y, Jiang W, Yang L, Zhu Y, Xue L, Gao S, Liu W, Chen T, Meng J. Key toxic components and sources affecting oxidative potential of atmospheric particulate matter using interpretable machine learning: insights from fog episodes. J Hazard Mater. 2023;465:133175. https://doi.org/10.1016/j.jhazmat.2023.133175.

Liu L. Introduction, heart failure: epidemiology and research methods. Amsterdam: Elsevier; 2018. p. 1–12. https://doi.org/10.1016/B978-0-323-48558-6.00001-3.7.

Liu J, Wang Y, Li PH, Shou YP, Li T, Yang MM, Wang L, Yue JJ, Yi XL, Guo LQ. Polycyclic aromatic hydrocarbons (PAHs) at high mountain site in North China: concentration, source and health risk assessment. Aerosol Air Qual Res. 2017;17:2867–77. https://doi.org/10.4209/aaqr.2017.08.0288.

Nemery B, Hoet PH, Nemmar A. The Meuse Valley fog of 1930: an air pollution disaster. Lancet. 2001;357(9257):704–8. https://doi.org/10.1016/S0140-6736(00)04135-0.

Nieberding F, Breuer B, Braeckevelt E, Klemm O, Song Q, Zhang Y. Fog water chemical composition on Ailaoshan Mountain, Yunnan Province, SW China. Aerosol Air Qual Res. 2018;18:37–48. https://doi.org/10.4209/aaqr.2017.01.0060.

Novakov A, Novakov T. Eyewitness: the chromatic effects of late nineteenth- century London fog. London: Interdisciplinary Studies; 2006.

Park J, Lim MN, Hong Y, Kim WJ. The influence of Asian Dust, Haze, Mist, and Fog on hospital visits for airway diseases. Tuberc Respir Dis. 2015;78(4):326–35. https://doi.org/10.4046/trd.2015.78.4.326.

Pérez-Díaz JL, Ivanov O, Peshev Z, Álvarez-Valenzuela MA, Valiente-Blanco I, Evgenieva T, Dreischuh T, Gueorguiev O, Todorov PV, Vaseashta A. Fogs: physical basis, characteristic properties, and impacts on the environment and human health. Water. 2017;9(10):807. https://doi.org/10.3390/w9100807.

Qi J, Li M, Zhen Y, et al. Characterization of bioaerosol bacterial communities during hazy and foggy weather in Qingdao, China. J Ocean Univ China. 2018;17:516–26. https://doi.org/10.1007/s11802-018-3307-2.

Rajeev P, Singh AK, Singh GK, Vaishya RC, Gupta T. Chemical characterization, source identification and health risk assessment of polycyclic aromatic hydrocarbons in ambient particulate matter over central Indo-Gangetic Plain. Urban Clim. 2021;35:100755. https://doi.org/10.1016/j.uclim.2020.100755.

Saikh SR, Das SK. Fog-induced alteration in airborne microbial community: a study over Central Indo-Gangetic Plain in India. Appl Environ Microbiol. 2023;89:e01367.

Samet JM. London fog—the biography. Am J Public Health. 2016;106(8):1352–3. https://doi.org/10.2105/AJPH.2016.303292.

Shammi M, Rahman MM, Tareq SM. Distribution of bioaerosols in association with particulate matter: a review on emerging public health threat in Asian megacities. Front Environ Sci. 2021;9:698215. https://doi.org/10.3389/fenvs.2021.698215.

Singh DK, Gupta T. Effect through inhalation on human health of PM1 bound polycyclic aromatic hydrocarbons collected from foggy days in northern part of India. J Hazard Mater. 2016;306:257–68. https://doi.org/10.1016/j.jhazmat.2015.11.049.

Sokhi RS, Moussiopoulos N, Baklanov A, Bartzis J, Coll I, Finardi S, Friedrich R, Geels C, Grönholm T, Halenka T, Ketzel M, Maragkidou A, Matthias V, Moldanova J, Ntziachristos L, Schäfer K, Suppan P, Tsegas G, Carmichael G, Franco V, Hanna S, Jalkanen J-P, Velders GJM, Kukkonen. Advances in air quality research – current and emerging challenges. Atmos Chem Phys. 2022;22:4615–703. https://doi.org/10.5194/acp-22-4615-2022.

Tan F, Wang W, Qi S, et al. Air pollutants and outpatient visits for cardiovascular disease in a severe haze-fog city: Shijiazhuang, China. BMC Public Health. 2019;19:1366. https://doi.org/10.1186/s12889-019-7690-4.

Tanaka H, Honma S, Nishi M, Igarashi T, Teramoto S, Nishio F, Abe S. Acid fog and hospital visits for asthma: an epidemiological study. Eur Respir J. 1998;11(6):1301–6. https://doi.org/10.1183/09031936.98.11061301.

Tian L, Bi Y, Ho SC, et al. One-year delayed effect of fog on malaria transmission: a time-series analysis in the rain forest area of Mengla County, south-west China. Malar J. 2008;7:110. https://doi.org/10.1186/1475-2875-7-110.

Verma PK, Sah D, Satish R, Rastogi N, Kumari RM, Lakhani A. Atmospheric chemistry and cancer risk assessment of polycyclic aromatic hydrocarbons (PAHs) and nitro-PAHs over a semi-arid site in the Indo-Gangetic plain. J Environ Manag. 2022;317:4797. https://doi.org/10.1016/j.jenvman.2022.115456.

Villeneuve PJ, Leech J, Bourque D. Frequency of emergency room visits for childhood asthma in Ottawa, Canada: the role of weather. Int J Biometeorol. 2005;50:48–56. https://doi.org/10.1007/s00484-005-0262-6.

Vishwakarma YK, Ram K, Gogoi MM, Banarjee T, Singh RS. Size-segregated characteristics of bioaerosols during foggy and non-foggy days of winter, meteorological implications, and health risk assessment. Environ Sci Adv. 2024;2024(3):1163–72. https://doi.org/10.1039/d4va00108g.

Wasilevich EA, Rabito F, Lefante J, Johnson E. Short-term outdoor temperature change and emergency department visits for asthma among children: a case-crossover study. Am J Epidemiol. 2012;176(suppl_7):S123–30. https://doi.org/10.1093/aje/kws326.

Wei W, Qi J, Yin Y, Gong J, Yao X. Characteristics of inhalable bioaerosols on foggy and hazy days and their deposition in the human respiratory tract. Environ Pollut. 2022;307:119593. https://doi.org/10.1016/j.envpol.2022.119593.

Xiang J, Hansen A, Liu Q, Tong MX, Liu X, Sun Y, Cameron S, Hanson-Easey S, Han GS, Williams C, Weinstein P, Bi P. Association between malaria incidence and meteorological factors: a multi-location study in China, 2005-2012. Epidemiol Infect. 2018;146(1):89–99. https://doi.org/10.1017/S0950268817002254.

Yan Y, Li X, Liu Y, Zhao J. Effects of foggy conditions on drivers' speed control behaviors at different risk levels. Saf Sci. 2014;68:275–87. https://doi.org/10.1016/j.ssci.2014.04.013.

Zhang YS, Ma GX, Yu F, Cao D. Health damage assessment due to PM2.5 exposure during haze pollution events in Beijing-Tianjin-Hebei region in January 2013. CAMPH. 2013;93:2707–10.

# Chapter 9
# Governance of Side Effects: The Case of Impact of Fog on Economy

**Jurgen van der Heijden**

## 9.1 Introduction

Fog is hazardous for vehicle and air traffic. When combined with air pollution, fog becomes even more detrimental to health than usual. These are two reasons to limit traffic during fog, yet this may harm the economy. But is that truly the case? Couldn't a healthy city benefit the economy as well? City administrators can weigh these factors, but this remains abstract without a coalition to clearly show the benefits a healthy city could have on traffic safety, health, and the economy. Coalitions advocating for more traffic are often better organized than those advocating for a healthy city. How can the latter coalition be more effectively organized, and if that fails, how can administrators make balanced decisions between safety, health, and economic interests when fog occurs in the city?

An often-overlooked fact in these considerations is that interests are not inherently opposed. Interests such as traffic, health, housing, food, energy, well-being, culture, nature, and education are not inherently in conflict. Housing is not opposed to nature, traffic is not opposed to water, and education is not opposed to safety. Conflicts arise when people pursue these interests by setting goals and using resources that may clash. This is a characteristic of interests—they can lead to conflict—but this differs from inherent opposition between two or more interests.

Another aspect often ignored is the tendency to overlook side effects. For example, it's common to build a residential area without considering that it could also serve as a small clean energy hub or incorporate ample greenery. These are positive side effects, and clean energy helps mitigate the negative externalities the neighborhood would have on the outside world if it continued to rely on fossil fuels. Likewise, green spaces help reduce the negative impacts the outside world, such as extreme

J. van der Heijden (✉)
TNO (Dutch Organisation for Applied Scientific Research), The Hague, The Netherlands
e-mail: jurgen.vanderheijden@tno.nl



P. Saxena et al. (eds.), *Fog: Formation, Chemistry, Dynamics, Impact and Dissipation*,
https://doi.org/10.1007/978-981-96-4422-3_9

weather and heat, can have on the neighborhood. Unless legally required, there's a strong inclination to ignore negative side effects, often due to the cost of addressing them.

Even more remains in the background, because positive side effects can mitigate negative ones. To achieve a healthy city, it could be valuable to foster a coalition between clean energy, greenery, and public health. Prioritizing clean energy and greenery could reduce air pollution and mitigate its consequences. The positive side effects of clean energy and greenery promote health and reduce the negative impacts of external factors on public health. Governance of side effects is an emerging, relatively unknown concept. This chapter explores whether it could help city administrators balance economic and health considerations.

The next step is to better outline the problem of air pollution during fog in the city and the balance administrators must strike between the economy and public health. From this, a theory on the potential of side effects will be developed and how administrators might consider this in their decision-making. Before applying this to the case of urban air pollution and fog, the theory will first be tested on a different case from the agricultural sector. The fiber cultivation case has already provided many insights, and these lessons could inform the application of this theory to air pollution and fog. Following this, an analysis of the economic, financial, and legal implications of the theory will be conducted. Finally, the conclusion will assess the potential value of these insights and offer recommendations for both theory and practice.

## 9.2 Problem

Other chapters in this book portray fog in the city as a societal problem when air pollution levels are high. Based on this premise, this chapter focuses on a different issue—the effectiveness and efficiency of measures to reduce air pollution. Here, it is not a technical problem, as addressed in other chapters, but a governance problem. What control options does society have, especially through legal, economic, and financial means, to reduce air pollution, particularly in cases of urban fog?

The environmental department of a municipality is responsible for combating air pollution. However, it cannot independently implement economic measures to encourage companies to use clean energy, as such decisions involve the economic department. Nor can it promote green spaces directly, as the green space department would handle that. Furthermore, businesses rarely join forces to reduce air pollution on their own; they are under constant pressure to keep costs down and usually wait for government initiatives. Thus, the division of functions within and between governments and companies undermines the effectiveness and efficiency of society's regulatory control over air pollution.

When governments and companies do manage to collaborate, it is usually because they both see mutual benefits—government funds are better spent, and companies profit, often through reduced costs. However, various established practices, many of which have become legally entrenched, hinder this collaboration. For example, it is legally challenging to partially redirect government funds intended for combating air pollution toward initiatives like clean energy or green spaces. Representatives and financial controllers may rightly raise questions about this. Law and tradition create barriers for governments and companies to pursue solutions that might be most beneficial for society.

The way governments and companies operate can result in a standoff, leaving them unable to adequately address issues like air pollution. Moreover, even the most successful interventions tend to be undermined by recurring factors that cause environmental problems to continue growing. For example, while China's environmental policies were initially very effective, economic growth eventually outweighed these efforts (Lakra and Avishek 2022). The result is environmental degradation, while regeneration is essential. The following Crystal Bath case study demonstrates the potential of an area to regenerate itself.

### 9.2.1 Case Study: Crystal Bath

Crystal Bath is a water storage area between Enschede and Hengelo, the two easternmost cities in the Netherlands. The increasing urban development in both cities means rainwater cannot be adequately retained, leading to flooding risks. Since Enschede is located at a slightly higher elevation, this exacerbates the problem for Hengelo. The available space between the two cities allows for surface water storage in Crystal Bath, where natural landscapes help retain water, and the water, in turn, benefits nature. The interplay of water, soil, and nature optimizes Crystal Bath's capacity for water retention and environmental sustainability (RVO Netherlands Enterprise Agency 2016, p. 50).

This interaction between water, soil, and nature also enables Crystal Bath to purify water, as natural elements and soil contribute to the water purification process. Water storage, nature conservation, and water purification reinforce each other. Additionally, they provide mutual protection. For example, if urban development puts pressure on the area, three strong, interconnected interests and a coalition of stakeholders would work together to protect it legally against urbanization pressures.

Because Crystal Bath is near lower-quality urban neighborhoods in Hengelo, the combination of nature and recreational opportunities enhances the quality of life in the area. Nature enables recreational activities, and income from these activities can be reinvested in the area's quality. This continuous enhancement of Crystal Bath positively impacts the surrounding area. Such influence mitigates negative impacts, such as improving low-quality residential neighborhoods, purifying water, and providing flood protection during heavy rainfall. Crystal Bath and its surroundings experience regeneration, with increased quality through recreation as the latest step.

The practice of economics is in the tradition that the market and government organize resources to pursue interests and not that interests reinforce each other. Similarly, law is structured to protect specific interests rather than to facilitate mutual protection between interests. Examples like Crystal Bath show how an integration of interests can provide solutions to societal problems that lie outside the traditional realms of economics and law. Could such an approach be feasible for addressing urban air pollution during fog?

Crystal Bath is a multifunctional solution, combining water storage with nature conservation, water purification, and recreation. The four involved parties, each responsible for one of these interests, coordinate efforts not only for their own benefit but also for positive effects on other interests. All four stakeholders benefit from cost-sharing and contribute mutual benefits, such as nature's ability to retain and purify water. These positive spillover effects strengthen the area and counteract negative effects on the surroundings caused by policies or other developments, such as urbanization and climate change.

The combined impact of urban development and climate-driven extreme weather has a negative effect on Enschede and especially on Hengelo. By reinforcing Crystal Bath through positive spillover effects, this area compensates for the negative effects on the surrounding area. This model suggests that the problem of urban air pollution during fog could similarly be addressed. This would begin with reducing air pollution and then, by fostering and combining positive side effects that regenerate the city, build resilience that mitigates or absorbs negative effects.

### *9.2.2 Case Study on Air Pollution During Fog in the City*

Fog benefits nature, and nature, in turn, benefits people. However, fog becomes harmful to people and nature when it mixes with air pollution. This is especially true in cities, where high populations lead governments to set legal limits on air pollution. At the same time, it is possible to reinforce other interests, such as clean energy and expanded green spaces. Both help reduce air pollution as a side effect: clean energy replaces fossil fuels, which pollute the air, and green spaces can purify the air.

Focusing on side effects is not traditionally practiced, but the Crystal Bath case shows that this approach can be both legally sound and economically advantageous. If funds allocated to combat air pollution are directed toward promoting clean energy, nature, or both, this strengthens the side effect of reducing air pollution. This secondary effect can complement the impact of direct antipollution measures, such as installing filters at the source. Legally, pollution levels remain more within regulatory limits, and economically, it keeps pollution control within budget. When this is achieved, society as a whole benefits.

In cities, sources of air pollution include industry, traffic, and aviation. These activities may continue if pollution control improves, but this approach focuses only on adaptation, not mitigation. Meanwhile, controlling the interests of industry, traffic, and aviation remains challenging. Legal measures to limit air pollution, including during fog, can be effective, but gains are often offset by the growth of industry, traffic, and aviation. Can a strong coalition of alternative interests help ensure that the law is better enforced while also offering solutions that make areas more resilient to pollution that cannot be easily eliminated?

For example, if clean energy and green spaces continue to grow, regeneration can occur, with pollution reduction as a byproduct. Fog then becomes less of an issue and may even have benefits. In a clean, green city, fog is not problematic; instead, it becomes a natural element of the city, supporting more and better greenery and regeneration. Strong green spaces are resilient enough to tolerate some pollution, partly through the benefits of fog (Chang and Schemenauer 2021). Additionally, green spaces are known to provide a range of functions, beyond just improving air quality.

Green spaces are often edible, help with cooling, can generate energy through surplus biomass, retain and purify water, and much more. These are economic functions that support environmental regeneration and protect community interests. Can these benefits compete with dominant, degenerative production practices, such as those that generate excessive traffic? In agriculture, there is a growing number of companies using regenerative practices to compete against harmful traditional methods. These companies do not contribute to air pollution, such as nitrogen emissions, at the same level; they may emit nitrogen, which is unavoidable, but they sequester more than they release.

Some companies produce more clean energy than they consume, while others discharge cleaner process water than they take in. Regeneration is possible but only sustainable if it remains economically feasible. Can society foster this approach through legal, economic, and financial means? These considerations lead to the primary research question addressed in this chapter. Although focused on the specific case of air pollution during fog in urban areas, the findings may inform broader pollution issues.

**Primary Question**

Can regeneration resolve the issue of air pollution during fog in the city?

**Sub-Questions**

1. How can regeneration be made economically feasible?
   (a) Can fog contribute economically?
   (b) How can financial barriers be overcome?
2. How can the law promote regeneration?
   (a) Should the law also address financial barriers?

**Method**

This chapter uses various practice-oriented studies to gain insight into the interoperability of side effects. Two ongoing research projects are particularly relevant:

1. Communitas, Home—COMMUNITAS (www.communitas-project.eu). As part of this research, the economic model under development will be further refined in Sect. 9.6.1.

2. Atelier, ATELIER—Positive Energy Districts (www.smartcity-atelier.eu). The financial analysis in Sect. 9.6.2 builds upon this project.

Results from these and other studies will undergo a qualitative analysis of documents and cases in this chapter. The case study on fiber cultivation will help assess the advantages of interoperability. This case raises the question of how to conduct a cost–benefit analysis of a series of reinforcing side effects. A theory regarding the impact of multiple side effects on urban regeneration will be applied to the case of air pollution during fog.

## 9.3 Theory

### *9.3.1 Government Instruments*

According to the Dutch College of Government Architects and Advisors (CRa), building within neighborhoods can serve as a tool for "a broad range of social investments. (... these) become a driver to address major societal challenges of our time, from the climate crisis to loneliness" (College van Rijksadviseurs (CRa) 2022, p. 12). "Examples include: strengthening economic vitality and social justice, sustainable use of limited space, improvement of the living environment, and acceleration of the energy and mobility transition" (College van Rijksadviseurs (CRa) 2022, p. 7). In addition to the intended effect—housing—there are positive side effects: housing construction also becomes a tool for other objectives.

The CRa's approach reflects a completely different view on tools compared to traditional interventions in governance, law, and economics. In conventional fields, tools include people, regulations, money, markets, and communication. Their application affects areas like housing, energy, and mobility. There has never been the idea that these areas themselves could serve as tools for change in each other. For example, climate adaptation is an effective instrument for advancing the energy transition. Trees can be strategically planted in neighborhoods to cool homes, reducing the need for air conditioning. Green roofs can cool solar panels, increasing their efficiency.

Conversely, the energy transition can also support climate adaptation, primarily because clean energy has no $CO_2$ emissions. Another example is that solar panels in agriculture provide shade for crops and animals during extreme heat. At the same time the crops help cool the panels, as on green roofs. Combining these actions reduces costs and yields more benefits than addressing each need separately. Society benefits when energy measures become the most effective means for climate adaptation, and adaptation measures the best means for the energy transition (Pascaris and Pearce 2020). However, parties must be willing to adopt these methods.

Returning to the CRa's perspective, housing construction can become a tool for a variety of policy areas when responsible policymakers recognize that housing has the greatest impact on their domain. Through their investments, they can "order" housing as a tool to advance their own goals. In turn, housing developers gain additional resources, not to cover regular housing costs, but to deliver higher quality overall. As a result, society receives greater quality for the same expenditure. In December 2024, the City of Amsterdam embraced this way of ordering and spending public funds on that (van Erkel 2024, p. 51).

For the government, "procurement" here means that results in one policy area are purchased from another. This approach fosters cross-sectoral collaboration, which may sometimes clash with the deeply ingrained, centuries-old tradition of sectoral focus. By emphasizing procurement, each sector retains its value while still enabling cooperation. For example, the effect of energy transition policy is amplified when delivered successfully through the climate adaptation sector. Realizing these effects often requires the involvement of businesses, which can expand the notion of government instruments.

### *9.3.2 Businesses*

Government procurement from businesses is common practice, with companies typically providing what they excel at, such as building homes or creating nature reserves to support climate adaptation. However, companies can do more, as seen in cases where they offer side benefits to the government to enhance policy impact. For example, businesses may support a district heating network the government seeks to develop.

A company that supplies industrial waste heat to a district heating network benefits itself by reducing its $CO_2$ emissions, which is advantageous for regulatory compliance. Additionally, it contributes to affordable energy costs for local residents, including its own employees. If the company relies on a prosperous local customer base, it helps itself when the surrounding community remains well-off, partly due to manageable energy prices. This fosters a favorable business environment.

A favorable business environment benefits the government, as it seeks to attract and retain economic activity. The company also aids the government's efforts to combat energy poverty, accelerate the energy transition, and potentially improve water management. If process water is recycled in a heating network rather than being discharged, it significantly reduces environmental impact, supporting the government's environmental protection goals.

There are many reasons why the government might procure services from a company to collaborate on a district heating network. One key reason is that this process helps the company become more sustainable, aligning with government policy to promote such efforts. In this case, the government is not providing a subsidy but instead achieving mutual benefits through procurement. By procuring services, the government commissions the company to support public policy on heat while also directing the company to adopt more sustainable practices.

A company that pursues its own objectives and can assist the government with the side effects of that goal can employ these effects to make government policies more effective. The government would do well to encourage this, as it also helps the company better align with governmental aims. The government is not simply purchasing a product that contributes to, for example, energy transition or climate adaptation policies but is also advancing its policy to transform the industry.

By purchasing these side effects from companies, which adapt themselves to government policies, the government can increase its effectiveness while saving on subsidies and enforcement costs. How can the government encourage this? In this arrangement, companies are not selling their main products but rather selling their side effects, provided these are beneficial to the government. Essentially, these companies are doing what the CRa advocates: using resources not only for the intended purpose—the product for which the company exists in the market—but also considering what more these resources can achieve. Perhaps, there are byproducts that are marketable, or there may be side effects that can be sold to the government.

### 9.3.3 Finances

The government has more instruments at its disposal than it realizes. In fact, any policy area can serve as a tool to achieve results in another policy area. One government agency could, in this way, purchase outcomes beneficial for a different agency's policy objectives. Companies may also produce side effects that benefit a government policy area. However, purchasing cross-sector results or side effects encounters financial barriers.

The biggest financial barrier is not the money itself but the accountability required in spending it. For example, there may be funds available for the energy transition, but if they are earmarked for solar fields, it might not be easy to allocate this money to solar panels on rooftops. Although rooftop panels are technically solar fields, will financial controllers accept this if solar fields are the stated requirement? This depends on the conditions tied to the funding, which are often set on a sectoral basis (Scott 2020). This approach limits the cross-sector allocation of funds.

Another obstacle, particularly with public funds, is that many budgets must be spent within a specific period, or else the funds are no longer available. Various projects, in which governments and companies could align their budgets differently, may only yield results after a long time. In both cases, the solution could be setting up a fund, for example, by a municipality or even a group of governments. Such a fund would, of course, be subject to conditions, but in this case, these conditions would specifically focus on cross-sector spending and allow long-term investment.

The political value of a fund should not be underestimated. For instance, it may be challenging for a local councilor to explain choosing a district heating network, as it provides not only heating and cooling but also contributes to social cohesion if users have a stake in it and helps reduce energy poverty. In making such a choice, the councilor responsible for energy sustainability also addresses welfare, which falls under the jurisdiction of a colleague responsible for social affairs. Would social affairs contribute funding to the district heating network?

The councilor for social affairs would face some scrutiny if they supported a district heating network, especially if funding earmarked for poverty reduction were invested in it. Such a move may not easily win the council's approval. However, if the money comes from a fund with a broad objective, such as promoting broad prosperity in the long term, both councilors might succeed in gaining the council's support.

### 9.3.4 *Fog and Fibers*

A water retention area can impact nature, water purification, and recreation. Housing developments have a wide array of social effects. Could such a solution for societal issues also address urban air pollution during foggy weather? Could a series of interventions in the city, such as increasing clean energy and green spaces, be the best way to reduce this problem? In pursuit of an answer, the next section discusses the case of fiber cultivation, which provides an example of a long chain of effects and side effects that help reduce a series of problems.

## 9.4 Fiber Cultivation, from Land to Property

The goal of the Dutch national IDOBB-program (Integral Sustainable Building with Biobased Materials) is to transition from traditional, abiotic construction toward largely biobased construction by using wood, biobased fibers, and reclaimed biobased materials. The underlying principle of the IDOBB is that both agriculture and construction need to change and can support each other in doing so. In other words, agriculture can solve issues with solutions from the construction sector and vice versa. However, this synergy goes further. Transforming agriculture and construction are just two of twelve goals, which also include climate objectives, for example. Many of these goals are shared concerns of the government, and companies in agriculture and construction provide side benefits to the government to address these goals. This is why various government agencies collaborate with the IDOBB (2023) (see Building Balance).

Solutions from agriculture and construction often also address climate issues and vice versa. This integration goes further: for example, solutions to water quality problems also contribute to agriculture and climate objectives. Integration here means solving problems by applying solutions from other issues. The IDOBB also illustrates how to organize these different problems so that their solutions truly support each other. This is done by using a single tool to tackle these issues: fiber cultivation.

The IDOBB offers a glimpse into what might be an entirely different economy from what is currently common. What's unique is that this single resource, fiber cultivation, serves twelve goals, yet there's more to it. It's not just that each of these twelve goals has a desirable impact and that the result is the sum of these twelve parts. Instead, these goals are interconnected, which allows for mutual reinforcement. Not all twelve interact directly, but each can often fit into a model that facilitates mutual reinforcement.

### 9.4.1 Increasing Returns

With twelve goals rather than ten or eight, the longer the chain, the greater the returns. The twelve goals are: (1) climate mitigation and adaptation, (2) circular use of resources, (3) construction (new builds and renovations), (4) reduced nitrogen, (5) water quality, (6) biodiversity, (7) spatial quality, (8) clean energy, (9) agriculture, (10) landscape, (11) building materials industry, and (12) soil and subsidence. Each goal is discussed below, explaining how it affects at least one other goal or vice versa. The IDOBB's scaling plan illustrates how fiber cultivation initiates a chain of objectives.

### 9.4.2 Increasing Intra-action

The twelve goals contribute to twelve impacts, layered together. Additionally, the impact is greater than the sum of the parts due to a unique form of synergy known as intra-action (van der Heijden 2022). An example of standard synergy is giving someone a tool or building a road that connects houses. Intra-action, however, takes it further. An example of intra-action is placing solar panels over agricultural fields. The panels protect crops from heat and heavy rain, while the crops cool the panels, enhancing their performance. Intra-action is synergy with additional multipliers. Many of the twelve goals exhibit intra-action among themselves.

In fiber farming, farmers and others utilize the intra-action present in relationships, such as those between water and soil or between nature and climate, biodiversity, and landscape. For instance, enhancing biodiversity inherently improves landscape quality. In other words, investing in biodiversity is also an investment in landscape and spatial quality, adding value through a multiplier effect. The landscape manager, often a governmental entity, benefits from this. As goals accumulate, more intra-action occurs among various objectives.

The crops used for biobased building materials are very low-impact in terms of fertilization, pesticide use, and their deep root systems. This contributes to improved soil and water quality and increased above- and below-ground biodiversity, which is also intra-action. Additionally, circularity, where organic material returns to the soil at the end of its life cycle, feeds the soil with organic matter essential for moisture retention, nourishing soil life, and providing a resilient environment for crops.

Intra-action and regeneration are unique characteristics of this chain. A series of problems is solved, with each solution reinforcing the others. This is the opposite of the long-standing approach where solutions often led to new problems through negative side effects, which are then off-loaded onto society now and later, resulting in degeneration. The government usually steps in to counteract this but rarely completely. While the chain does not solve everything, it can build resilience to address a great deal, fostering regeneration.

### 9.4.3 *Decreasing Investment Costs by Sharing*

Fiber crop farming offers an alternative to intensive livestock farming, which harms nature, the environment, landscapes and soil, because of subsidence. Fiber crop farming includes investing in healthier soil, which benefits water management authorities (government entities responsible for water quality and quantity), because healthier soil helps purify and retain water. Drinking water companies located nearby also benefit from improved water quality. When these organizations plan their next investments, they can take the farmer's positive impact on water quality and quantity into account. If this impact is substantial, they might even co-invest in the farmer's land management, reducing investment costs for all involved parties.

Landowners near the farmer, such as other farmers or conservation managers, could join this soil health-focused management model. If they share investment costs, and water management authorities and drinking water companies participate as well, costs will decrease for all parties, including the farmer who initiated the effort. The farmer might grow not only fiber crops but also rotate crops with sugar beets, for example. The sugar producer, a buyer of these beets, could co-invest as well, as they would receive higher-quality beets and contribute to a soil that captures $CO_2$—a benefit the producer can account for in their $CO_2$ emissions and compensation records.

Like the sugar producer, manufacturers of building materials could also co-invest, especially in fiber crops. By investing directly, they gain influence over quality standards and can also account for part of the $CO_2$ reduction in their records. As the number of participants grows, each subsequent investment further reduces costs because expenses are shared among more stakeholders. A new entrant incurs fewer costs than they would if they had to finance everything alone. By compensating those who already invested (from which the new entrant benefits), earlier participants gain from their initial investment.

### 9.4.4 Decreasing Usage Costs by Sharing

The same principle of shared investment applies to usage costs. If two or more parties share land for fiber crop farming, water retention and purification, $CO_2$ storage, and ecological and landscape enhancement, they also share usage costs. As with investment, costs decrease as more participants join, new entrants have lower costs, and compensating earlier investors is reasonable. Beyond cost-sharing, parties can support each other in reducing expenses.

In construction, producing abiotic materials requires substantial energy and emits considerable $CO_2$, whereas biotic material production is more energy-efficient and stores $CO_2$. Additionally, biobased materials often provide good insulation, saving energy, while wood and fibers are lighter than concrete and steel. This allows for lighter construction tools, reduced transport weights, and up to a 20% reduction in trips. These advantages lower usage costs, particularly energy-related expenses.

### 9.4.5 Decreasing Risk, Thus Lowering Financing Costs

Regeneration builds resilience in an area, where one goal strengthens another—such as healthier soil that can sequester more $CO_2$ and resist subsidence. Resilience also means that one objective can protect others from external threats; for example, healthy soil can purify water, protecting nature and agriculture from pollution. Fiber farming, as proposed by IDOBB, promotes regeneration and resilience, reducing overall risks.

Shared land use, as described above, improves resilience against issues like heat and flooding, far more than land used year after year for intensive agriculture. Resilience is built on the land and also among stakeholders; if one party struggles for a season while others remain stable, that party can recover the following year. As risks decrease, so do financing costs. Companies manufacturing fiber-based building materials benefit from this reduced risk, even without direct investment in agriculture.

There's a high likelihood that unmet climate goals will lead to industrial emission restrictions, similar to those for nitrogen. Since approximately 40% of the industry's revenue is tied to the construction sector, there's a high risk of construction facing stricter nitrogen regulations and halting activity. For this reason, increased use of biobased materials is essential. The same holds for sustainability goals, as the $CO_2$ footprint of renovation and sustainability projects is as significant as that of new construction, given the sheer number of homes and buildings involved.

### 9.4.6 *Decreasing Transaction Costs*

Farmers, water managers, drinking water providers, conservationists, landscape managers, and building materials producers all seek their own optimal outcomes within the collaborative model of the IDOBB-program. This requires coordination, which takes time, but as they become familiar with each other, transaction costs decline. New entrants join an established community with valuable experience from which they can learn, reducing costs. Since an IDOBB-community has formed, new members are welcome, allowing initial members to recover some of their costs. Moreover, they can benefit from the new participant's contributions to the group's overall success. Could such a collaborative approach also address air pollution in urban fog conditions?

## 9.5 Case Study: Air Pollution During Fog in the City

Fog combined with air pollution has severe consequences for human health and the environment. Additionally, fog creates unsafe traffic conditions, leading to accidents and delays (Pérez-Díaz et al. 2017, pp. 14–15). However, there are also positive effects of fog, particularly for nature. Fog contributes to various natural processes that support growth and decomposition. Many of these contributions can be considered ecosystem services that benefit people, such as providing cooling and water that can be collected for human use, especially for agriculture (Lakra and Avishek 2022, pp. 338–343).

One factor contributing to the unhealthy combination of fog and air pollution is climate change, which also results in periods of insufficient fog. This lack of fog is detrimental to nature and its ecosystem services, with negative consequences for agriculture, among other areas (Lakra and Avishek 2022, pp. 340–342). Reversing these effects by controlling fog is almost impossible, as fog is influenced by a complex interplay of natural elements, climate, geology, atmosphere, and other factors beyond human control (Pérez-Díaz et al. 2017, p. 10).

However, people can influence the causes of air pollution and climate change, which in turn affect the unhealthy combination of air pollution and fog and the decrease in fog occurrence. Switching to clean energy and increasing green spaces are interventions that humans are capable of implementing. Transitioning from fossil fuels to clean energy has the added benefit of reducing air pollution. In addition to directly targeting air pollution, policies could focus on encouraging the shift away from fossil fuels.

In fact, aiming to reduce air pollution as a side effect may be more effective than policies that directly combat this pollution, such as source-based filters. With clean energy, that pollution source no longer exists. Clean energy also brings other benefits, such as stabilizing energy prices and promoting energy democracy. Urban greenery acts as a filter to prevent contact with pollution once it has spread. While other filters are available (e.g., for air intake in ventilation systems), greenery has additional benefits, such as cooling, water retention, biodiversity support, fruit production, and biomass provision (Vranken et al. 2023).

The extensive range of additional benefits from clean energy and urban greenery makes them self-sustaining antipollution measures that may require no budgetary allocation for combating air pollution. However, this does not mean that the air pollution budget could not be used to accelerate the introduction of clean energy and green spaces. Beyond economic impact, this wide range of benefits also has legal implications. The measures for more clean energy and more greenery serve multiple interests at once. Precisely, because they serve a coalition of interests, they cannot be easily dismissed. By relying solely on a legal rule to protect a single interest against air pollution, you may find yourself in a weaker position.

Crystal Bath, housing construction, and fiber crop cultivation are examples of single measures that achieve multiple goals. In the case of fiber crop cultivation, there are twelve goals, prompted by alternative agriculture and construction needs. In the case of air pollution during urban fog, clean energy and increased urban greenery can serve as such measures independently or together. When used separately, each has side effects on the city, including reducing air pollution. When used together, they can also reinforce each other. For example, greenery helps lower temperatures, reducing the need for air conditioning.

Biomass can also be fermented, especially when it no longer serves another purpose, like converting fibers into building materials. The business cases for clean energy and urban greening can enhance each other, amplifying both measures' impact on air pollution. Together, they form the foundation of a strong coalition for a healthy city, but city officials must weigh multiple factors before making this choice. This decision-making process also includes the interests of parties reliant on traffic that contributes to air pollution. How can officials balance allowing less traffic during fog in the city? Since it concerns building a healthy city, expanding the coalition to include health would be beneficial, though the health sector often proves to be a challenging partner for such alliances.

The fire department engages significantly in prevention but will always need to invest in firefighting. Healthcare operates similarly, but with a key difference. The fire department does not profit from more fires; its sole motivation for investing in firefighting is preparedness. It reports to the government and the community, who encourage prevention alongside readiness. In healthcare, however, there is profit from providing more care, and a significant part of the accountability is to shareholders, who benefit from increased care, not from prevention.

Coalitions aiming to invest in a healthy city are missing a key partner in healthcare, as the incentives are misaligned. Moreover, any incentive to work on prevention, to the extent that it exists, is weak. For instance, if everyone in a city were insured by the same provider, that provider would have a vested interest in promoting a healthy city (Vranken et al. 2023). There are examples of insurers contributing to urban greening. However, if a park is created, perhaps only 15% of the neighborhood residents are insured by the contributing provider. This setup does not strongly motivate the insurer to invest further.

Without healthcare as a strong partner in the healthy city coalition, the balance remains somewhat abstract between activities that pollute and those that promote health. City officials may recognize the importance of health, but there is no support from the healthcare sector itself. Without this, the coalition for a healthy city cannot make a compelling case. Would healthcare representatives see the benefit of a broad coalition like IDOBB, recognizing that their sector would ultimately thrive more in a coalition focused on health and sustainability rather than one driven by profit through increased care provision?

Agricultural and construction companies also face pressure to specialize in a single product and ignore any side effects. Yet leading companies participate in the IDOBB coalition, because they understand that this path offers a more sustainable future. This approach involves generating a series of positive side effects that mitigate negative ones. Without this regenerative approach, construction and agriculture cannot survive in the long term. Many companies still see short-term gains in the old model, as do numerous parties in healthcare, especially since demand is expected to grow both in the short and long term.

However, healthcare providers cannot continue to see rising revenue indefinitely if society can no longer afford the costs or provide enough staff. Looking ahead, even in the short term, a coalition like IDOBB—or in their case, a coalition focused on a healthy city—could be helpful. This would involve efforts in clean energy and urban greening to preempt certain healthcare challenges. City officials would then be offered financially stronger solutions to make their city more sustainable. Their decision-making between economic growth and health would thus shift, impacting areas like transportation, aviation, and air quality.

The essence of what is happening in a healthy city is the development of positive side effects. These offset negative side effects within the area and its surroundings. Positive effects include enhanced energy quality and natural spaces, with beneficial outcomes for climate, well-being, health, and quality of life, including air quality. Negative effects include climate change, soil, water, and air pollution, and reduced quality of life due to urbanization. If positive side effects outweigh the negatives, regeneration occurs. Could urban development reduce the negative effects of air pollution during foggy conditions by eliminating them and introducing positive factors instead?

A positive balance can significantly reduce government spending, as the government typically bears the costs of negative side effects. Ultimately, this also saves taxpayer money, though the question remains how well governments and citizens perceive this benefit. Most governments and citizens focus on single-value business cases rather than multi-value ones, let alone regeneration. For instance, the idea that risks and transaction costs can decrease is counterintuitive, and many financiers need time to adjust to this perspective. Normally, they drop out when they have to bear too many costs and risks and receive too little return in relation to others—a split incentive. However, within a coalition such as Crystal Bath or fiber cultivation, they are prepared to invest.

## 9.6 Analysis

Resources allocate resources—a realization drawn from the Crystal Bath case, the CRa's stance on housing development, the fiber cultivation case, and also the healthy city case. A traditional perspective in economic theory holds that markets allocate resources, with the government stepping in where the market cannot or is not permitted to act. However, without changing this framework, it can also be observed that resources themselves can allocate resources. For instance, Crystal Bath's water retention area naturally fosters the growth of nature, which in turn supports additional activities like water purification and recreation.

The CRa even regards resource allocation by resources as a strategic tool. In the CRa's view, housing development can serve as a mechanism to allocate sustainable energy use and climate adaptation measures. Similarly, in IDOBB's fiber cultivation plan, fiber cultivation serves as a means to allocate various resources, such as reducing nitrogen emissions and limiting soil subsidence. In the healthy city case, clean energy and greenery help allocate resources that contribute to cleaner air, among other benefits.

When resources allocate other resources, it can be more efficient than allocation solely by the market or government. The difference lies in that the market and government tend to allocate each resource independently. They function in a monofunctional manner, whereas resource allocation by resources is, by definition, multifunctional; it involves two or more resources, each with its unique goals. These goals serve underlying interests such as housing, health, water safety, and water quality. When resources allocate one another, they mutually reinforce these interests.

A single resource that strengthens multiple interests, creating multi-value outcomes, is an emerging phenomenon in economics. How can economists make this concept actionable, both for themselves and for financiers? The next two subsections will address these points, followed by a discussion of the legal perspective. When two or more interests are linked through the application of a single resource, they inherently support each other through this interconnectedness. The final subsection will attempt to translate this concept into a practical framework for legal professionals.

### *9.6.1 Economy*

The multi-value creation of combining clean energy and greening efforts involves mutual reinforcement across a range of activities, such as water storage, cooling, biogas production, and improved air quality. This results in a unique business case compared to a typical business case. For example, take water purification: usually, resources like space and equipment are dedicated solely to purifying water. In the case of Crystal Bath, however, these resources are also used for other benefits, such as water retention and nature management.

Many investors and other stakeholders are accustomed to deploying resources exclusively for a single purpose, not multiple ones. A standard business case involves a single product with its associated costs and revenues, assuming that costs decrease and revenues increase as scale grows. In a multi-value business case, however, scope plays a significant role. For example, the scope of Crystal Bath broadens from water retention to include nature management, among other goals. Moreover, by expanding the area, scale can also increase, so scaling up is still possible.

The biggest barrier to multifunctional projects lies in the fact that production is predominantly organized for single-purpose objectives. Organizations, their cultures, and policies are entirely oriented toward single-purpose operations, creating substantial hurdles to integrated working. This is also reflected in the use of financial instruments. One way to overcome these hurdles is by rethinking these instruments and the methods of applying them. To do so, it's first necessary to understand the business case, as otherwise, the precise target of funding remains unclear. A way to achieve this understanding is by breaking down the business case into its components.

When the scope broadens—for instance, from greening to also include clean energy—the costs and benefits change as follows:

(a) Additional revenue is generated by a layering of activities that otherwise would not exist.
(b) Cooling from greenery increases energy production efficiency, intra-action, while shade from solar panels enhances green growth, also intra-action.
(c) The energy company can benefit from the green company's investments in greenery, which provides cooling and biomass for biogas production. Similarly the green company can benefit from the energy company's investments in solar fields that promote green growth, such as green roofs. Both companies can compensate each other for these investments, but the amount is less than if each company invested individually, reducing the additional investment costs.
(d) Both companies benefit from managing an area that combines healthy green spaces with efficient solar energy production. They can share operational costs by jointly contracting a company to manage the area, reducing additional operational expenses.
(e) The resilience of the business case increases, which lowers financing costs due to reduced risk.
(f) Although collaboration between the two companies may initially increase transaction costs, they learn from each interaction, decreasing transaction costs with each subsequent step.

Table 9.1 illustrates a scenario where the green company is already active and the energy company joins.

With these six elements, a financier can understand what the multi-value business case entails and therefore know exactly what they are investing in. It is possible to go further by illustrating how the multi-value business case can be regenerative. The combination of greening and clean energy demonstrates this, as it enhances the area's resilience against deterioration from heat, drought, and heavy rainfall. This

**Table 9.1** Comparative benefits of green and energy company

| | Stacked revenue | Intra-action | Additional investment costs | Additional operational costs | Additional financing costs | Additional transaction costs |
|---|---|---|---|---|---|---|
| Green company activities | a | b | c | d | e | f |
| Energy company activities | >a | >b | <c | <c | <e | <f |

[a] Increases with additional layering
[b] Increases when additional layering reinforces previous layers through intra-action
[c] Decreases due to shared costs
[d] Decreases due to shared costs
[e] Decreases due to increased resilience
[f] Decreases due to shared learning

resilience extends further by contributing to health, well-being, and prosperity, helping to prevent the decline of social quality, for example.

The shift in focus on resilience is based on outcomes and intended impact, i.e., multi-value creation. This also includes "de-risking" the business case through greater integration, as well as the addition of another partner. This concept may seem counterintuitive, as an extra partner might initially feel like added risk. However, the largest barriers are different—namely, the culture and politics of organizations. To overcome these, a clear understanding of the multi-value business case structure, including regenerative potential, is helpful. Additionally, insight into the use of financial instruments for multi-value and regenerative projects can contribute to overcoming cultural and political obstacles. The following section provides insights that can help address these barriers (Bril and Schramade 2023).

### 9.6.2 *Finance*

Existing financial instruments are typically designed to finance a project benefiting a single stakeholder using that stakeholder's own financial resources. For instance, if a stakeholder can secure a loan for their project, that loan becomes the financial instrument. Integration, however, calls for funding structures that also involve other stakeholders, such as when they jointly take out a loan. To facilitate and arrange this, investment platforms and funds specifically aimed at integration are emerging.

An investment platform is not a financial instrument like a loan; it is a method to organize funding. Likewise, a fund is not an instrument but can use instruments like loans to disburse funds. Like the investment platform, a fund enables the collaboration and integrated investment essential for integration. Here, first instruments and then methods such as the investment platform and fund are discussed, examining each for their potential to support integrated project financing.

#### 9.6.2.1 Instruments

In the Netherlands, a significant portion of land is municipally owned. When selling land to a developer, for example, for the construction of a residential area, the municipality has various instruments to enforce an integrated approach. Some of these are contractual, while others arise from environmental law (van der Heijden et al. 2025). Beyond property ownership, municipalities have funding allocated for a range of objectives. An integrated approach brings together multiple goals, such as greening public spaces with the aims of climate adaptation, recreation, well-being, and health. Normally, these four goals have separate budgets.

A single budget for a single project aligns with typical government accountability practices. Innovatively, however, funds can be pooled and barriers removed to make an integrated project feasible. A key example is combining projects, like installing a district heating system alongside sewer replacement. This approach saves substantial funds on labor and minimizes public disruption, as the ground doesn't need to be excavated twice. In procurement, municipalities can request project integration for both construction and maintenance.

In addition to pooling budgets for an integrated project, it is also possible to expand a single budget to encompass an integrated project. For example, funds allocated for sewer replacement can be repurposed for neighborhood greening. This greening helps with water retention, allowing only partial sewer replacement. In this way, green infrastructure serves as an alternative means to achieve the same goals as sewer infrastructure, fitting within the purpose of sewer funding.

A municipality may also participate in another entity's integrated project, whether public or private. In return, the municipality expects a result, such as increased green space if a water authority initiates a project aimed at enhancing natural areas. Conversely, a water authority may participate in a municipal greening project, expecting better water retention in return and contributing funding. This method of attracting funds is known as "outcome payment" and is an innovative concept well-suited for integrated projects.

Alongside traditional public and private financial institutions, individuals are increasingly contributing as financiers. Local citizen initiatives are examples of integrated projects that raise funds from individuals through crowdfunding. Citizen initiatives are also noteworthy examples of innovative cost savings through integration. For instance, residents who maintain public green spaces instead of the government simultaneously enhance well-being, thereby reducing the government's efforts and costs in this area. Such integration reduces expenses, like combining projects does, and highlights cost savings as a significant financial benefit of integration (van der Heijden et al. 2025).

#### 9.6.2.2 Investment Platforms

Undertaking integrated investments is almost unimaginable without a platform to handle the preliminary work. Such a platform visualizes the synergies and possible intra-action between values and seeks an optimal outcome for both financiers and society. Knowledge sharing and development within the platform further enhance this optimal outcome. Within this optimum, the involved parties share their knowledge and costs, reducing transaction costs and risks. To assess the risk of a party exiting, it's helpful to understand what collaboration yields for this party compared to proceeding alone. The smaller the difference, the greater the risk.

It's evident that the investment platform facilitates blended finance. This is the strategic use of public funds, philanthropic resources, community contributions, and other sources to mobilize a wide range of additional private commercial capital for sustainable development projects. The goal is to address market and often government failures by attracting private investments in sectors or cases that would otherwise be considered too risky and thus financially unattractive. By combining different capital sources, the risk burden can be distributed among the various parties, aligning it with the benefits each stakeholder can gain.

#### 9.6.2.3 Funds

Members of an investment platform who wish to collaborate on an integrated project pool their money. They must do this in a coordinated way to maximize cost reductions and the additional quality they can achieve together. The idea is that there's a difference between individually channeling all funds toward the integrated project and doing so in a structured way through a fund. This involves a structure with both long-term and short-term funding, resulting in varying costs and conditions. This latter approach is considered more efficient. A fund can offer a solution to better allocate money, with more accountability, for an integrated project.

An investment platform can focus on project financing without fund formation but can also serve as a precursor to a fund. Unlike a platform, a fund is a formally registered entity with professional management. A fund can pool various funding streams, blending finance from both public and private sources, which can collectively support the realization of an integrated project. There are still few practical examples of funds aimed at integration. Existing funds often have broad spending purposes but do not necessarily integrate the projects they finance. An example is the Landscape Investment Portfolio.

The Landscape Investment Portfolio includes a series of activities that, when implemented in a coordinated way, can generate shared value and regenerate an ecologically degraded and economically impoverished landscape. The portfolio approach is based on the idea that spatially coordinated and sequential investments at the landscape scale will promote synergies and intra-action. This allows each individual project or enterprise to achieve a higher return, a lower risk profile, or more social, economic, and ecological benefits (Commonland 2023).

Funding the preliminary phase is crucial, since it's during this phase that integration can be added to the plan. This phase often requires more time and thus more money than single projects. For instance, parties must reach agreements on project details and cost-sharing. As a result, smart integrated plans can be developed in the preliminary phase, which will pay off during the realization and management phases. This phase can be separately organized and funded by an investment platform but can also be included in financing by a fund (van der Heijden et al. 2025).

### *9.6.3 Law*

Resources allocate resources; consider the fiber cultivation case, where a clean soil rich in biodiversity also stores a large amount of $CO_2$. This soil, in turn, can contribute to water quality. The interests of clean soil, biodiversity, $CO_2$ storage, and water quality reinforce one another. In doing so, they also protect one another: firstly, because reinforcement itself signifies protection. After all, if biodiversity, for example, increases, there is simply more of it, which alone means protecting species.

Secondly, representatives of clean soil, biodiversity, $CO_2$ storage, and water quality interests can come together within a coalition in which they collaborate. They coordinate on how a measure to improve soil quality, for example, can also benefit the other interests. Since they rely on one another, they also support each other's needs. If water quality needs protection from pollution risks, this also threatens the other interests. Together as a coalition, they can defend against this risk, which in turn strengthens the protection of each interest. In this way, a healthy city with clean energy and greenery can protect the interest of air quality.

When resources allocate resources, interests protect interests—at least if we consider both individual and collective interests that are not oppositional, as discussed in the Introduction. When resources like clean energy and greenery allocate each other, a more balanced protection against air pollution in the city during fog can develop. This protection is more balanced than one that solely follows from laws setting limits on this pollution. Interests are not only protected individually by the law, which often leaves them weak on their own. As part of a coalition, they stand stronger and reveal their true value, which can also carry legal weight.

Automobile traffic and aviation are also part of an interest coalition. A city council that must balance the interests of this coalition against health will fare better when health is part of a coalition that includes clean energy and green spaces. Not only does the council gain a more holistic view of the interconnections among interests, but it can also see what these mean for the city. The importance of health for the city becomes a central consideration for the council, presented, for example, as the "healthy city." The coalition's vision of a healthy city also includes its economic significance, which could even outweigh the economic importance of automobile traffic and aviation (Gössling et al. 2018). Halting both during fog could prove more economically harmful than allowing them to operate.

## 9.7 Conclusion

Can regeneration solve the problem of air pollution during fog in the city? That is the main question of this study, with the first sub-question being how regeneration can be economically feasible. The answer to this sub-question lies in the interplay of side effects, such as clean energy and greening efforts. Table 9.1 shows how this can lead to revenue generation and cost savings. Fog itself can contribute economically in this context, which is part of this first sub-question. Even within the city, fog has qualities that can enhance nature and even mitigate pollution to some extent.

Another part of the first sub-question is how to overcome financial obstacles in the interaction of side effects, as these are almost always funded from different budgets. Various financial instruments can help overcome this obstacle, such as pooling and flexible budgeting. Additionally, there are specific methods, notably an investment platform and a fund designated to finance the interplay of side effects. Both the platform and the fund can utilize blended finance, which fits well with managing diverse side effects.

The second sub-question is how the law can stimulate regeneration. The answer is that legal professionals, such as judges, must recognize that the protection of interests comes not only from legal frameworks but also from the interests themselves. They may not immediately realize this, because situations like these often escape their attention. For instance, if a neighborhood's noise level remains below the standard due to extensive greenery, the green space protects the interest of silence, but legal professionals may not see this since no standards are being violated. They may only notice this if regulations are breached, such as when funds intended for noise reduction are allocated to green spaces instead.

Another part of the second sub-question is whether the law should remove financial obstacles to promote regeneration. Financial obstacles include funds earmarked strictly for noise reduction that cannot be allocated to green spaces. While such allocations might not be mandated by laws, they are often established in lower-level regulations, such as budget planning and the allocation of funds within those budgets. Removing such obstacles could encourage regeneration, and legal professionals can advocate for this if they recognize the issue.

The protection of interests by other interests is both a legal and economic matter. The stronger the economic protection, the stronger the legal protection. This starts with the layering of interests: the more interests involved, the greater their economic weight, and the stronger the legal protection. This can go even further, as an economic dynamic may develop among the layered interests, though this intra-action does not always happen.

In other words, an economic dynamic within layered interests can create more economic weight and, therefore, more legal pressure, leading to better protection against degeneration. However, achieving such a layering of interests requires collaboration among often very different parties. This is not typical in most countries, where parties are accustomed to keeping their interests separate. Financially, they may face challenges with accountability when funding streams are mixed.

Answering the two sub-questions leads to the conclusion that regeneration can solve the problem of air pollution during fog in the city, which is the main research question. Reaching this conclusion requires an understanding that resource allocation is not solely conducted by the market and government. However, the prevailing view in many Western-oriented countries is that the market allocates resources, with the government stepping in only when the market cannot or may not do so. There appears to be no third way, yet this alternative emerges in projects like fiber agriculture, with many other examples.

Recognizing this third way, alongside the market and government, can mean that sectors like aviation may need to position themselves as part of a coalition of interests rather than solely allocated by market and government. This coalition can more accurately represent its true value, which includes noise, $CO_2$ emissions, and other outputs and their impact on society. Determining that value lies not only with market players and government but also with the participants in this coalition of interests.

Members of the coalition of interests surrounding aviation, along with city administrators, may believe that the economic loss from halting flights during fog is too high (Kulkarni et al. 2019). They may feel that the safety risks and pollution outweigh the economic loss, as a healthy city faces greater risks from reduced safety and increased pollution. Others may disagree. Shifting the balance by including side effects does not determine the outcome, but it makes the assessment more balanced by considering not just isolated conflicting interests but how these can strengthen each other in long chains of interests.

**Competing Interests** Section 9.6.1 builds upon work funded by the European Union, COMMUNITAS, Grant Agreement No. 101096508. Section 9.6.2 builds upon work funded by the European Union, ATELIER | Positive Energy Districts—Grant Agreement No. 864374.

## References

Bril H, Schramade W. Strengthening investment portfolios through resilience – a primer. 2023. https://ssrn.com/abstract=4667697.

Chang SC, Schemenauer RS. Fog deposition. In: Foken T, editor. Springer handbook of atmospheric measurements. Cham: Springer; 2021.

College van Rijksadviseurs (CRa). Bouw in de buurt, Leidraad voor Nederland. 2022.

Commonland. Towards financing large-scale holistic landscape restoration in Europe, United Nations Decade on Ecosystem Restoration 2021-2030 & Commonland. 2023.

Gössling S, Choi A, Dekker K, Metzler D. The social cost of automobility, cycling and walking in the European Union. Ecol Econ. 2018;158:65–74. https://doi.org/10.1016/j.ecolecon.2018.12.016.

IDOBB. Building balance, Advies Interdepartementaal Opschalingsplan Biobased Bouwen (IDOBB) 2023–2030, 19-6-2023, V 1.0. 2023.

Kulkarni R, Jenamani RK, Pithani P, Konwar M, Nigam N, Ghude SD. Loss to aviation economy due to winter fog in New Delhi during the Winter of 2011–2016. Atmos. 2019;10:198.

Lakra K, Avishek K. A review on factors influencing fog formation, classification, forecasting, detection and impacts, Rendiconti Lincei. Sci Fisiche Nat. 2022;33:319–53.

Pascaris ASCS, Pearce JM. A first investigation of agriculture sector perspectives on the opportunities and barriers for agrivoltaics. Agronomy. 2020;10(12):1885.

Pérez-Díaz JL, Ivanov O, Peshev Z, Álvarez-Valenzuela MA, Valiente-Blanco I, Evgenieva T, Dreischuh T, Gueorguiev O, Todorov PV, Vaseashta A. Fogs: physical basis, characteristic properties, and impacts on the environment and human health. Water. 2017;9:807.

RVO Netherlands Enterprise Agency. Reinventing multifunctionality, combining goals, sharing means, linking interests, the Netherlands Ministry of Economic Affairs. 2016.

Scott I. Governing by silos. In: Oxford research encyclopedia of politics. Oxford: Oxford University Press; 2020. https://doi.org/10.1093/acrefore/9780190228637.013.1414.

van der Heijden GMA. New materialism and climate change adaptation. In: Enthoven G, Rudnicki S, Sneller R, editors. Towards a science of ideas, an inquiry into the emergence, evolution and expansion of ideas and their translation into action. Wilmington: Vernon Press; 2022.

van der Heijden J, Rambharos S, Dorst H. Investing in co-benefits: a financial-economic and political explanation. In: Okitasari M, Hudson P, Suzuki M, editors. Equitable and sustainable cities – economics and disruptive technologies. Singapore: Springer; 2025.

van Erkel F. Naar een volwaardige en volwassen positie van energiegemeenschappen in de energietransitie in Amsterdam, Versterken van publiek-civiele samenwerking naast en met overheid en markt. Amsterdam: Gemeente Amsterdam en APEC; 2024.

Vranken A, Bijnens E, Horemans C, Leclercq A, Kestens W, Karakaya G, Vandenthoren L, Trimpeneers E, Vanpoucke C, Fierens F, Nawrot T, Cox B, Bruyneel L. Association of air pollution and green space with all-cause general practitioner and emergency room visits: a cross-sectional study of young people and adults living in Belgium. Environ Res. 2023;236:116713.

# Chapter 10
# Role of Satellites and Sensors in Fog Predictability

**Pradhi Rajeev**

## 10.1 Introduction

A fog episode refers to a period during which fog persists over a specific region, significantly reducing visibility and impacting various socio-economic activities (Lakra and Avishek 2022). Fog is defined as a cloud which is close to ground (Stull 2017). Fog formation is a complex meteorological phenomenon resulting from the interplay of various atmospheric conditions and processes, including temperature, humidity, wind patterns, and topography. Understanding the dynamics of a fog episode involves analyzing the meteorological factors and mechanisms that contribute to its development, maintenance, and dissipation. Fog is classified into different types based on their occurrence such as radiation fog, advection fog, upscale fog, and evaporation fog. Radiation and advection fog formation is due to condensation, when the surface temperature falls below the dew point leading to formation of dense layer of fog. Radiation fog occurs under clear skies and calm winds when Earth's surface cools rapidly through radiative heat loss, typically during the night. This type of fog is common in valleys and low-lying areas where cold air can pool. Advection fog forms when warm, moist air moves over a cooler surface, prevalent along coastlines. Upslope fog occurs when moist air is forced to ascend along a sloping terrain, cooling adiabatically as it rises. Evaporation fog, also known as steam fog, forms when cold air passes over a warmer water surface, causing water vapor to evaporate into the air and subsequently condense as the air cools. This type of fog is often observed over lakes and rivers during autumn and winter (Stull 2017).

Once formed, fog can be maintained or dissipated by various atmospheric processes such as temperature inversion, wind speed, and moisture availability. A

P. Rajeev (✉)
Department of Civil and Environmental Engineering, Indian Institute of Technology Patna, Patna, India
e-mail: pradhi@iitp.ac.in



P. Saxena et al. (eds.), *Fog: Formation, Chemistry, Dynamics, Impact and Dissipation*,
https://doi.org/10.1007/978-981-96-4422-3_10

temperature inversion, where a layer of warmer air sits above cooler air near the surface, can trap fog and prevent it from dissipating. Low wind speeds help maintain fog by limiting the mixing of air layers, which would otherwise disperse the fog. Conversely, fog dissipation occurs when the sun's radiation warms the Earth's surface, increasing the air temperature and reducing relative humidity. Additionally, stronger winds can mix the air layers, diluting the fog and enhancing evaporation.

Numerous studies and observational programs of fog formation have been conducted to improve the understanding of fog formation, growth, and dissipation (Bergot and Lestringant 2019; Fuzzi et al. 1998; Ghude et al. 2017; Meyer et al. 1986; Price et al. 2018; Smith et al. 2023). An interesting study have reported that increase in irrigation activities led to enhancement of fog events over Northern India (Smith et al. 2024). However, fog prediction is still a challenge despite the utilization of advanced technology, higher resolution data, and improved numerical weather prediction (NWP) models. This is generally owing to the complex mesoscale circulations, as well as small-scale and microphysical processes (Bergot et al. 2015; Gultepe et al. 2007, 2017; Price et al. 2018).

## 10.2 History

The first fog occurrence and prediction study was performed in England by meteorologist named Taylor (1917). He introduced the fog forecasting through fog prediction diagram by plotting dew point depression temperature against dry bulb temperature (Taylor 1917). A notable fog episode can be analyzed by examining a specific event, such as the London Fog of December 1952. This episode, known as the Great Smog, was a severe air pollution event exacerbated by cold weather, which led to the formation of a dense fog mixed with pollutants from coal burning. The temperature inversion trapped the fog and pollutants near the surface, resulting in extremely low visibility and significant health impacts.

During this episode, the high concentration of particulate matter (PM) and sulfur dioxide in the air contributed to the formation of sulfuric acid droplets within the fog, creating a toxic mixture. The Great Smog persisted for several days, causing widespread disruption and leading to thousands of respiratory illnesses and fatalities. This event highlighted the critical importance of understanding the interplay between meteorological conditions and air quality in fog formation and maintenance. Detailed analysis of specific fog episodes, such as the Great Smog of London, provides valuable insights into the mechanisms driving these phenomena and underscores the need for accurate forecasting and effective mitigation strategies to minimize their impact on human activities and health.

The initial study on fog occurrence and the atmospheric phenomenon leading to fog occurrence was studied in Kolkata and Hyderabad (Chakravortty 1948; Chandiramani 1958). Gupta, 1987, has given the "Technique of Elimination" for the fog forecasting in India. This study has predicted the fog episodes of winter 1984–1985 using the elimination technique and compared it with the prediction diagram and wet bulb depression tendency forecasting technique (Gupta 1987). This study has also identified the specific conditions under which these forecasting techniques will perform optimally.

The mathematical fog models for the formation, growth, and dissipation of radiation and advection fog were based on the heat balance (heat transfer coefficient), surface potential temperature, and dew point temperature. This has been discussed in detail in Stull (2017). However, predicting fog is a complex task that relies on a variety of technological tools and methodologies. The fog formation and decay occur due to intricate interactions among microphysical, thermodynamical, and dynamical processes. Previous studies have reported that the key factors influencing fog formation include the moist air cooling, vegetation, wind, heat and moisture transport and mixing, advection, and topographic parameters. He also highlighted that the impact of each factor is influenced by atmospheric stability, location, and season. Advances in meteorological science have integrated multiple data sources and computational techniques to enhance the accuracy of fog predictions.

## 10.3 Fog Prediction Techniques

The real-time monitoring of atmospheric conditions can be done by utilizing satellites and sensors, enabling meteorologists to monitor the formation and dissipation of fog as it happens. This real-time capability is vital for issuing timely warnings and updates, particularly for transportation sectors like aviation and shipping. Satellite data is integrated with information from ground-based observations and numerical weather prediction models. This integration creates a comprehensive picture of current and forecasted weather conditions. By combining satellite imagery with other data sources, meteorologists can improve the accuracy of fog predictions (Fig. 10.1). Satellites help identify patterns and trends in atmospheric conditions that lead to fog formation. Historical satellite data can be analyzed to understand the typical precursors and behaviors of fog in different regions. This analysis enhances

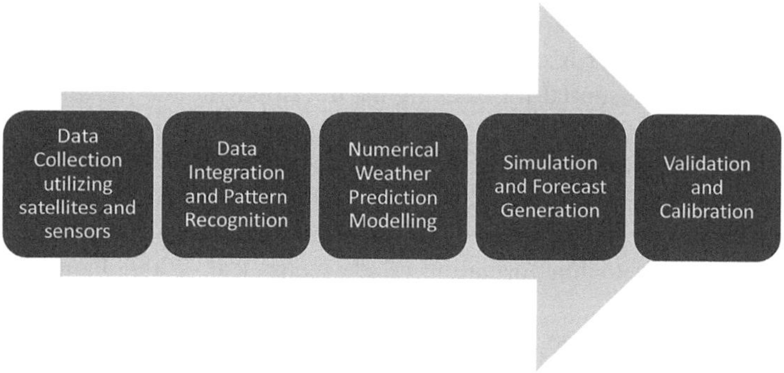


**Fig. 10.1** Flowchart depicting fog prediction process by utilizing satellites and sensors data

predictive models and improves the understanding of fog dynamics. Advances in satellite technology have led to higher resolution sensors, allowing for more detailed observations of foggy conditions. The finer spatial resolution helps in distinguishing between fog and other low-lying clouds, providing clearer insights into fog extent and density (Guidard and Tzanos 2007). The assimilation of meteorological parameters from various measurements such as aircraft measurements (AMDAR) and surface synoptic observations (SYNOP) generally improves the forecasting of fog (Bari 2019).

Numerical weather prediction (NWP) models are sophisticated computational tools that simulate the atmosphere's behavior based on physical laws. These models incorporate data from various sources, including satellites, radar, and ground stations, to forecast weather conditions (Singh et al. 2018). Specific models, such as the high-resolution rapid refresh (HRRR) model, are tailored to predict short-term weather phenomena like fog by providing high-resolution forecasts updated hourly. Machine learning (ML) and artificial intelligence (AI) techniques, such as neural networks (ANNs, CNN, RNN), fuzzy logics, and decision trees, have become increasingly important in fog prediction (Bari et al. 2023; Lakra and Avishek 2022). By analyzing vast datasets from historical weather patterns, these algorithms can identify complex relationships and patterns that traditional models might miss. ML models, such as neural networks and decision trees, can enhance the accuracy of fog predictions by continuously learning from new data.

### 10.3.1 Satellite-Based Observations

Satellites equipped with advanced sensors play a crucial role in fog prediction. They provide real-time data on atmospheric conditions, including temperature, humidity, and cloud cover. The integration of satellite technology into meteorology has significantly enhanced the ability to forecast fog, particularly in areas where ground-based observations are sparse. Geostationary satellites, like the Geostationary Operational Environmental Satellites (GOES) series, continuously monitor large areas, offering high-frequency observations that are essential for tracking fog development and dissipation. Geostationary satellites remain fixed over a specific point on the Earth's equator, providing continuous monitoring of the same area. This constant observation is essential for tracking the development and movement of fog in real-time. Polar-orbiting satellites travel in a north–south orbit around the Earth, covering the entire globe as the planet rotates beneath them. These satellites, such as NOAA-20, EUMETSAT, FY-3D, and Suomi NPP, provide high-resolution data and global coverage, which are crucial for understanding broader atmospheric patterns that contribute to fog formation (Badarinath et al. 2009; Pauli et al. 2020; Wang et al. 2023).

Various types of satellite sensors are used, such as infrared, visible, and microwave sensors, to measure the atmospheric conditions. Infrared sensors on satellites measure the temperature of the Earth's surface and atmosphere. They can detect the cooling of the surface at night, which is a precursor to fog formation. By analyzing the infrared radiation emitted by the ground and clouds, these sensors help identify temperature inversions and the presence of low-lying clouds, which are indicative of fog. Price and Stokkereit (2020) have reported that thermal infrared imagery can provide better information leading to precision in fog prediction. Visible light sensors capture images of the Earth in the visible spectrum. These sensors help identify the presence and extent of fog during daylight hours by detecting the reduced visibility caused by fog. High-resolution imagery from satellites, such as those in the MODIS (Moderate Resolution Imaging Spectroradiometer) program, provides detailed views of fog-covered areas. Microwave sensors penetrate clouds and provide data on atmospheric moisture levels. They are particularly useful in monitoring the amount of water vapor in the atmosphere, which is a critical factor in fog formation. Satellites equipped with microwave sensors, such as AMSU (Advanced Microwave Sounding Unit), contribute valuable data to fog prediction models (Lu et al. 2023). Recently, brightness temperature difference thresholding method is dominantly used in fog identification (Lakra and Avishek 2022). The details of various types of satellites with the types of sensors used for prediction are given in Table 10.1.

### *10.3.2 Ground-Based Observations*

Ground-based radar systems, particularly those that operate at microwave frequencies, can detect moisture and other atmospheric particles. These systems help meteorologists observe the fine-scale structures of fog and its movement. Doppler radar is useful for detecting changes in wind patterns that can contribute to fog formation. Automated weather stations scattered across various locations collect critical data, such as temperature, humidity, and visibility. These stations are equipped with a variety of sensors, including thermometers, hygrometers, anemometers, and barometers. The data collected helps determine the local conditions that can lead to fog, such as high humidity and low wind speeds. These stations provide real-time updates and feed information into larger forecasting systems. LIDAR (Light Detection and Ranging) systems use laser pulses to measure the distance to an object or surface. When used in meteorology, LIDAR can map the density of fog and aerosols in the atmosphere, providing detailed vertical profiles of the fog layer. Ceilometers are specialized LIDAR systems used to measure the height of cloud bases. They emit a laser beam upwards and measure the time it takes for the light to be reflected by the cloud base. This data helps determine the presence of low-lying clouds that can contribute to fog formation. Doppler radar systems measure the velocity of particles in the atmosphere. They can detect the movement of moisture-laden air masses and provide data on wind patterns. Changes in wind patterns and the movement of moist air are important indicators of fog formation. Doppler radar is also useful for monitoring fog dispersal. Particle Counters measure the concentration of particulate matter (PM) in the air. High levels of particulates can lead to the

**Table 10.1** Satellites and sensors used for fog prediction

| S. No. | Satellite | Organization | Instruments used | Remarks |
|---|---|---|---|---|
| 1. | Geostationary Operational Environmental Satellites (GOES) | NOAA (National Oceanic and Atmospheric Administration) | Advanced Baseline Imager (ABI); Geostationary Lightning Mapper (GLM) | Continuous and high-frequency monitoring for tracking the fog development and dissipation<br>Detection of temperature differences through infrared imagery |
| 2. | Suomi National Polar-orbiting Partnership (NPP) | NOAA and NASA | Visible Infrared Imaging Radiometer Suite (VIIRS); Cross-track Infrared Sounder (CrIS) | High-resolution imagery across the visible and infrared spectrum<br>Detailed atmospheric temperature and moisture profiles |
| 3. | Meteorological Operational Satellites (Metop Series) | EUMETSAT (European Organisation for the Exploitation of Meteorological Satellites) | Advanced Very High-Resolution Radiometer (AVHRR); Infrared Atmospheric Sounding Interferometer (IASI) | AVHRR data helps in identifying fog cover, particularly in remote areas<br>IASI's detailed profiles aid in understanding the vertical distribution of temperature and moisture, which is critical for predicting fog formation |
| 4. | Cloud-Aerosol LIDAR and Infrared Pathfinder Satellite Observations (CALIPSO) | NASA and CNES (French Space Agency) | Cloud-Aerosol LIDAR with Orthogonal Polarization (CALIOP); Imaging Infrared Radiometer (IIR) | Precise vertical profiles of fog layers, aiding in the understanding of fog depth and density<br>Helps distinguish between different types of aerosols and cloud particles, improving fog detection accuracy |
| 5. | Terra and Aqua Satellites | NASA | Moderate Resolution Imaging Spectroradiometer (MODIS); Atmospheric Infrared Sounder (AIRS) | MODIS data is used to monitor fog extent, cloud properties, and surface conditions, which are crucial for fog analysis<br>AIRS provides high-resolution atmospheric profiles, enhancing the understanding of conditions conducive to fog formation |

(continued)

**Table 10.1** (continued)

| S. No. | Satellite | Organization | Instruments used | Remarks |
|---|---|---|---|---|
| 6. | Global Precipitation Measurement (GPM) Mission | NASA and JAXA (Japan Aerospace Exploration Agency) | Dual-frequency Precipitation Radar (DPR); GPM Microwave Imager (GMI) | Provides data on precipitation and moisture, which are key factors in fog formation Enhances understanding of the relationship between precipitation events and subsequent fog development |

formation of fog by providing nuclei for water vapor to condense upon (Stolaki et al. 2015). This study has reported the change in the number of cloud condensation nuclei (CCN) will directly affect the fog droplet formation (Stolaki et al. 2015). By monitoring particle concentrations, meteorologists can better predict when and where fog might form. Visibility sensors measure the distance at which objects can be clearly seen. These sensors are often installed at airports and along highways to provide real-time data on visibility conditions. A drop in visibility is a direct indicator of fog presence and intensity. Combining data from various sources enhances the robustness of fog prediction systems. Advanced data fusion techniques integrate satellite imagery, radar data, and ground observations to create a comprehensive picture of current atmospheric conditions (Bari et al. 2023; Guidard and Tzanos 2007). This integrated approach allows for more accurate and timely fog forecasts, which are essential for transportation safety and public awareness.

## 10.4 Conclusion

The technology used for fog prediction is multifaceted, leveraging advancements in remote sensing, computational modeling, AI, and data integration. By utilizing these tools, meteorologists can provide more accurate and timely forecasts, helping to mitigate the impact of fog on transportation and public safety. As technology continues to evolve, the precision and reliability of fog predictions are expected to improve further. The use of diverse sensors for fog prediction exemplifies the complexity and interdisciplinary nature of meteorological science. By leveraging the strengths of different sensor technologies, meteorologists can monitor and predict fog with greater precision, thereby enhancing public safety and operational efficiency in sectors affected by fog, such as aviation and transportation. As sensor technology continues to advance, the accuracy of fog predictions is expected to improve, offering even more reliable forecasts in the future. Satellites are indispensable tools in the prediction of fog, offering extensive coverage, real-time data, and detailed insights into atmospheric conditions. The use of various sensors on geostationary and polar-orbiting satellites enables continuous monitoring and accurate forecasting of fog. As satellite technology continues to advance, the precision and reliability of fog predictions will further improve, enhancing safety and operational efficiency across multiple sectors impacted by fog. As reported in various studies that data integration from AMDAR, SYNOP, satellite, and ground-based stations will collectively help in improving the fog prediction.

## References

Badarinath KVS, Kharol SK, Sharma AR, Roy PS. Fog over Indo-Gangetic Plains—a study using multisatellite data and ground observations. IEEE J Sel Top Appl Earth Obs Remote Sens. 2009;2:185–95. https://doi.org/10.1109/JSTARS.2009.2019830.

Bari D. A Preliminary Impact Study of wind on assimilation and forecast systems into the one-dimensional fog forecasting model COBEL-ISBA over Morocco. Atmosphere (Basel). 2019;10 https://doi.org/10.3390/atmos10100615.

Bari D, Lasri N, Souri R, Lguensat R. Machine learning for fog-and-low-stratus nowcasting from Meteosat SEVIRI satellite images. Atmosphere (Basel). 2023;14:953.

Bergot T, Lestringant R. On the predictability of radiation fog formation in a mesoscale model: a case study in Heterogeneous Terrain. Atmosphere (Basel). 2019;10 https://doi.org/10.3390/atmos10040165.

Bergot T, Escobar J, Masson V. Effect of small-scale surface heterogeneities and buildings on radiation fog: Large-eddy simulation study at Paris–Charles de Gaulle airport. Q J Roy Meteorol Soc. 2015;141:285–98. https://doi.org/10.1002/qj.2358.

Chakravortty KC. India Meteorological Department. Sci Note. 1948;10:133–40.

Chandiramani WG. Incidence of fog and low stratus clouds over Begumpet airport during winter months. Mausam. 1958;9:345–8.

Fuzzi S, Laj P, Ricci L, Orsi G, Heintzenberg J, Wendisch M, Yuskiewicz B, Mertes S, Orsini D, Schwanz M. Overview of the Po Valley fog experiment 1994 (CHEMDROP). Contrib Atmos Phys. 1998;71:3–19.

Ghude SD, Bhat GS, Prabhakaran T, Jenamani RK, Chate DM, Safai PD, Karipot AK, Konwar M, Pithani P, Sinha V, Rao PSP, Dixit SA, Tiwari S, Todekar K, Varpe S, Srivastava AK, Bisht DS, Murugavel P, Ali K, Mina U, Dharua M, Rao YJ, Padmakumari B, Hazra A, Nigam N, Shende U, Lal DM, Chandra BP, Mishra AK, Kumar A, Hakkim H, Pawar H, Acharja P, Kulkarni R, Subharthi C, Balaji B, Varghese M, Bera S, Rajeevan M. Winter fog experiment over the Indo-Gangetic plains of India. Curr Sci. 2017;112:767–84.

Guidard V, Tzanos D. Analysis of fog probability from a combination of satellite and ground observation data. Pure Appl Geophys. 2007;164:1207–20.

Gultepe I, Tardif R, Michaelides SC, Cermak J, Bott A, Bendix J, Müller MD, Pagowski M, Hansen B, Ellrod G, Jacobs W, Toth G, Cober SG. Fog research: a review of past achievements and future perspectives. Pure Appl Geophys. 2007;164:1121–59. https://doi.org/10.1007/s00024-007-0211-x.

Gultepe I, Milbrandt JA, Zhou B. In: Koračin D, Dorman CE, editors. Marine fog: a review on microphysics and visibility prediction BT—marine fog: challenges and advancements in observations, modeling, and forecasting. Cham: Springer; 2017. p. 345–94. https://doi.org/10.1007/978-3-319-45229-6_7.

Gupta RK. On the techniques of forecasting fog/stratus over the Dundigal airfield of Hyderabad. Mausam. 1987;38:401–6.

Lakra K, Avishek K. A review on factors influencing fog formation, classification, forecasting, detection and impacts. Rend Lincei Sci Fis e Nat. 2022;33:319–53. https://doi.org/10.1007/s12210-022-01060-1.

Lu H, Ma Y, Zhang S, Yu X, Zhang J. Daytime Sea fog identification based on multi-satellite information and the ECA-TransUnet model. Remote Sens (Basel). 2023;15 https://doi.org/10.3390/rs15163949.

Meyer MB, Lala GG, Jiusto JE. FOG-82: a cooperative field study of radiation fog. Bull Am Meteorol Soc. 1986;67:825–32. https://doi.org/10.1175/1520-0477(1986)067<0825:FACFSO>2.0.CO;2.

Pauli E, Andersen H, Bendix J, Cermak J, Egli S. Determinants of fog and low stratus occurrence in continental Central Europe–a quantitative satellite-based evaluation. J Hydrol. 2020;591:125451.

Price J, Stokkereit K. The use of thermal infra-red imagery to elucidate the dynamics and processes occurring in fog. Atmosphere (Basel). 2020;11 https://doi.org/10.3390/atmos11030240.

Price JD, Lane S, Boutle IA, Smith DKE, Bergot T, Lac C, Duconge L, McGregor J, Kerr-Munslow A, Pickering M, Clark R. LANFEX: a field and modeling study to improve our understanding and forecasting of radiation fog. Bull Am Meteorol Soc. 2018;99:2061–77. https://doi.org/10.1175/BAMS-D-16-0299.1.

Singh A, George JP, Iyengar GR. Prediction of fog/visibility over India using NWP model. J Earth Syst Sci. 2018;127:26. https://doi.org/10.1007/s12040-018-0927-2.

Smith DKE, Dorling SR, Renfrew IA, Ross AN, Poku C. Fog trends in India: relationships to fog type and western disturbances. Int J Climatol. 2023;43:818–36. https://doi.org/10.1002/joc.7832.

Smith DKE, Reka S, Dorling SR, Ross AN, Renfrew IA, Jayakumar A, Anurose TJ, Parde AN, Ghude SD, Rumbold H. Forecasts of fog events in northern India dramatically improve when weather prediction models include irrigation effects. Commun Earth Environ. 2024;5:141.

Stolaki S, Haeffelin M, Lac C, Dupont J-C, Elias T, Masson V. Influence of aerosols on the life cycle of a radiation fog event. A numerical and observational study. Atmos Res. 2015;151:146–61. https://doi.org/10.1016/j.atmosres.2014.04.013.

Stull R. Meteorology for scientists and engineers. 3rd ed. University of British Columbia; 2017.

Taylor GI. The formation of fog and mist. Q J Roy Meteorol Soc. 1917;43:241–68. https://doi.org/10.1002/qj.49704318302.

Wang Y, Qiu Z, Zhao D, Ali MA, Hu C, Zhang Y, Liao K. Automatic detection of Daytime Sea fog based on supervised classification techniques for FY-3D satellite. Remote Sens (Basel). 2023;15 https://doi.org/10.3390/rs15092283.

# Chapter 11
# Cost-Effective Measures of Fog Remediation and Control: Mitigation of Health Impacts and Safety Risks

**Atisha Sood, Gautam Pathare, Srishti Sinha, and Anil Kumar Gupta**

## 11.1 Introduction

Fog is a meteorological phenomenon that creates complex hazards across multiple sectors, affecting public health, transportation systems, and economic productivity. Defined as a mass of suspended water droplets or ice crystals near the Earth's surface, fog occurs when the air near the ground cools, leading to condensation. This reduction in visibility intensifies risks across densely populated and industrial areas, especially in regions experiencing frequent fog events (Mcaleenan and Oloke 2010; Gultepe et al. 2007). The impacts of fog are diverse, spanning from health complications to transportation delays and economic losses (Fig. 11.1).

**Health Risks** Fog significantly exacerbates respiratory and cardiovascular health risks, particularly in urban areas where pollutants like particulate matter (PM2.5) and nitrogen oxides are already high. Fog traps these pollutants close to the ground, increasing exposure for vulnerable populations, including the elderly and individuals with chronic respiratory conditions. Studies have documented increases in hos-

A. Sood (✉)
Centre for Disaster & Health, National Institute of Disaster Management, Ministry of Home Affairs, GoI, Delhi, India
e-mail: soodatisha@gmail.com

G. Pathare
HighCAP Project, National Institute of Disaster Management, Ministry of Home Affairs, GoI, Delhi, India

S. Sinha
DFPD-DMPD Project National Institute of Disaster Management, Ministry of Home Affairs, GoI, Delhi, India

A. K. Gupta
Integrated Centre for Adaptation, DRR and Sustainability (ICARS, GOI-DST Joint COE), IIT, Roorkee, India



P. Saxena et al. (eds.), *Fog: Formation, Chemistry, Dynamics, Impact and Dissipation*,
https://doi.org/10.1007/978-981-96-4422-3_11

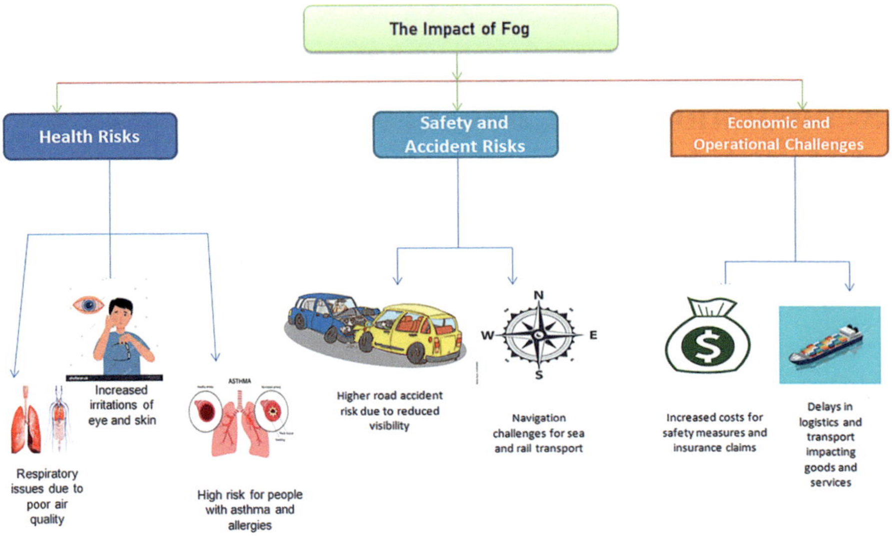


**Fig. 11.1** The impacts of fog

pital admissions and respiratory-related health complications during prolonged fog episodes (Jayanthi and Vishnuvarthan 2024; Ward-Caviness et al. 2021).

**Safety and Accident Risks** Reduced visibility due to fog poses serious safety risks, especially in transportation sectors such as roadways, railways, aviation, and maritime. Road accidents are notably more frequent in foggy conditions, which strain healthcare systems and emergency services. Aviation and maritime operations are similarly affected, with fog frequently causing delays, route diversions, and even collisions, leading to broader disruptions in logistics and supply chains (Dadwal and Dadwal 2021).

**Economic and Operational Challenges** Beyond health and safety risks, fog presents significant economic challenges, particularly in sectors reliant on transportation and timely logistics. Fog-related accidents and delays disrupt supply chains, hinder productivity, and increase operational costs across industries. For instance, in agricultural regions, prolonged fog can damage crops by creating persistently high humidity conditions, contributing to mold growth and yield reduction.

Addressing these diverse risks requires a comprehensive and multifaceted approach. Effective fog control involves implementing engineering solutions, providing appropriate personal protective equipment (PPE), and fostering public awareness. Engineering controls, including advanced fog detection and warning systems, are critical for improving safety, especially in transportation settings. PPE considerations may include high-visibility clothing, respiratory protection, and specialized eyewear to improve safety in low-visibility conditions (Salvador 2023). Moreover, promoting a safety culture within organizations, where both management and employees engage in risk mitigation, enhances the overall effectiveness of these interventions.

This chapter examines various cost-effective fog remediation and control measures, with an emphasis on minimizing health and safety risks. We explore both traditional and innovative techniques, including vegetative buffers, terrain modifications, and technological solutions like infrared heating and fog nets. By combining these approaches with policy frameworks and collaborative strategies, this chapter aims to provide a holistic guide for mitigating the adverse effects of fog, emphasizing sustainability and community resilience in fog-prone areas.

## 11.2 Health and Safety Risks Associated with Fog

Fog contributes to a wide range of health and safety hazards, impacting diverse sectors such as transportation, manufacturing, agriculture, and urban living (Table 11.1). The risks associated with fog are often underestimated, yet they have significant implications for public health, safety, and economic stability. This section delves into the specific health and safety concerns caused by fog, including respiratory health risks, transportation accidents, and workplace hazards.

### *11.2.1 Health Risks from Fog Exposure*

Fog exacerbates health risks by trapping airborne pollutants close to the ground, creating an environment where individuals, especially those in urban areas, experience prolonged exposure to harmful particulates and gases. Pollutants such as particulate matter (PM2.5), nitrogen dioxide (NO2), and sulfur dioxide (SO2) are commonly suspended in fog, leading to an increase in respiratory and cardiovascular health issues (Jayanthi and Vishnuvarthan 2024; WHO 2021).

Individuals with pre-existing conditions, including asthma, chronic obstructive pulmonary disease (COPD), and cardiovascular diseases, are particularly vulnerable. Studies have shown that hospital admissions for respiratory and cardiac issues spike during prolonged fog events, as inhalation of pollutants in fog-laden air

**Table 11.1** Health risks associated with fog exposure

| Health condition | Impact of fog exposure |
|---|---|
| Respiratory issues | Aggravation of asthma, bronchitis, and COPD |
| Cardiovascular strain | Increased risk of heart attack and stroke |
| Reduced visibility (safety risk) | Accidents and injuries, particularly in traffic |
| Mental health | Increased anxiety and stress due to poor air quality |

exacerbates symptoms (Dadwal and Dadwal 2021). Children, the elderly, and outdoor workers are also at a heightened risk, as they are more likely to experience adverse effects from these pollutants (Ogwu et al. 2024).

Notably, fog-related health risks extend beyond respiratory issues. Some studies indicate that chronic exposure to fog-related pollutants can have long-term effects on lung function and contribute to systemic inflammation, potentially leading to other chronic illnesses (Mcaleenan and Oloke 2010). Consequently, addressing the health implications of fog is essential for public health planning, especially in urban centers with high pollution levels.

### *11.2.2 Transportation Safety Hazards*

Reduced visibility during foggy conditions presents one of the most severe safety risks, particularly in the transportation sector. Road traffic accidents are significantly more common in fog, with low visibility leading to collisions, pileups, and increased fatalities. Research suggests that foggy conditions contribute to a substantial percentage of highway accidents, with some regions reporting up to a 30% increase in collision rates on foggy days (Salvador 2023).

**Road Transportation** Drivers face challenges such as limited sight distance and delayed reaction times, which increase the likelihood of accidents. Furthermore, the difficulty of perceiving distance in dense fog causes misjudgment in vehicle spacing, leading to rear-end collisions. Highways, with faster-moving traffic and minimal room for maneuver, are particularly susceptible to fog-related accidents (Mcaleenan and Oloke 2010).

**Aviation and Maritime Operations** The impact of fog extends to aviation and maritime sectors, where it disrupts flight schedules, delays cargo shipments, and increases operational risks. In aviation, fog limits pilots' visibility during takeoffs and landings, often resulting in delayed or canceled flights. Maritime navigation also suffers, as fog obscures coastal landmarks and navigational aids, increasing the risk of collisions and groundings (Dadwal and Dadwal 2021).

These transportation-related fog hazards lead to broader economic implications. Delays in cargo shipments impact supply chains, while accidents increase insurance costs and place additional strain on emergency response and healthcare systems.

### *11.2.3 Workplace Hazards in Foggy Environments*

In addition to transportation risks, fog creates hazards in various workplace settings. Outdoor workers, particularly those in construction, agriculture, and manufacturing, face increased risks due to reduced visibility and exposure to pollutants. For instance, construction sites may see an increased likelihood of falls, collisions, and other accidents during foggy conditions. Similarly, in agricultural settings, fog can contribute to mold growth, reducing crop quality and yield (Jayanthi and Vishnuvarthan 2024).

For industries reliant on outdoor operations, fog can also impede workers' ability to recognize hazards, navigate safely, and operate machinery. To mitigate these risks, employers are encouraged to implement safety measures such as enhanced lighting, high-visibility clothing, and structured work schedules that minimize exposure during dense fog periods (Salvador 2023).

In conclusion, the health and safety risks associated with fog are complex and multifaceted. Addressing these challenges requires a combination of environmental, technological, and policy-based solutions tailored to the specific risks in transportation, public health, and workplace safety.

## 11.3 Current Methods of Fog Control and Their Limitations

Efforts to control and dissipate fog have traditionally relied on methods aimed at enhancing visibility and reducing the density of fog particles. These methods, while effective in certain conditions, often come with limitations related to cost, environmental impact, and practicality in large-scale applications. This section reviews common fog control methods, including heating, chemical dispersal, and physical barriers, examining their effectiveness and challenges.

### *11.3.1 Heating as a Fog Control Technique*

Heating is one of the oldest methods used to dissipate fog by evaporating water droplets, making it particularly useful in environments where clear visibility is essential, such as airport runways. Common heating methods include burning fossil fuels to generate heat, using infrared radiation, and deploying heat exchangers (Kunkel et al. 1974). The application of heat raises the air temperature locally, causing water droplets to evaporate and reduce fog density *(*Mcgowan, 1971*)*.

Although heating can be effective in confined or high-priority areas, such as runways, it is not viable for large-scale applications. Heating methods require a significant energy input, making them both costly and environmentally unsustainable for extensive use, such as along highways or across urban landscapes. Studies indicate that continuous heating over long stretches, like highways, would be prohibitively expensive and environmentally taxing due to the reliance on fossil fuels or nonrenewable energy sources.

Additionally, the effectiveness of heating is influenced by environmental conditions. In colder regions, the temperature difference required to dissipate fog is larger, further increasing energy demands and operational costs. Therefore, while heating remains a valuable tool for specific settings, it is not suitable as a general solution for fog dissipation across large areas.

### 11.3.2 Chemical Dispersal for Fog Mitigation

Chemical dispersal involves introducing hygroscopic substances, such as calcium chloride or urea, to absorb moisture from fog droplets, causing them to coalesce and precipitate. This approach is often used on highways and in seaports, where quick fog dissipation is crucial to prevent accidents. However, while chemical dispersal can clear fog temporarily, it has notable drawbacks (Knapp 1972).

Environmental concerns are significant, as chemicals used in fog dispersal can infiltrate soil and water systems, introducing pollutants that affect local ecosystems. Research has shown that repeated use of chemicals like calcium chloride can alter soil composition, harm plant life, and pose risks to aquatic organisms if these chemicals enter water bodies. The frequent need for reapplication, due to dilution from wind or precipitation, also increases costs and environmental impact.

Moreover, the effectiveness of chemical dispersal is heavily dependent on atmospheric conditions, such as wind speed and direction. These variables can reduce the concentration of the chemicals, necessitating repeated applications to maintain visibility, which adds to both financial and environmental costs. Due to these limitations, chemical dispersal is generally reserved for specific high-risk areas where other fog mitigation methods are unavailable.

### 11.3.3 Physical Barriers and Structural Solutions

Physical barriers, such as fog fences, windbreaks, or terrain modifications, represent a passive approach to managing fog. These barriers aim to disrupt airflow patterns, prevent fog from forming in critical areas, or redirect it away from high-traffic zones. For instance, fog fences are often installed along highways or around airports to block or divert fog, while vegetative windbreaks can be effective in agricultural or rural settings (Veldhuizen and Ledbetter 1971).

While physical barriers can be beneficial in specific, well-defined areas, they are limited in scope and effectiveness. Barriers must be strategically placed to match local wind patterns and are less effective in regions where fog flows in multiple directions. Physical barriers do not reduce fog density above ground level, limiting their utility in areas requiring high visibility at varying altitudes, such as airports.

Physical barriers also require substantial initial investments and regular maintenance, especially in large-scale applications. Additionally, these structures can disrupt local ecosystems by impeding animal movement or altering natural habitats. Consequently, while physical barriers offer a sustainable, low-maintenance option in certain environments, they are often insufficient for broad applications and are generally used as part of a larger fog control strategy (Lee et al. 2020).

### *11.3.4 Broader Limitations Across Fog Control Techniques*

Each of these fog control methods comes with inherent challenges. Heating and chemical dispersal are expensive and environmentally harmful when scaled beyond small, targeted areas. Physical barriers, though environmentally safer, lack adaptability and are limited in effectiveness under variable wind and fog conditions. This lack of scalability and environmental impact hinders the ability of current methods to address fog-related issues in a sustainable, cost-effective manner (Gultepe et al. 2007).

The need for environmentally friendly alternatives is pressing. Current methods are often impractical for long-term or large-scale deployment, particularly in areas with frequent or heavy fog. Research into new, sustainable technologies for fog mitigation that can overcome these limitations is critical. Solutions that prioritize energy efficiency, minimal environmental impact, and adaptability to different weather conditions could serve as more effective, scalable approaches to fog control (Brighente et al. 2022).

This section has highlighted the need for innovation in fog mitigation technologies. As we explore alternative fog control measures in the next section, it becomes clear that a comprehensive strategy combining multiple techniques may be essential to manage fog efficiently and sustainably.

## 11.4 Cost-Effective Fog Remediation Techniques

Developing cost-effective fog remediation techniques is essential to ensure safety and maintain productivity across fog-prone regions. This section explores both natural and low-cost technological solutions, focusing on methods that offer efficient fog reduction with minimal environmental impact. By utilizing passive approaches like vegetative buffers and fog nets, along with emerging low-cost technologies, these strategies present practical options for communities and industries affected by fog.

## 11.4.1 Natural and Passive Techniques

### 11.4.1.1 Vegetative Buffers

Vegetative buffers, consisting of strategically planted trees and shrubs, provide a sustainable method to reduce fog density in targeted areas. These buffers intercept moisture-laden air, causing fog droplets to settle on plant surfaces rather than dispersing into surrounding areas. Research indicates that well-designed vegetative buffers can reduce fog density by up to 30% in adjacent zones, contributing to improved visibility and safer conditions on nearby roads or farms.

A combination of coniferous and deciduous trees has been shown to be particularly effective in capturing fog. The needle-like structure of coniferous species, for example, allows for high water interception rates, which helps dissipate fog naturally. The initial costs for vegetative buffers primarily include planting and minor maintenance, making this approach a low-cost, long-term solution, particularly beneficial in rural and semi-urban areas.

### 11.4.1.2 Terrain Modifications

Altering terrain features is another passive technique for reducing fog formation, especially in areas where fog regularly accumulates. Modifications, such as creating gentle slopes and avoiding abrupt elevation changes, allow for smoother airflow, which can reduce the likelihood of temperature inversions that contribute to fog formation. Terrain modifications often include drainage channels designed to redirect cold air, preventing it from settling in low-lying areas and forming dense fog.

Although terrain modifications require initial infrastructure investment, these changes can often be integrated into existing development or land-use projects, minimizing additional costs. Studies have shown that terrain adjustments in fog-prone valleys can reduce fog persistence by as much as 40%, improving safety and operational efficiency in affected regions.

## 11.4.2 Low-Cost Technological Solutions

### 11.4.2.1 Salt Dispersion

Salt dispersion is a technique that involves spraying a mist of salt or other hygroscopic compounds into foggy areas. Salt particles absorb moisture from fog droplets, causing them to coalesce and fall out of suspension, thus clearing the fog. This method has proven effective on highways and in port areas where visibility needs to be restored quickly. A study demonstrated that salt dispersion could improve visibility by up to 50% within minutes, significantly reducing the risk of accidents in critical zones (Zhu et al. 2023).

Despite its effectiveness, salt dispersion requires careful application due to potential environmental impacts. Prolonged or excessive use of salt can lead to soil salinization, affecting nearby vegetation and water sources (Qadir et al. 2018). Therefore, this technique is best suited for short-term use in high-traffic or emergency areas where rapid fog clearing is essential, with environmental considerations taken into account to prevent long-term damage.

#### 11.4.2.2 Infrared Heating Devices

Infrared heating devices offer a targeted approach to fog dissipation by raising air temperature in specific areas, causing fog droplets to evaporate. These devices are often installed in high-traffic intersections and highway rest stops to improve visibility and reduce accident risks. Studies indicate that strategically placed infrared heaters can improve visibility by up to 60%, particularly in dense fog conditions, making this method both effective and energy-efficient (Li et al. 2024; Wo et al. 2023).

The cost-effectiveness of infrared heating systems comes from their energy efficiency and targeted deployment. While initial installation costs can be substantial, these systems are more economical over time due to low operational costs. Recent studies show that infrared heating is particularly advantageous in urban areas where space is limited, and large-scale heating methods are impractical.

#### 11.4.2.3 Fog Nets

Fog nets, traditionally used for water harvesting, are increasingly recognized for their potential to reduce fog density in high-risk areas. These nets are made from fine, mesh-like materials that capture fog droplets as they pass through, effectively clearing the air in surrounding regions. In addition to improving visibility, fog nets also provide the added benefit of water collection, which can be especially valuable in arid and semi-arid areas (Ding et al. 2023; Feld et al. 2016).

Fog nets have proven to be a low-cost and low-maintenance solution suitable for rural and coastal areas where more expensive technologies are not feasible. Research indicates that fog nets can capture up to 10 L of water per square meter daily under heavy fog conditions, making them a dual-purpose tool for both fog mitigation and water supply.

### *11.4.3 Implementation Considerations and Future Directions*

Each fog remediation technique discussed here has unique benefits and limitations. Vegetative buffers and terrain modifications are best suited to rural or semi-urban environments where natural infrastructure can play a role in fog control. In contrast, technological solutions such as salt dispersion and infrared heating devices are more practical for rapid fog dissipation in high-traffic or emergency areas where immediate visibility improvements are critical.

Combining these approaches may yield superior results. For instance, integrating fog nets and vegetative barriers in coastal areas or valley regions could enhance visibility while providing environmental benefits such as soil stabilization and water conservation. The need for customized fog mitigation strategies that align with local environmental conditions and economic resources is evident.

Future research should focus on refining these methods, particularly in developing eco-friendly, low-impact technologies. Advances in AI-based fog prediction systems and internet of things (IoT)-enabled fog monitoring devices could further enhance the effectiveness of these approaches by providing real-time data on fog formation and dispersion patterns, enabling proactive fog management strategies (Lakra and Avishek 2022).

The cost-effective fog remediation techniques, especially those leveraging natural or low-cost technologies, offer promising solutions for improving safety and visibility in fog-prone areas. Continued innovation and integration of multiple approaches are essential for building resilient and sustainable fog control frameworks that balance efficiency with environmental responsibility.

## 11.5 Mitigating Health Impacts: Practical Measures and Interventions

The health impacts of fog, particularly in polluted urban areas, necessitate proactive measures to protect public health. Fog, by trapping pollutants such as particulate matter (PM2.5) and nitrogen oxides, increases the risk of respiratory and cardiovascular issues, especially for vulnerable populations. This section explores cost-effective interventions that communities and individuals can implement to minimize health risks associated with fog exposure, including air quality improvements, protective gear, public awareness campaigns, and early warning systems (Fig. 11.2) (WHO 2021).

### *11.5.1 Air Quality Improvements*

#### 11.5.1.1 Low-Cost Air Purifiers in Public Spaces and Homes

Installing affordable air purifiers in high-risk public spaces and homes is an effective intervention to reduce indoor exposure to fog-trapped pollutants. These devices, especially those equipped with high efficiency particulate air (HEPA) filters, can capture fine particulates and significantly improve indoor air quality. Studies show

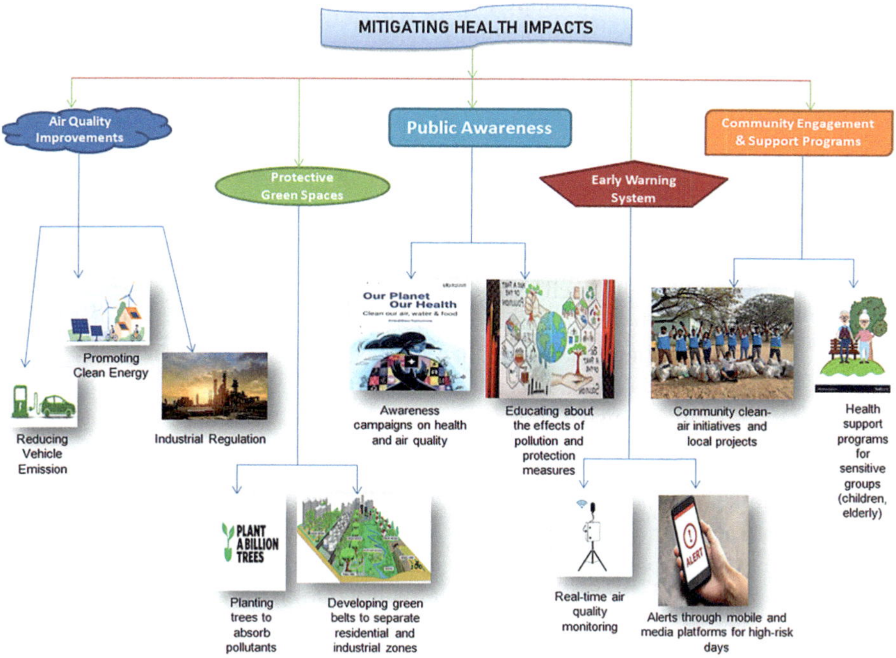


**Fig. 11.2** Mitigating health impacts of fog

that HEPA-filtered purifiers reduce particulate matter indoors by over 60%, offering a vital shield against fog-related air pollution (Dadwal and Dadwal 2021).

In public areas like schools, hospitals, and community centers, air purifiers create safer environments for vulnerable groups, including children and the elderly. In residences, air purifiers are particularly beneficial in rooms where individuals spend the most time, such as bedrooms and living spaces. Municipal initiatives that subsidize or provide air purifiers to low-income households can make this intervention more accessible, helping reduce the health burden in economically challenged areas (Ward-Caviness et al. 2021).

#### 11.5.1.2 Strategies for Effective Use

Proper maintenance is essential to ensure purifiers function optimally. Filters should be replaced regularly, and purifiers should be strategically placed in high-traffic or frequently used areas. Public education on using air purifiers effectively during fog episodes is also crucial, as many users may not be aware of how to maximize their impact.

## *11.5.2 Protective Gear and Public Awareness*

### 11.5.2.1 Distribution of Respiratory Masks

Respiratory masks, particularly those rated N95 or higher, can protect against fine particulate matter in fog. Public health agencies can distribute these masks in areas frequently affected by dense fog, prioritizing distribution to high-risk groups such as the elderly, individuals with respiratory conditions, and outdoor workers. Studies have shown that N95 masks filter out approximately 95% of airborne particulates, significantly reducing inhalation of pollutants during heavy fog events (Salvador 2023).

### 11.5.2.2 Public Awareness Campaigns

Effective public awareness campaigns are crucial for encouraging the use of protective gear and promoting behaviors that minimize fog exposure risks. Campaigns should educate the public on the risks associated with fog, the proper use of respiratory masks, and ways to improve indoor air quality during fog episodes. Using various communication channels—including social media, local radio, and community outreach programs—ensures that critical information reaches a broad audience.

Key messages in these campaigns could include recommendations to limit outdoor activities during dense fog, keep windows closed, and use indoor air purifiers. By increasing public awareness, communities can better protect vulnerable populations and reduce the overall health impact of fog events.

## *11.5.3 Early Warning Systems*

Early warning systems that alert communities about incoming fog events allow individuals and organizations to take preventive measures in advance. These alerts, similar to those used for severe weather events, can be disseminated through SMS notifications, mobile apps, social media, and digital signs in public places. Such systems rely on meteorological data and visibility sensors to detect fog formation and issue warnings.

Advanced fog prediction models, leveraging AI and machine learning, can significantly enhance the accuracy and timeliness of these warnings. For example, AI algorithms can analyze real-time weather patterns, topography, and other environmental factors to predict fog formation up to 24 h in advance. By integrating these alerts with health advisories, communities can reduce fog exposure, particularly for individuals who are highly susceptible to respiratory issues (Lakra and Avishek 2022).

**Components of an Effective Fog Warning System**

- Tiered Alert System: A color-coded alert system (e.g., yellow, orange, red) can indicate varying fog intensity levels, allowing individuals to respond appropriately.
- Guidance for Action: Each alert level should be accompanied by recommendations, such as limiting outdoor activities, wearing protective masks, and using air purifiers indoors.
- Accessibility of Alerts: Using diverse dissemination methods (SMS, mobile apps, and local news channels) ensures the information reaches different segments of the population, including those without internet access.
- Collaboration with Local Health Agencies: Alerts can be coordinated with health services to prepare for increased patient influx during severe fog episodes.

### *11.5.4 Community Engagement and Support Programs*

#### 11.5.4.1 Education on Indoor Air Quality Maintenance

Community workshops and seminars on air quality improvement techniques during fog events can empower residents to make informed choices. Programs can cover topics such as the effective use of air purifiers, tips for maintaining indoor air quality, and the importance of ventilation management. For instance, educating residents on sealing doors and windows during fog episodes to prevent pollutants from entering can help maintain safer indoor air environments.

#### 11.5.4.2 Supporting Vulnerable Populations

Programs that prioritize support for vulnerable populations, such as the elderly and those with pre-existing health conditions, are essential. Services may include the provision of respiratory masks, installation of air purifiers in communal facilities like senior centers, and transportation assistance to reduce outdoor exposure. Community health workers and volunteers can play a vital role in delivering these services and ensuring that high-risk groups have access to protective resources.

## 11.6 Enhancing Transportation and Workplace Safety

Fog poses severe safety risks across transportation sectors and work environments, particularly those with outdoor operations. Addressing these risks requires practical interventions designed to improve visibility, navigation, and safety protocols. This section outlines low-cost technologies, training programs, and safety protocols that reduce accident risks and enhance operational safety in foggy conditions.

## *11.6.1 Traffic Management Systems*

### 11.6.1.1 Radar and LiDAR-Based Vehicle Guidance

In low-visibility conditions, radar and LiDAR (Light Detection and Ranging) systems help drivers navigate safely by providing real-time data on nearby objects. These systems detect obstacles and communicate vital information to drivers, reducing the risk of collisions. LiDAR systems, for instance, use lasers to measure distances and map surrounding objects, which is particularly useful on highways and in urban settings with heavy traffic. Studies show that radar-guided navigation systems can reduce accident rates by up to 30% in foggy conditions.

Many countries are now integrating these systems into their transportation infrastructure, connecting radar and LiDAR technology to traffic management centers that can inform drivers of changing conditions through roadside digital displays. This approach is cost-effective, because it builds on existing traffic management systems, allowing easy updates and integration with other safety features.

### 11.6.1.2 Smart Traffic Lights and Adaptive Speed Control

Smart traffic signals and adaptive speed control systems adjust signal timing and speed limits based on visibility and weather conditions. For instance, when fog sensors detect low visibility, these systems can automatically lower speed limits and extend red-light times to reduce accident risks at intersections. Road signs with dynamic displays communicate these adjustments to drivers, helping them respond appropriately. Cities that have implemented adaptive traffic lights report improved safety in low-visibility conditions due to reduced driver error (Salvador 2023).

## *11.6.2 Enhanced Lighting Solutions*

### 11.6.2.1 LED Roadway Lights and Reflective Markings

LED-based roadway lights offer better illumination and visibility compared to traditional lighting. They are energy-efficient and cost-effective, making them an ideal choice for fog-prone regions. Research indicates that LED lights improve visibility by up to 40% in low-visibility conditions, significantly reducing accident rates on foggy days.

Reflective road markings are another simple, cost-effective solution. High-quality reflective paint enhances lane visibility, guiding drivers and reducing the likelihood of lane departures or collisions. Combining LED lights with reflective markings provides layered visual cues that are essential in dense fog conditions. Roadway lights placed at regular intervals, alongside reflective road markings, create a well-lit path for vehicles to follow, enhancing navigation during fog events.

#### 11.6.2.2 Fog-Specific Vehicle Lights

Vehicles equipped with fog lights provide additional illumination focused closer to the ground, where fog tends to be denser. These lights use a broad, flat beam to illuminate the road directly in front of the vehicle without reflecting off the fog. Public awareness campaigns on proper fog light usage, combined with vehicle safety checks to ensure functioning fog lights, contribute to safer driving practices during fog events.

### *11.6.3 Safety Protocols for Outdoor Workers*

#### 11.6.3.1 Fog-Specific Safety Training

For industries with outdoor operations, such as construction, agriculture, and transportation, fog-specific safety training programs are essential. These programs teach workers how to navigate and operate safely in low-visibility conditions, including the proper use of visibility-enhanced gear and emergency protocols.

Training sessions include situational awareness exercises and communication protocols for use in foggy conditions. Workers are trained to rely on visibility-enhanced gear and follow strict safety protocols when operating machinery or vehicles during fog events. Training programs also emphasize hazard identification, including recognizing potential accident risks and ensuring proper spacing between individuals and equipment on-site (Salvador 2023).

#### 11.6.3.2 Visibility-Enhanced Protective Gear

High-visibility clothing, such as reflective vests and jackets, is essential for outdoor workers during fog events. Reflective gear enables workers to be seen from a distance, reducing the risk of accidents in low-visibility settings. In addition to standard reflective clothing, anti-fog safety glasses and masks with proper filtration capabilities are also recommended for workers, particularly those exposed to pollutants and particulate matter present in fog.

#### 11.6.3.3 Adjusting Work Schedules Based on Fog Conditions

Flexible work scheduling is another cost-effective measure for protecting outdoor workers. Supervisors can arrange work hours around fog conditions, prioritizing high-visibility periods for outdoor tasks and limiting outdoor exposure during peak fog hours. This approach minimizes the risk of fog-related accidents and can be especially beneficial in high-risk sectors, such as construction and heavy manufacturing. Some organizations use portable weather stations to monitor fog density in real time, enabling supervisors to make informed scheduling decisions based on current visibility levels.

### *11.6.4 Portable Fog Monitoring Devices*

#### 11.6.4.1 Use of Portable Sensors for On-Site Fog Monitoring

Portable weather stations equipped with fog sensors offer a practical solution for monitoring fog conditions in real time. These devices can measure humidity, temperature, and visibility levels, providing valuable data to supervisors and safety personnel in construction and manufacturing sectors. Workers can be informed immediately of changing fog conditions, enabling prompt adjustments to safety protocols or work schedules. Portable fog sensors are cost-effective and easy to install, making them suitable for various work environments.

#### 11.6.4.2 Mobile Apps for Safety Alerts

Mobile applications integrated with portable fog monitoring devices can provide instant alerts to workers and managers when fog density reaches dangerous levels. These alerts can prompt users to take safety precautions, such as wearing visibility-enhanced gear or temporarily halting certain tasks. In construction and agricultural settings, where workers may be dispersed across large areas, these apps facilitate real-time communication and promote a culture of safety.

## 11.7 Case Studies on Cost-Effective Fog Control

Implementing fog control measures requires tailored solutions based on geographic, environmental, and economic factors. This section presents case studies from various regions that have successfully adopted cost-effective fog control measures, showcasing the adaptability and efficacy of different methods in real-world scenarios.

### *11.7.1 Chile's Fog Water Harvesting in the Atacama Desert*

The Atacama Desert in Chile, one of the driest places on Earth, faces frequent fog events that offer both challenges and opportunities. The local communities have effectively addressed these challenges through fog water harvesting using fog nets.

This technique, initially developed for water collection in arid environments, has proven to be an efficient method for both fog mitigation and water supply (Albornoz et al. 2023, 2024; Del Río et al. 2018).

#### 11.7.1.1 Fog Nets as Dual-Function Tools

Fog nets installed in the desert capture fog droplets, which condense on the mesh material and are collected as freshwater. This setup not only clears the fog from surrounding areas, reducing its density, but also provides a valuable water source for local communities. Studies have shown that these fog nets can collect up to 10 L of water per square meter daily, making them a sustainable, low-cost option for fog remediation in rural and semi-urban settings.

#### 11.7.1.2 Community Impact and Sustainability

Beyond water collection, fog nets have improved visibility in the area, contributing to safer navigation for travelers and locals. The dual-purpose design demonstrates the potential of fog nets as a solution that combines environmental sustainability with practical fog control. The project's success in the Atacama Desert has inspired similar initiatives in other arid regions worldwide.

### *11.7.2 Northern India's Highway Safety Measures*

In Northern India, dense winter fog frequently disrupts transportation, contributing to accidents and fatalities on highways. The government and local agencies have implemented several cost-effective interventions to improve visibility and enhance road safety in these fog-prone areas.

#### 11.7.2.1 Reflective Road Markings and Fog Lights

Highways in the region are now equipped with reflective road markings that improve lane visibility during low-light and foggy conditions. Additionally, government regulations mandate fog lights for vehicles, providing better illumination directly on the road surface. Studies indicate that reflective road markings, combined with fog lights, have reduced accident rates by approximately 25% on major highways during peak fog season (Dadwal and Dadwal 2021).

#### 11.7.2.2 Fog Sensors and Adaptive Traffic Systems

In high-traffic areas, fog sensors have been installed to detect visibility levels in real time. These sensors are linked to digital traffic signs that adjust speed limits according to visibility conditions. Such adaptive systems help reduce the risk of accidents by encouraging drivers to slow down in dense fog. This approach has proven both effective and cost-efficient, as it leverages existing traffic infrastructure and requires minimal maintenance.

### *11.7.3 China's Urban Fog Monitoring and Safety Interventions*

China's rapid urbanization has increased the frequency of fog events in cities, exacerbated by high pollution levels. In response, several urban areas have implemented innovative fog control and monitoring systems to protect public health and improve transportation safety (Fu and Chen 2017).

#### 11.7.3.1 Smart Fog Detection Systems

Urban areas in China, including Beijing and Shanghai, have invested in AI-powered fog detection systems that predict fog formation and issue public alerts. These systems use real-time data from weather stations and satellite imagery, combined with machine learning algorithms, to forecast fog density and duration. Public alerts are then disseminated via mobile apps, digital billboards, and local news channels, helping residents take preventive measures.

#### 11.7.3.2 Public Health Campaigns and Air Quality Management

In addition to fog detection, public health campaigns encourage residents to stay indoors and use air purifiers during high-fog days. This proactive approach has been shown to reduce hospital admissions related to respiratory issues by 15% during peak fog periods. By addressing both fog density and public health education, China's urban areas demonstrate a comprehensive, multipronged approach to fog management.

## 11.8 Challenges and Limitations

Fog remediation and control present several challenges, particularly when considering cost-effective measures. The comparison of fog remediation technologies is shown in Table 11.2. The primary challenge lies in the complexity of fog formation, which is influenced by a combination of meteorological, geographical, and environmental factors. As such, the development of universally effective solutions remains elusive. Some key limitations include:

1. Environmental and Geographical Variability: Fog conditions vary significantly across different regions due to factors such as local weather patterns, altitude, and proximity to water bodies. This variability makes it difficult to implement one-size-fits-all remediation strategies, as measures effective in one area may not be suitable for others.
2. Technological Limitations: While technologies for fog dispersion and control exist, they often come with high operational and maintenance costs. For example, the use of fog-dispersing aerosols, which may be effective in certain settings, can be prohibitively expensive for large-scale or long-term deployment, particularly in low-resource environments. Moreover, technological solutions may not always achieve the desired reduction in visibility or health impacts.
3. Health Impact Variability: The health effects of fog, particularly on respiratory and cardiovascular systems, can differ depending on the composition of pollutants present in the fog. While some fogs may contain high concentrations of harmful particulate matter (PM), others may not. This variability complicates the identification of appropriate mitigation measures, as interventions may need to be tailored to specific pollutant profiles.

**Table 11.2** Comparison of fog remediation technologies

| Technology | Effectiveness | Cost | Environmental impact | Suitability |
|---|---|---|---|---|
| **Aerosol dispersal systems** | High (immediate fog dispersion) | High | Potential air pollution (short term) | Urban areas, limited budget regions |
| **Fog netting** | Moderate (reduces fog concentration) | Moderate | Minimal (low impact) | Mountainous regions, low-budget areas |
| **Fog draining systems** | High (reduces fog buildup) | Moderate to high | Minimal (uses natural elements) | Coastal regions, high rainfall areas |
| **Climate engineering (geoengineering)** | High (modifies fog formation) | Very high | Potential long-term environmental risks | Large-scale applications, regions prone to persistent fog |
| **Public health campaigns** | Indirect (reduces exposure) | Low | Minimal (no direct environmental impact) | All regions, especially in high-risk populations |

4. Economic Constraints: While cost-effective measures are essential, the implementation of fog control strategies often requires upfront investments in infrastructure, technology, and maintenance. For low-income communities or developing regions, the financial burden can be a significant barrier to adoption. Moreover, ongoing funding for fog-related research and development remains limited, affecting progress in the field.
5. Public Awareness and Behavioral Change: Even with the introduction of effective fog mitigation measures, public awareness and behavior play a crucial role in their success. People must be educated about the risks of fog exposure and encouraged to adopt protective behaviors, such as wearing masks or limiting outdoor activities during fog events. Lack of awareness can limit the impact of any interventions.
6. Policy and Regulatory Challenges: In many regions, there is limited policy focus on fog-related health risks, which hinders the development and enforcement of regulations for fog control. Additionally, the lack of standardized guidelines for fog remediation means that many interventions are implemented on an ad hoc basis, further limiting their effectiveness (WHO 2021).

## 11.9 Future Directions and Research Opportunities

While there are notable challenges in fog remediation and control, significant opportunities exist to enhance the effectiveness of strategies aimed at reducing the health impacts and safety risks of fog. To address the limitations mentioned above and improve the overall outcomes, several future directions and recommendations are proposed:

1. *Development of Region-Specific Solutions*: Given the geographical and environmental variability in fog formation, future research should focus on developing region-specific fog control strategies. Localized studies on fog characteristics, pollution sources, and meteorological conditions can provide insights into tailored interventions (Ogwu et al. 2024). Collaborative efforts between climate scientists, public health experts, and environmental engineers are crucial for devising context-appropriate solutions.
2. *Advancement of Cost-Effective Technologies*: Continued research into cost-effective technological solutions is necessary. Innovations in fog dispersal technology, such as low-cost aerosol systems or passive fog dispersal mechanisms, could provide more affordable options for fog control, particularly in resource-constrained settings. Moreover, integrating fog control systems with other environmental management practices, such as air quality monitoring and pollution control, could enhance their overall impact.
3. *Integration of Public Health and Environmental Monitoring*: Establishing systems for real-time monitoring of fog conditions, air quality, and health outcomes can help guide timely interventions. Wearable health monitors, combined with

air quality data, could offer valuable insights into the health effects of fog and facilitate more targeted and effective mitigation measures. Such systems would also enable the evaluation of the effectiveness of control measures in real time, ensuring that interventions remain adaptive and responsive.

4. *Strengthening Policy and Regulatory Frameworks*: To effectively manage fog-related health risks, comprehensive policy and regulatory frameworks must be developed at local, national, and international levels. Governments should prioritize fog management as part of broader climate change and air quality policies. Standardized guidelines for fog prevention and control could assist in the implementation of evidence-based strategies, while public health initiatives can ensure that the most vulnerable populations are protected (WHO 2021).
5. *Public Awareness and Education*: Effective fog remediation goes beyond technological solutions—it requires active community engagement and awareness. Public health campaigns focused on educating populations about the risks of fog exposure and appropriate preventive actions are essential. Governments, NGOs, and local institutions should collaborate to disseminate information on safe practices during fog events, such as staying indoors, using masks, and limiting outdoor activities, particularly for sensitive groups such as children, the elderly, and individuals with pre-existing health conditions.
6. *Cross-Disciplinary Research and Collaboration*: Fog control and mitigation require the integration of knowledge across multiple disciplines, including atmospheric sciences, public health, environmental engineering, and urban planning. Future research should focus on interdisciplinary collaboration, fostering dialogue among experts in these fields to develop holistic approaches to fog management. This collaboration should also extend to policymakers, industry stakeholders, and local communities to ensure that solutions are practical, scalable, and sustainable.
7. *Investing in Long-Term Monitoring and Data Collection*: For a deeper understanding of the long-term impacts of fog exposure on human health and safety, consistent data collection and monitoring are vital. Investing in long-term studies will help build a comprehensive evidence base, facilitating the development of targeted mitigation strategies that account for seasonal variations, long-term health effects, and the evolving nature of fog-related hazards.

## 11.10 Conclusion

Fog, while a natural atmospheric phenomenon, poses significant health and safety risks, particularly when compounded by environmental pollutants and geographical factors. The complexities of fog formation and its variable impacts across different regions present unique challenges for effective mitigation. However, through innovative and cost-effective measures, as well as an integrated approach combining technology, policy, public health, and community engagement, the negative impacts of fog can be significantly reduced.

The key to successful fog remediation lies in the development of region-specific solutions that are adaptable to local conditions. Investment in low-cost, scalable technologies, along with the integration of real-time monitoring systems, will facilitate timely interventions and enable ongoing improvements in fog management. Furthermore, increasing public awareness of the risks of fog exposure and encouraging protective behaviors will empower individuals and communities to better respond to fog events.

Effective policy frameworks and cross-disciplinary collaboration are essential to ensuring the sustainability and success of fog mitigation efforts. Governments, research institutions, and local communities must work together to create a robust support system that prioritizes both immediate fog control measures and long-term public health protection.

In conclusion, while challenges remain, the future of fog mitigation holds promise. With continued research, technological advancement, and collaborative efforts, fog-related health risks can be minimized, paving the way for healthier, safer environments across the globe. The efforts made today will shape the resilience of communities in the face of fog, mitigating its harmful effects and contributing to a more sustainable future.

## References

Albornoz F, Escobar R, Carter V, Del Río C, Vásquez L. Fog water collection for local greenhouse vegetable production in the Atacama Desert. Sustainability. 2023;15(22):15720. https://doi.org/10.3390/su152215720.

Albornoz F, Carter V, Del Río C, Vilches I. Soilless production of lettuce (Lactuca sativa) in the Atacama Desert using fog water: water quality and produce mineral composition. Horticulturae. 2024;10(9):957. https://doi.org/10.3390/horticulturae10090957.

Brighente A, Conti M, Mohammadnia F. Physical layer security challenges and solutions for beyond 5G fog computing networks. Springer; 2022. p. 59–77. https://doi.org/10.1007/978-3-031-08254-2_4.

Dadwal SS, Dadwal D. Management of health and safety risks at large events. igi global. 2021:726–41. https://doi.org/10.4018/978-1-7998-8960-1.ch034.

Del Río C, Larraín H, Zanetta N, Rivera D, Siegmund A, Wolf N, Osses P, Garcia J-L, Lambert F, Lobos F, Cereceda P. ENSO influence on coastal fog-water yield in the Atacama Desert, Chile. Aerosol Air Qual Res. 2018;18(1):127–44. https://doi.org/10.4209/aaqr.2017.01.0022.

Ding L, Yu Y, An L, Li X, Dong S. Bionic surfaces for fog collection: a comprehensive review of natural organisms and bioinspired strategies. ACS Appl Bio Mater. 2023;6(12):5193–209. https://doi.org/10.1021/acsabm.3c00859.

Feld SI, Spencer BR, Bolton SM. Improved fog collection using turf reinforcement mats. J Sustain Water Built Environ. 2016;2(3) https://doi.org/10.1061/jswbay.0000811.

Fu H, Chen J. Formation, features and controlling strategies of severe haze-fog pollutions in China. Sci Total Environ. 2017;578:121–38.

Gultepe I, Tardif R, Michaelides SC, Cermak J, Bott A, Bendix J, et al. Fog research: a review of past achievements and future perspectives. Pure Appl Geophys. 2007;164:1121–59.

Jayanthi A, Vishnuvarthan R. A study on health and safety measures towards manufacturing industry. Int J Sci Res Eng Manag. 2024;08(06):1–5. https://doi.org/10.55041/ijsrem36230.

Knapp FW. Chemical fog dispersal with ground equipment for mosquito control. J Med Entomol. 1972;9(1):112–3. https://doi.org/10.1093/jmedent/9.1.112.

Kunkel BA, Weinstein AI, Silverman BA. An evaluation of some thermal fog dispersal experiments. J Appl Meteorol. 1974;13(6):666–75. https://doi.org/10.1175/1520-0450(1974)013.

Lakra K, Avishek K. A review on factors influencing fog formation, classification, forecasting, detection and impacts. Rendiconti Lincei Scienze Fisiche e Naturali. 2022;33(2):319–53.

Lee C, Said AO, Stoliarov SI. Passive mitigation of thermal runaway propagation in dense 18650 lithium ion cell assemblies. J Electrochem Soc. 2020;167(9):090524. https://doi.org/10.1149/1945-7111/ab8978.

Li M, Xie S, Tian G, Chen G, Guo Z. Biomimetic leaf-shaped wedge structure with mixed wettability for fog harvesting. ACS Appl Mater Interfaces. 2024;16(32):42931–41. https://doi.org/10.1021/acsami.4c08254.

Mcaleenan C, Oloke D. ICE manual of health and safety in construction. Thomas Telford. 2010; https://doi.org/10.1680/mohs.40564.

Mcgowan W. Fog Dispersal Technol. 1971; https://doi.org/10.2514/6.1971-1709.

Ogwu MC, Imarhiagbe O, Ikhajiagbe B, Osawaru ME. Toward understanding the impacts of air pollution. In: The handbook of environmental chemistry. Berlin/Heidelberg: Springer; 2024. https://doi.org/10.1007/698_2024_1107.

Qadir M, Jiménez G, Dodson L, Smakhtin V, Farnum R. Fog water collection: challenges beyond technology. Water. 2018;10(4):372. https://doi.org/10.3390/w10040372.

Salvador D. Welding safety and health: occupational hazards and risk mitigation. Int J Adv Res Sci Commun Technol. 2023:993–7. https://doi.org/10.48175/ijarsct-11904.

Veldhuizen H, Ledbetter J. Cooling tower fog: control and abatement. J Air Pollut Control Assoc. 1971;21(1):21–4. https://doi.org/10.1080/00022470.1971.10469490.

Ward-Caviness CK, Danesh Yazdi M, Moyer J, Weaver AM, Cascio WE, Di Q, et al. Long-term exposure to particulate air pollution is associated with 30-day readmissions and hospital visits among patients with heart failure. J Am Heart Assoc. 2021;10(10):e019430.

WHO global air quality guidelines: particulate matter (PM2.5 and PM10), ozone, nitrogen dioxide, sulfur dioxide and carbon monoxide [internet]. Geneva: World Health Organization; 2021. 6, Implementation of the guidelines. Available from: https://www.ncbi.nlm.nih.gov/books/NBK574587/

Wo Z, Su Y, Ma H, Shi C, Sun H, Zhang X. A nc-titania modified cellulose Janus membrane for unidirectional water penetration and fog collection. J Alloys Compd. 2023;968:172207. https://doi.org/10.1016/j.jallcom.2023.172207.

Zhu S, Liu Y, Bai T, Shi X, Li D, Feng L. High-efficient and robust fog collection through topography modulation. Surf Coat Technol. 2023;468:129747. https://doi.org/10.1016/j.surfcoat.2023.129747.

# Chapter 12
# Policy Implications and Mitigation Strategies for Combating Fog Problem

**Rimsha Arshad and Salman Tariq**

## 12.1 Introduction

Despite substantial reductions in air pollutant concentrations in the developed nations of the Western Hemisphere due to pollution restrictions, levels are anticipated to rise in Asia. Rapid development, insufficient monitoring, and regulation of pollutants, coupled with specific natural and climatic phenomena such as Asian brown haze and extended periods of winter fog, render the atmospheric chemistry of South Asia exceedingly complex. Data regarding airborne quantities of harmful chemicals in developed nations is rather plentiful. Nevertheless, data availability is significantly diminished in less-developed nations and those characterized by largely arid climates. Air pollution in South Asian countries escalates due to rising populations, economic expansion, and urbanization.

Since the late 1990s, many towns in eastern India and northeastern Pakistan have been subjected to severe fog for several months during winter (Hameed et al. 2000). In the atmospheric boundary layer, fog is a physical phenomenon that is essential to atmospheric and environmental processes and has a significant impact on human life. The formation affects radiative transfer in the atmosphere by enhancing solar radiation reflection at the fog layer's surface and concurrently diminishing radiative cooling at the surface level. This modification in the energy equilibrium influences numerous ecosystem functions. The attenuation of solar radiation caused by fog restricts the light available to vegetation, hence affecting photosynthesis and other biological functions. Furthermore, the emergence of liquid water droplets on plant surfaces during fog events can promote the deposition of airborne nutrients and pollutants, which are either absorbed by plants or deposited into the soil, thereby

R. Arshad · S. Tariq (✉)
Department of Space Science, University of the Punjab, Lahore, Pakistan
e-mail: salman.spsc@pu.edu.pk



P. Saxena et al. (eds.), *Fog: Formation, Chemistry, Dynamics, Impact and Dissipation*,
https://doi.org/10.1007/978-981-96-4422-3_12

modifying nutrient cycles and potentially resulting in ecosystem imbalances (Isil et al. 2022; Saikh et al. 2024; Saikh and Das 2023; Weathers et al. 2020).

Fog can be classified into two groups, namely radiation fog and scattering fog. The radiation fog is associated with the cooling of the ground through radiation. It manifests when the air is cold enough and reaches saturation. This fog typically manifests at night and throughout the evening hours. Particle sizes exhibit slight variation around 4 μm, while the liquid water content ranges from 0.01 to 0.1 $g/m^3$. The scattering fog is generated by the dynamics of humid and warm air masses over cooler marine or terrestrial surfaces. It is defined by a liquid water concentration exceeding 0.2 $g/m^3$ and a particle diameter of 20 μm (Gebhart et al. 2005). In northern India, the majority of fog generation is attributed to radiative cooling; nevertheless, advection fog has also been documented (Dutta and Chaudhuri 2015; Sawaisarje et al. 2014). Radiation fog typically develops near the ground under skies with no clouds in stagnant air associated with an anticyclone (Li et al. 2023). The primary mechanism is radiative cooling; however, the counteracting effects of upward soil heat flux, along with warming influences (Ba et al. 2024) and moisture losses due to dew deposition and turbulent mixing in the stable boundary layer, significantly influence the probability and timing of radiation fog formation (Roach et al. 1976; Syed et al. 2012). Bushahab et al. (2009) investigated the impact of soil moisture and vegetation cover on soil thermal conductivity, which contributes to the development of radiation fog.

Radiation fog is more probable under conditions of low wind speed, high relative humidity, and low surface temperature, resulting in elevated static stability (Liu et al. 2011; Liu et al. 2012; Niu et al. 2010; Sachweh and Koepke 1995). Under these conditions, the mixing of air with the open troposphere is restricted due to the presence of a trapping boundary-layer inversion. Such conditions ultimately lead to diminished visibility and elevated pollution concentrations in metropolitan areas, as pollutants are unable to disperse (Tiwari et al. 2011). The impact of urbanization and human activities on fog formation is highly intricate.

Singh (2014) and Collier (1970) noted that fog is more prevalent in urban areas than at airports, which are often situated distant from urban air pollution. Yan et al. (2020) also noted that urban regions may have a twofold increase in winter fog relative to neighboring rural areas. In urban areas, the prevalence of persistent fog is ascribed to the effects of air pollution, which introduces a significant number of additional nuclei, leading to an augmented quantity of smaller fog droplets (WHO 2024). The polluted cloud can contain eight times as many droplets of half the size, resulting in twice the surface area and twice the optical depth, thereby causing greater obscurity compared to natural clouds (Toon et al. 2000). Additionally, the creation of fog droplets under subsaturated circumstances may transpire in contaminated air. In locations where aerosol concentrations have diminished, dense fog has also dissipated (Igawa et al. 2017; LaDochy 2005). Sachweh and Koepke (1995) indicated that heightened urban construction density correlates with a decrease in the average number of fog days, which they attributed to the urban heat island effect and moisture deficit. The thin and fragmented fog cover over the urban area may disintegrate fog due to radiative heating of the surface following sunrise. Soil characteristics significantly influence the production of radiation fog. Clay and silt loams, prevalent soils in the plains, possess a greater capacity for moisture retention than sandy soils due to their increased porosity and reduced thermal conductivity (Abu-Hamdeh 2001; Rosenberg et al. 1983).

This restricts latent heat transfer to the surface, permitting radiative cooling to prevail and facilitate the development of surface inversions and fog. Fog substantially diminishes horizontal visibility, posing considerable issues for transportation systems. Reduced visibility due to fog poses significant risks for aviation, road transportation, and sea navigation, frequently resulting in accidents and disruptions (Abdel-Aty et al. 2011; Du et al. 2021; Pagowski et al. 2004; WMO 2020). The economic ramifications of fog-induced transit delays are significant, as these interruptions can influence trade, logistics, and daily commutes. The overall economic detriment resulting from fog events can be comparable to that of tornadoes or, in certain instances, winter storms and hurricanes (Syed et al. 2012; Yao et al. 2020).

Notwithstanding these limitations, fog may also have significant advantages under specific environmental conditions. In semi-arid and subtropical areas, where freshwater resources are limited, fog collection systems have been established to capture moisture from fog for human consumption and agricultural purposes. Fog nets offer a sustainable water source by trapping minute water droplets present in fog, rendering them a significant resource in regions with scarce rainfall (Ali 2024; Domen et al. 2014; Klemm and Lin 2016).

It can be concluded that fog's varied effects on radiation, ecosystems, and human endeavors highlight its significance in atmospheric science and environmental studies. Comprehending and regulating fog-related phenomena is essential for alleviating its adverse impacts, especially in transportation and air quality, as well as for exploiting its potential advantages in arid areas.

There is extensive evidence in scientific literature demonstrating that both the frequency and intensity of fog events have undergone significant changes over time. Numerous studies across various geographic locations indicate a general decline in fog occurrence, with the majority of monitoring stations worldwide reporting a decrease in fog events. This trend is often attributed to regional climatic changes, such as the urban heat island effect and alterations in dominant regional circulation patterns, which contribute to rising air temperatures and, consequently, a reduction in relative humidity. As fog formation depends critically on high humidity levels, this reduction in relative humidity, induced by temperature increases, makes it less likely for fog to form under typical atmospheric conditions (Dione et al. n.d.; Guo et al. 2020; Wang et al. 2021; Zhang et al. 2023).

The influence of changes in cloud condensation nuclei (CCN) concentrations on fog trends has also been suggested as another contributing factor. CCN are tiny particles in the atmosphere that facilitate cloud droplet formation, and their presence plays a role in fog formation as well. However, the precise physical mechanisms linking CCN concentrations to changes in fog patterns have not been thoroughly explored in the literature. The central question remains whether a decrease in CCN concentrations could lead to a corresponding decrease in fog occurrence, and if so, how significant this effect might be compared to the impact of temperature increases.

Understanding the relative influence of CCN concentrations and temperature changes on fog formation is critical, and this study aims to address these questions using deliquescence and Mie scattering calculations across different air masses. By systematically modulating CCN concentrations and temperature in these models, the study seeks to quantify their effects on visibility and fog formation. The findings will contribute to ongoing discussions on whether global climate change could lead to a global decrease in fog events (Li et al. 2023).

In Pakistan, winter precipitation predominantly arises from eastward-moving midlatitude cyclones originating in the Mediterranean region. These systems are frequently identified as closed cyclonic circulatory systems at sea level and are commonly referred to as Western Disturbances (WDs). Precipitation is augmented in the northern region due to obstruction and orographic conditions created by the high mountain ranges in this area. Western disturbances also induce precipitation in the northern plains of Pakistan. Upon rainfall, a substantial quantity of water vapor is introduced into the lower atmosphere through the evaporation of precipitation. Upon the passage of a low-pressure system, the advancing edge of a high-pressure system generally establishes clear and statically stable air conditions, resulting in the development of robust surface-based inversions. This increases surface-relative humidity and promotes fog formation.

Lahore (31°35'N, 74°21'E; Fig. 12.1), with a population of around ten million, is the second largest city in Pakistan (Husain et al. 2007). Hameed et al. (2000) and

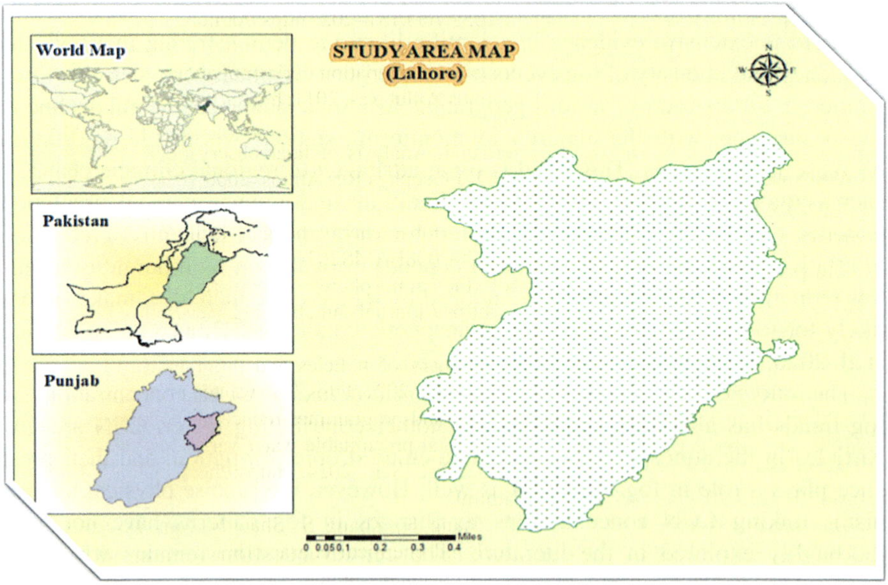


**Fig. 12.1** Study area map of Lahore. (Source: author's drawing)

Rattigan et al. (2002) quantified sulfate ($SO_4^{2-}$), nitrate ($NO_3$), sulfur dioxide ($SO_2$), and specific trace components in air samples obtained after a significant pollution event in Lahore. The World Health Organization (WHO) classified Lahore's concentrations of total suspended particles as among the highest globally during a global monitoring program (Smith et al. 1996). Among the 63,000 yearly fatalities in Lahore, approximately 1250 are attributable to air pollution, which contributes to numerous chronic illnesses, particularly asthma.

Several studies have been conducted to understand the fog problem in Lahore and its implications. For instance, Majeed et al. (2023) investigated the phenomenon of smog in Lahore, often referred to as the "fifth season," due to its annual recurrence and severe socioeconomic disruptions. Their research utilized ground-based and satellite observations to analyze the increase in air pollution sources, such as crop residue burning and vehicular emissions, which contribute to the intensity and frequency of smog events (Majeed et al. 2023).

Another study by Jabeen et al. (2021a, b) assessed the health impacts of winter smog on various occupational groups in Lahore. The research employed a questionnaire survey and multivariate logistic regression models to examine the effects of pollution on individuals working in both outdoor and indoor environments. The findings highlighted the significant health risks posed by smog, particularly for middle-aged individuals with a history of respiratory ailments (Jabeen et al. 2021a).

In Bangladesh, India, and Pakistan, 82 locations—the majority of which were in cities—were examined for the presence of fog between 1976 and 2010 (Syed et al. 2012). Notable fog patterns were predominantly observed in a band extending from Pakistan to India to Bangladesh, nearly parallel to the Himalayan mountains. The trends in this region are favorable, characterized by an increase in fog. The authors propose potential causes including land-use alterations (e.g., irrigation), global changes in temperature, humidity, and regional circulation patterns, as well as variations in atmospheric aerosol concentrations. Yasmeen et al. (2012) investigated the impact of aerosols on winter fog in Pakistan, focusing on Punjab and Sindh. They found that fog frequency increased from 2000 to 2010, particularly in dense fog-prone areas like Sialkot and Lahore, with visibility dropping significantly. The study highlighted that higher aerosol concentrations, especially dust and biomass burning, contributed to enhanced fog formation, posing environmental and health risks.

The primary causes are the elevated concentration of pollution sources, such as numerous small industrial operations, and the absence of effective environmental regulations. To comprehensively elucidate the origins of the contaminants and the underlying chemical processes, the study undertook exhaustive research from the winter season 2010 to 2024. The main aims of this study were: (1) to chemically describe aerosols during fog events and in the absence of fog, (2) to examine the influence of fog on the chemical composition of ambient air, and (3) to identify the principal sources of pollutants.

The current study provides a significant contribution to the literature by addressing critical gaps in the existing research on fog formation, its impacts, and mitigation strategies. While earlier studies, such as Lakra and Avishek (2022) have offered in-depth insights into the complexity of fog forecasting due to multilevel factors, there has been a clear absence of comprehensive frameworks that translate this understanding into actionable mitigation policies. Our study bridges this gap by moving beyond the theoretical discussion of fog factors to propose targeted strategies that involve policy design, community engagement, and the integration of innovative technologies to minimize the adverse socioeconomic and environmental impacts of fog, particularly in heavily affected regions like Lahore.

Moreover, while Saraswati (2022) highlights the environmental and societal risks posed by fog, it underscores the need for detailed characterization of fog's impacts but stops short of outlining specific strategies to mitigate these risks. Our study fills this void by proposing a holistic approach that not only addresses the environmental degradation caused by fog but also offers practical solutions to safeguard public health and the economy. Additionally, Gultepe et al. (2007) provide valuable insights into the scientific understanding of fog but calls for future research that moves beyond exploration and toward the development of actionable perspectives. This study answers that call by presenting both long-term and short-term policy interventions, thus advancing the existing literature from a theoretical understanding of fog to practical application in mitigation efforts.

This research fills several critical gaps in the existing fog-related literature. Firstly, while earlier works have focused extensively on the formation, classification, and forecasting of fog, they have not sufficiently explored the socioeconomic and policy-oriented dimensions of fog mitigation. The current study addresses this gap by providing a detailed analysis of the economic, health, and social challenges posed by fog in Lahore and suggesting comprehensive mitigation strategies. These strategies encompass policy reform, technological innovation, and community-based actions, providing a broader scope that ensures the issue is tackled from multiple angles. Secondly, much of the prior research, particularly the works reviewed, emphasizes the environmental risks associated with fog but does not delve into the localized impacts or region-specific challenges, such as those experienced in urban centers like Lahore. This study narrows this gap by focusing specifically on the regional context, offering location-specific insights into how fog affects air quality, economic productivity, transportation, and public health.

Additionally, this research advances the existing literature by integrating both scientific and societal approaches. While past studies have largely remained within the domain of environmental science or meteorology, this study interlinks environmental challenges with policy, economics, and health, advocating for interdisciplinary approaches in fog mitigation. The introduction of modern forecasting technologies, public awareness campaigns, and the emphasis on regulatory reform demonstrate a progressive step toward not only understanding but also combatting fog-induced challenges, thus filling an essential gap in the practical application of fog-related research. Through its regional focus and multifaceted policy suggestions, this study expands the scope of fog research and provides a blueprint for future studies that seek to combine scientific insights with real-world solutions.

## 12.2 Study Area

Lahore, the capital of the Punjab province in Pakistan, is geographically positioned at approximately 31.5497°N latitude and 74.3436°E longitude. As a significant cultural and economic center, Lahore boasts a population that benefits from its strategic location and historical significance. The city's climate is predominantly characterized by extreme seasonal variations. During summer months, temperatures often surpass 40 °C (104 °F), whereas winter months bring a significant drop in temperature, sometimes nearing the freezing point. The monsoon season, typically spanning from July to September, results in substantial rainfall, contributing to the city's greenery and agricultural productivity.

The meteorological conditions in Lahore exhibit a combination of high temperatures and elevated humidity levels, particularly during the summer season. The geographical and topographical features of the city greatly influence its weather patterns, leading to diverse climatic conditions throughout the year. Fog is a frequent phenomenon during the winter months, posing challenges to visibility and transportation. The interaction of fog with aerosols and particulate matter prevalent in the urban atmosphere exacerbates the situation, leading to significant implications for air quality and public health. An in-depth understanding of these meteorological conditions is essential for devising effective fog mitigation strategies and enhancing the quality of life for Lahore's residents.

## 12.3 Data, Empirical Models, and Techniques

At present, there exists only one study on Lahore using AERONET data conducted by Alam et al. in which aerosol optical and radiative properties have been discussed during only the summer and winter seasons of 2010–2011. These authors have studied the aerosol optical and radiative properties by analyzing the measurements (level 1.5) for only 6 months (April–June, 2010 and December 2010 to February 2011). However, in terms of climatic studies, we need an improved understanding of aerosol properties and radiative forcing based on long-term observations that should include variations in all four seasons.

The main objective of the present work is to study the aerosol properties by analyzing the AERONET measurements over the megacity of Lahore for a relatively longer period (~14 years) by considering variations in just winter seasons. In this study, we have used all the available AERONET data (level 1.5) for the winter season (DJF) during 2010–2024 to analyze the aerosol optical depth (AOD), Angstrom exponent (440/870) (AE), and precipitable water. These datasets are used in this study to examine the effect of aerosol on winter fog formation during the period 2010–2024 in Pakistan (Lahore).

The datasets containing seasonal means of meteorological variables, such as relative humidity and air temperature at the 850 mb analysis level, have been obtained from the Physical Sciences Laboratory (https://psl.noaa.gov/).

## 12.4 Results and Discussion

Fog is an immense problem in Pakistan during the winter months, especially in the eastern region, notably Punjab (PMD 2024). This phenomenon occurs more frequently under relatively calm meteorological conditions. Syed Faisal Saeed and Younas categorize the fog in Punjab into two primary types: radiation fog and frontal fog. The research locations are located in the eastern region of Pakistan, characterized by a temperate–moist environment. In winter, the average temperature in this location fluctuates between 5 and 15 °C.

Table 12.1 indicates a statistically significant negative connection (−0.528) between aerosol optical depth (AOD) and Angstrom exponent (AE) ($p$ = 0.042). This suggests that a rise in AOD correlates with a decrease in AE, indicating that elevated aerosol concentrations are linked to bigger particle sizes. The relationship between AOD and temperature is minimal and lacks statistical significance (−0.028, $p$ = 0.919). This indicates that there is no significant linear correlation between AOD and temperature in this dataset. Gautam et al. (2022) evaluate the influence of aerosol optical depth (AOD) on meteorological factors such as temperature and humidity, offering insights into analogous correlations seen in the current dataset.

A robust negative correlation (−0.766) is present between temperature and relative humidity, which is statistically significant ($p$ = 0.000). This signifies that elevated temperatures correlate with diminished relative humidity, a prevalent inverse association noted in meteorological research. A small positive connection (0.282) exists between relative humidity and AE, however it is not statistically significant ($p$ = 0.308). This is a minor correlation between increased relative humidity and reduced particle sizes. A moderate positive correlation (0.541) exists between AOD and precipitable water, which is statistically significant ($p$ = 0.037). This indicates

**Table 12.1** Correlation coefficient (R) among each variable

| Variables | AOD | AE | Temperature (°C) | Relative humidity | Precipitable water |
|---|---|---|---|---|---|
| AOD | 1.000 | | | | |
| AE | −0.528 (0.042) | 1.000 | | | |
| Temperature (°C) | −0.028 (0.919) | −0.382 (0.159) | 1.000 | | |
| Relative humidity | 0.124 (0.659) | 0.282 (0.308) | −0.766 (0.000) | 1.000 | |
| Precipitable water | 0.541 (0.037) | −0.315 (0.252) | 0.433 (0.106) | −0.231 (0.406) | 1.000 |

*Note*: *P*-values are shown in brackets to describe whether the correlation is significant

**Table 12.2** Equations of the linear fit of AOD and AE with meteorological parameters, where y denotes the AOD and AE and x represents the value of the corresponding meteorological parameter

| Dependent variables | Temperature (°C) | Relative humidity | Precipitable water |
|---|---|---|---|
| AOD | $y = -0.003x + 0.699$ | $y = 0.006x + 0.261$ | $y = 0.579x + 0.683$ |
| AE | $y = -0.035x + 1.516$ | $y = 0.011x + 0.421$ | $y = -0.390x + 1.555$ |

that elevated AOD values correlate with increased precipitable water levels, suggesting a relationship between aerosol concentrations and atmospheric moisture content. Bharali et al. (2024) examine the relationship between aerosols and atmospheric moisture, particularly precipitable water, and its impact on fog formation and visibility, consistent with the positive association between aerosol optical depth and precipitable water in this study's data. The relationship between precipitable water and AE is weakly negative (−0.315) and not statistically significant ($p = 0.252$), indicating a little inclination for increased moisture levels to correlate with larger particle sizes.

In Table 12.2, the linear equation shows a weak negative relationship between AOD and temperature, as indicated by the small slope value of −0.003. This means that for each unit increase in AOD, the temperature decreases by a very small amount (0.003 °C). The intercept (0.699) tells us that when AOD is zero, the expected temperature is 0.699 °C. However, given the small magnitude of the slope, this relationship suggests that AOD has a minimal direct effect on temperature. This weak influence could be because temperature is primarily influenced by other meteorological factors like solar radiation and atmospheric circulation, with AOD playing a secondary role.

There is a positive relationship between AOD and relative humidity, with a slope of 0.006. For each unit increase in AOD, relative humidity increases by 0.006, indicating a very modest impact of aerosols on the humidity levels. The intercept (0.261) shows the expected relative humidity when AOD is zero. Aerosols, particularly hygroscopic particles, can attract water vapor, which may explain this positive relationship. Higher aerosol concentrations can provide more condensation nuclei, leading to increased moisture retention and higher humidity.

Moreover, the next equation reveals a much stronger positive relationship between AOD and precipitable water, with a slope of 0.579. For each unit increase in AOD, precipitable water increases by 0.579 units. The intercept (0.683) indicates that when AOD is zero, precipitable water is expected to be 0.683 units. This stronger correlation suggests that higher concentrations of aerosols, especially those acting as cloud condensation nuclei, are associated with increased moisture content in the atmosphere. The relationship is consistent with the idea that aerosols contribute to cloud formation and water vapor retention.

The relationship between AE and temperature is negative, with a slope of −0.035 (see Table 12.2). This indicates that for each unit increase in AE, the temperature decreases by 0.035 °C. The intercept (1.516) suggests that the temperature would be 1.516 °C when AE is zero. A negative relationship between AE and temperature could reflect the influence of smaller aerosol particles (associated with higher AE values) in scattering sunlight, thus reducing surface temperatures. Smaller particles tend to have a higher scattering efficiency, which can lower the temperature.

There is a small positive relationship between AE and relative humidity, with a slope of 0.011. This means that for each unit increase in AE, relative humidity increases by 0.011 units. The intercept (0.421) shows the expected relative humidity when AE is zero. The positive correlation indicates that higher AE values, which suggest smaller particles, may be associated with increased humidity. This could be because smaller particles, such as sulfates, are more hygroscopic and capable of attracting water vapor.

The last equation shows a relatively strong negative relationship between AE and precipitable water, with a slope of −0.390. For each unit increase in AE, precipitable water decreases by 0.390 units. The intercept (1.555) represents the precipitable water value when AE is zero. This negative correlation might suggest that smaller aerosols (higher AE values) do not support the retention of large amounts of water vapor as effectively as larger particles, possibly due to their different condensation properties.

Lahore, a prominent city in Punjab, is a significant field of research. This region is recognized for its agricultural productivity, facilitated by ample moisture from the Ravi River and a comprehensive irrigation canal system. Furthermore, western disturbances, which are meteorological systems starting from the west, augment the moisture in the region as they progress from west to east.

The development of fog in Punjab is affected by multiple elements, including the cooling of near-surface air, elevated humidity levels, and stable atmospheric conditions. These conditions are prevalent during the winter months, resulting in the frequent manifestation of fog. Aerosols and particulate debris in the atmosphere contribute to fog production by offering surfaces for water vapor condensation. Fog substantially influences daily life in Punjab, impacting transportation, vision, and general safety. It may result in vehicular collisions, interruptions in air travel, and disturbances in daily routines. Comprehending the elements that lead to fog development and its effects is essential for devising effective mitigation techniques.

Dejmal and Repal (2010) examined the optimal conditions for the formation of radiation fog, which encompass gentle winds, clear skies, elevated humidity, and minimal turbulence. The mean seasonal temperature and mean relative humidity (RH) have also been calculated from 2010 to 2024 for the DJF season during the study. The average temperature and relative humidity generally range between 6.3 °C and 9.6 °C and 69.8% and 79.3%, respectively, during the above-mentioned period. December and January are the coldest months, characterized by increased fog formation, particularly in eastern Punjab (Muslehuddin et al. 2004). The low temperature and high relative humidity result in a shallow occurrence of fog in the current study area, whereas the smallest fog frequency is attributed to elevated temperatures and reduced relative humidity (Yasmeen et al. 2012).

The temperature and relative humidity plots for Lahore from 2010 to 2024 in Figs. 12.2 and 12.3 provide critical insights into the atmospheric conditions contributing to winter fog formation. The temperature chart shows notable fluctuations, with higher temperatures in winters like 2017, 2019, and 2023, while winters such

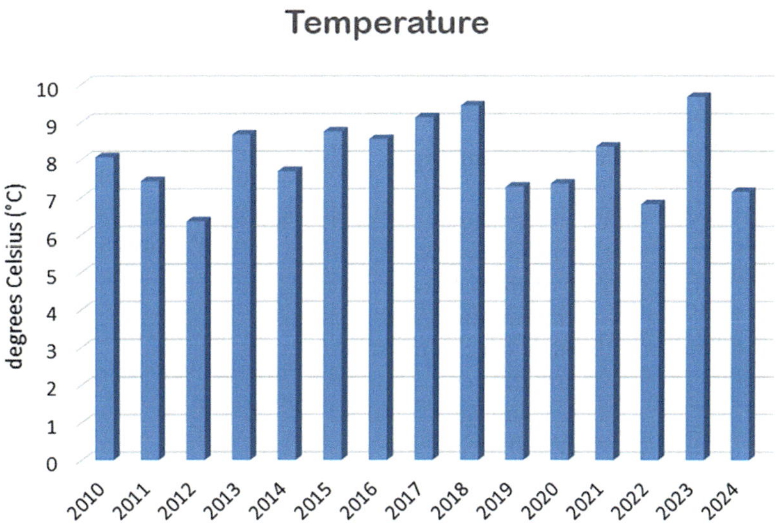


**Fig. 12.2** The trend of air temperature (°C) during the winter season of 2010–2024

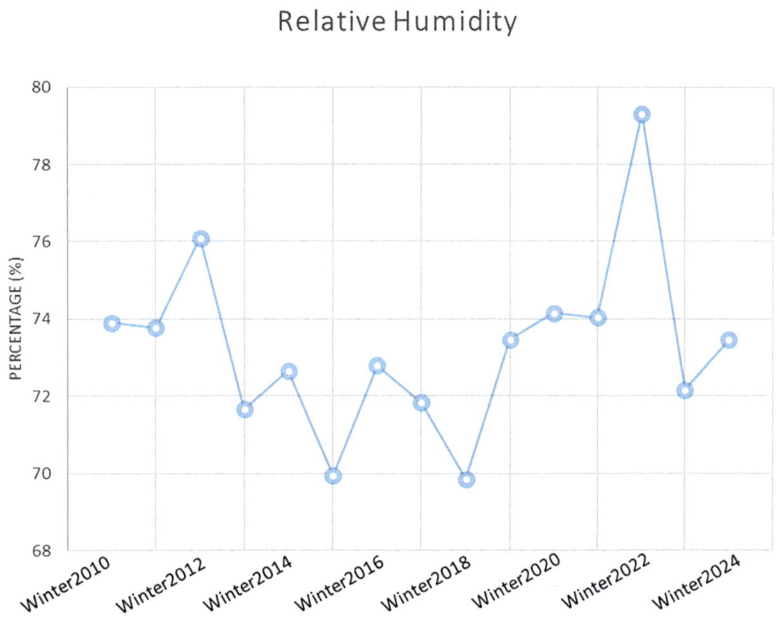


**Fig. 12.3** Variation in relative humidity (%) from winter 2010 to winter 2024

as 2014, 2016, and 2020 recorded comparatively lower temperatures. The relative humidity plot, on the other hand, exhibits more consistent behavior, but with significant peaks, especially in 2022, where the relative humidity reached approximately

80%, followed by a sharp decline in 2024. These variations in both temperature and humidity have direct implications for fog formation.

Higher temperatures tend to reduce the formation of fog by lowering relative humidity levels, while cooler temperatures, especially when coupled with higher humidity levels, create favorable conditions for fog. Winters with low temperatures and high humidity, like 2020 and 2022, correspond to seasons where dense fog events were likely more frequent, as fog formation requires high moisture content and cooler air temperatures to condense water vapor into fog droplets. Aslam (2012) examined fog hazards in Punjab utilizing 30 years of meteorological data from ten weather stations. Fog episodes were linked to substantial alterations in temperature and moisture, with average relative humidity above 70% during these periods.

The spike in relative humidity in 2022, along with relatively mild temperatures, suggests a season of intense fog formation, possibly exacerbated by local pollution sources and prevailing calm winds, typical for Lahore winters. In contrast, winters with lower humidity, like 2016 and 2024, would have experienced fewer and less intense fog events, as the reduced moisture content in the air limits the potential for condensation.

The overall trend shows a general correlation between lower temperatures and higher humidity, aligning with the classical fog formation mechanism. However, global factors like increasing urbanization, air pollution (such as aerosols), and changing weather patterns due to climate change could be influencing these trends, contributing to the variability in fog events. For instance, aerosol loading in Lahore, influenced by anthropogenic activities, has been shown to affect fog formation by serving as cloud condensation nuclei, further intensifying fog under appropriate conditions.

The data in Fig. 12.4 show a general upward trend in AOD values over the years, peaking in 2024. This suggests increasing aerosol concentrations, which align with increased industrial activities, vehicular emissions, and seasonal biomass burning in and around Lahore. Yasmeen et al. (2012) examined the influence of aerosols on winter fog in Pakistan, emphasizing the contribution of dust and biomass-burning aerosols to the intensification of fog formation.

Between 2010 and 2024, Lahore, Pakistan, saw significant fluctuations in aerosol optical depth (AOD) during the winter seasons, revealing critical insights into the region's fog patterns and air quality. Analyzing AOD data across this period provides a comprehensive view of the atmospheric behavior, influenced by both natural and anthropogenic factors. In Winter 2010, the AOD was notably low (~0.40), representing a period with fewer aerosols in the atmosphere. This likely corresponded to relatively clearer winter skies with less severe fog events. In Winter 2019, AOD remained high at ~0.75, similar to 2018, indicating persistent aerosol pollution. The severity of fog events during this season was a continuation of the trends observed in the previous years, with the combined effects of industrial activity, vehicular emissions, and transboundary pollution. Lahore faced some of the worst smog conditions, and public health suffered as a result of the poor air quality.

In Winter 2020, the AOD decreased marginally to approximately 0.60 relative to the preceding years (2018–2019), signifying a transient reduction in aerosol

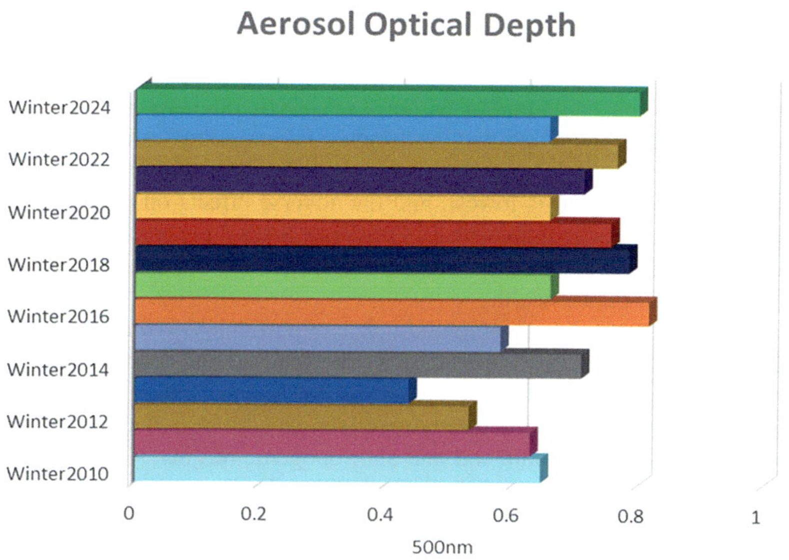


**Fig. 12.4** Winter season aerosol optical depth (500 nm) variations in Lahore over the period 2010–2024

concentrations. This reduction was primarily due to the statewide lockdowns and diminished human activities during the initial period of the pandemic. Industrial production diminished, automobile traffic significantly down, and building activities were restricted, resulting in a drop in particulate matter and other pollutants that usually contribute to the aerosol load. This reduction in AOD resulted in slightly improved air quality and less catastrophic fog conditions compared to previous years when industrial emissions were at their highest.

Nevertheless, despite the temporary decline in aerosol concentrations, by Winter 2021, the AOD increased once more to approximately 0.78, marking one of the highest levels seen during the research period. The abrupt increase in aerosol levels is probably attributable to the relaxation of COVID-19 limits, resulting in a revival of industrial operations, heightened emissions from transport, and a reversion to standard urban and agricultural activities. Moreover, although local airborne pollutants were temporarily reduced during the lockdown, cross-border emissions from agricultural residue combustion and other regional sources presumably continued, exacerbating the aerosol burden during this time. Furthermore, in the latter stages of the pandemic, when industries endeavored to recuperate losses, there may have been an escalation of emissions, especially from manufacturing sectors.

By Winter 2022, the AOD persisted at approximately 0.70, indicating sustained industrial and automobile emissions as the entire globe, including Pakistan, progressed toward post-pandemic recovery. Although transitory enhancements in air quality were noted during the first lockdown period, the subsequent rebound of aerosols post-lockdown demonstrates the profound impact of human activities, particularly in urbanized and industrial areas such as Lahore, on aerosol levels and the quality of the atmosphere.

Winter 2024 marked the highest AOD value observed during the study period, reaching ~0.85. This year some of the most severe fog and smog conditions were recorded in Lahore. The combination of industrial emissions, vehicular pollution, and transboundary aerosols created a perfect storm for extreme fog events. Visibility during this season was at its lowest, and the adverse effects on public health and transportation were particularly severe. The dense aerosol presence contributed to a prolonged period of poor air quality, with significant implications for the region's environmental and public health.

Bulbul et al. (2018) examined the influence of particulate matter on fog production in Islamabad, Pakistan, during the winter fog season. PM10 concentrations were observed, ranging from 123 μg $m^{-3}$ to 202 μg $m^{-3}$, with a mean value of 177 μg $m^{-3}$. The SEM-EDS analysis identified many metals, whereas morphological examinations revealed both soil and anthropogenic origins. A substantial association was identified between increased AOD levels and heightened PM10 concentrations on foggy days, as evidenced by sun photometers and MODIS satellite data.

Ali et al. (2020) observed a rising trend in AOD across multiple regions in Pakistan, with an annual increment between 0.0002 and 0.0047 $years^{-1}$. The rise was notably pronounced during the winter season, underscoring the influence of seasonal fluctuations on aerosol concentrations. Kabanov et al. (2020) conducted a comparative analysis of quasi-synchronous AOD data in adjacent regions from 2011 to 2017. Their findings revealed substantial seasonal fluctuations in AOD, with elevated values recorded in winter months attributed to intensified anthropogenic activity and stable atmospheric conditions.

Saxena et al. (2017) studied the seasonal effects on the variation of water-soluble inorganic species (WSIS) in PM10 and PM2.5 in Delhi, India. Mean concentrations of total WSIS in PM10 were 233.0 ± 124.6 μg $m^{-3}$ and in PM2.5 were 108.0 ± 86.5 μg $m^{-3}$. The study found higher concentrations of WSIS during the winter season (140.11 ± 90.67 μg $m^{-3}$ for PM10 and 74.41 ± 47.55 μg $m^{-3}$ for PM2.5) compared to other seasons.

Mkoma et al. (2009) conducted a study on the chemical characteristics of ambient daily PM10 aerosol samples collected at two sites in Tanzania: Morogoro and Dar es Salaam. During the wet season, the average PM10 mass concentration was found to be 27 ± 11 μg/$m^3$ in Morogoro and 51 ± 21 μg/$m^3$ in Dar es Salaam. Higher concentrations of organic carbon (OC), elemental carbon (EC), and water-soluble organic carbon (WSOC) were observed in Dar es Salaam compared to Morogoro. Significant species like black carbon, $NH_4^+$, non-sea-salt $SO_4^{2-}$, K, and Ni were predominantly present in the fine fraction at both sites. The PM10 mass in Morogoro comprised 48% organic matter (OM), 44% crustal matter, 4% sea salt, and 2% EC, while in Dar es Salaam, it was composed of 37% OM, 32% crustal matter, 9% sea salt, and 9% EC. Biomass burning and traffic were identified as major sources of OM in Dar es Salaam, with soil dust resuspension contributing significantly in Morogoro.

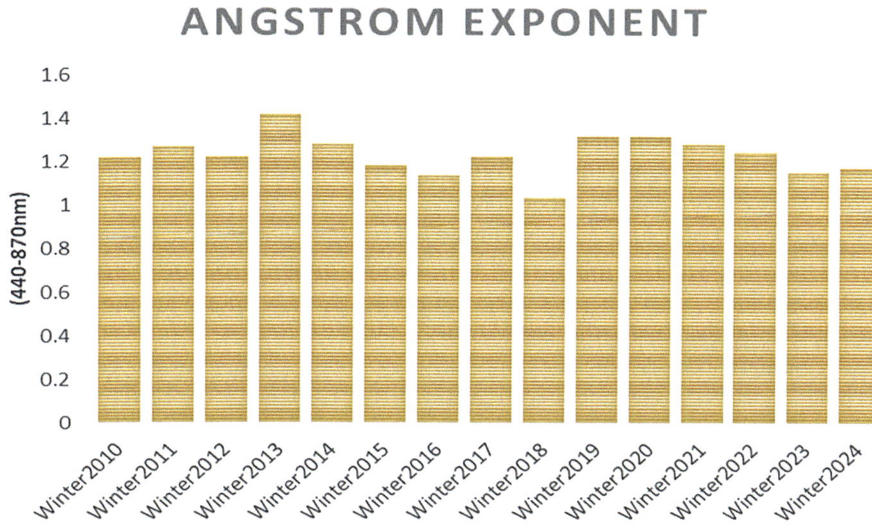


**Fig. 12.5** Angstrom exponent (400–870 nm) trends over Lahore during the winter seasons from 2010 to 2024

The Angstrom exponent (AE) plot in Fig. 12.5 for Lahore, Pakistan, spanning Winter 2010 to Winter 2024, offers critical insights into the features and variability of aerosol particle sizes during this timeframe. The AE is a vital metric in aerosol research, as it differentiates between fine particles (such as those from biomass combustion or pollution) and coarse particles (such as dust). Elevated AE values often signify a prevalence of tiny aerosols, whereas diminished AE values imply the existence of bigger particles such as dust.

During the study period, the AE values oscillate between around 0.9 and 1.5, signifying significant fluctuation in the types and sources of aerosols impacting Lahore during the winter seasons. The data indicates that Winter 2014 and Winter 2012 exhibit the highest AE values, approximately 1.5, suggesting these seasons were marked by an elevated concentration of fine-mode aerosols, perhaps attributable to urban pollution, biomass combustion, or other anthropogenic activities. The heightened AE throughout these years may be ascribed to augmented industrial activity, traffic emissions, or agricultural practices like crop residue incineration, prevalent in the region. The elevated AE values indicate that these tiny particles significantly influenced the atmospheric composition during these winters, contributing to haze and potentially more severe fog occurrences.

Conversely, Winters 2017 and 2023 demonstrate among the lowest AE values, decreasing to approximately 0.9. The reduced values indicate a greater prevalence of coarse-mode aerosols, predominantly consisting of dust particles. Lahore, situated in the Punjab region, is prone to dust storms, especially in winter when winds from adjacent desert areas can carry substantial quantities of dust. The diminished AE during these years suggests that dust aerosols exerted a more pronounced influence, which may decrease the likelihood of fine-particle-induced fog while still contributing to atmospheric opacity and diminished visibility owing to dust.

The mild AE values in previous years, including Winter 2016 and Winter 2018, suggest a combination of fine and coarse aerosols. The winters likely encountered a confluence of local pollution and regional dust transport, illustrating the intricate interplay of man-made and natural aerosol sources on Lahore's air quality.

The COVID-19 pandemic years (Winters 2020 and 2021) exhibit AE values of 1.3, marginally lower than the pre-COVID years, but still indicative of a substantial fine-particle impact. This minor drop may indicate diminished human activity during the pandemic, including less industrial production, transportation, and biomass combustion. Nonetheless, the prevalence of fine-mode aerosols persisted significantly, perhaps because of enduring pollution sources and seasonal agricultural burning.

The AE figure illustrates the dynamic characteristics of aerosol behavior in Lahore from 2010 to 2024. The elevated AE values in specific years indicate a heightened presence of fine-mode pollution, while the reduced AE values in other years imply a greater influence of natural sources such as dust. Comprehending these variances is essential for formulating appropriate mitigation measures to tackle air quality concerns and diminish the health effects of aerosols in the region.

Russell et al. (2010) determined that urban-industrial aerosols, marked by elevated AE values, predominate in winter owing to heightened heating and reduced dispersion rates. Liu et al. (2018) emphasized that black carbon, a fine particulate matter resulting from burning, markedly elevates AE values and promotes fog formation. This corroborates the notion that elevated AE values in winter result from intensified combustion activities.

Zhang et al. (2021) examined the long-term properties of aerosol absorption optical depth (AAOD) and Angstrom exponent (AAE) in the Tarim Basin from 2005 to 2018. AAE values were seen to be lower in winter than in summer, suggesting an increased prevalence of coarse-mode aerosols throughout the winter season. These findings correspond with the observation of elevated AE values during the winter months in the present investigation.

The graph in Fig. 12.6 depicting the precipitable water (PW) levels for Lahore from 2010 to 2024 reveals significant trends in atmospheric moisture content during the winter seasons. The PW, which quantifies total atmospheric water vapor, exhibits considerable variability during this period, with measurements spanning from 0.82 to 1.35 cm. The fluctuations in moisture levels significantly impact fog production, as atmospheric moisture is crucial for condensation and the creation of fog.

Between 2010 and 2024, there are significant fluctuations in PW over the winters. In Winter 2018, there was a significant rise in PW, reaching approximately 1.35 cm, suggesting elevated moisture levels that likely resulted in more intense and frequent fog occurrences during that season. The significant increase in PW indicates conducive conditions for fog development, as elevated water vapor levels enhance the likelihood of condensation, particularly when particles serve as cloud condensation nuclei (CCN). Likewise, there were moderate elevations in PW during

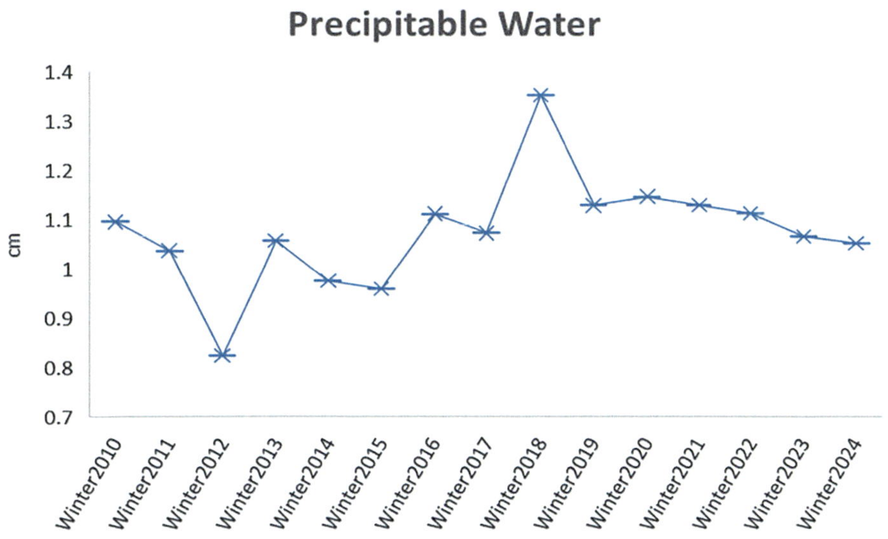


**Fig. 12.6** Precipitable water (cm) values over Lahore from winter 2010 to 2024

the winters of 2012 and 2016, which presumably contributed to intensified fog conditions owing to increased atmospheric humidity.

In contrast, the data indicate substantial reductions in PW during other years. Winter 2014 is notable for a decrease in PW below 1 cm, indicating drier air conditions that may have led to a reduction in the frequency or intensity of fog episodes. The persistent decline in PW since Winter 2018, reaching its nadir of approximately 0.9 cm in Winter 2024, indicates a progressive decrease in atmospheric moisture, possibly resulting in less conducive conditions for fog development in recent years. The reduction in moisture availability may be ascribed to alterations in meteorological circumstances, such as changes in regional circulation patterns or less precipitation, which restrict the accumulation of water vapor in the atmosphere.

The COVID-19 pandemic, which lasted from Winter 2020 to Winter 2022, had a nuanced impact on PW levels. Although worldwide lockdowns dramatically diminished human activity and thus reduced aerosol emissions, the effect on PW levels seems to have been very mild. A minor reduction in PW occurs during these winters, albeit the alterations are not significant. The results indicate a consistent decrease in PW from Winter 2018 onward, implying that wider atmospheric conditions, rather than solely diminished anthropogenic activity during the COVID period, were more influential in the observed patterns. The reduction in industrial and vehicular emissions during this period may have diminished the number of aerosols functioning as cloud condensation nuclei; however, this factor alone might not have been adequate to precipitate a significant decline in precipitable water, as local meteorological systems and regional climatic variables likely persisted in affecting moisture levels.

The variations in PW during these years underscore the intricate relationship between atmospheric moisture and aerosol levels in influencing fog conditions in Lahore. In years with enhanced PW, such as Winter 2018, the amalgamation of heightened moisture and increased aerosol concentrations likely resulted in thicker and more frequent fog occurrences. Conversely, in arid years such as Winter 2024, despite elevated aerosol concentrations, the deficiency of enough moisture would have constrained fog development. This suggests that aerosols are essential for fog formation, but their efficacy is maximized in the presence of sufficient atmospheric moisture.

Numerous research in the literature corroborates the findings of trends in precipitable water during winter. Wan et al. (2024) examine fluctuations in worldwide total precipitable water, highlighting an increase during winter months attributed to particular meteorological circumstances. Trenberth and Smith (2005) emphasize the importance of atmospheric water vapor, noting its considerable fluctuation with seasonal fluctuations, especially during winter. Landrigan et al. (2018) investigate the Asian winter monsoon and its effect on precipitable water levels, affirming a rise in atmospheric moisture during winter. Finally, K. E. Trenberth et al. (2009) analyze the global energy budget, highlighting the significance of water vapor and its seasonal fluctuations, particularly a considerable rise during winter.

This analysis provides significant policy implications for addressing the fog issue in Lahore. A crucial technique may involve diminishing aerosol emissions, especially those originating from biomass combustion and automobile pollution, which significantly contribute to fog formation. Enforcing more stringent emissions restrictions for industry and advocating for cleaner fuels and transportation techniques could mitigate the concentration of CCN in the atmosphere, thereby diminishing the occurrence and severity of fog episodes, particularly in years with elevated PW. Moreover, initiatives to enhance urban planning and alleviate the urban heat island effect may assist in regulating local temperatures and humidity, hence further diminishing fog occurrences in Lahore.

The imagery shown in Fig. 12.7 captured over Lahore on 21 December 2023 reveals a dense layer of whitish-gray fog enveloping the region. This phenomenon is a common occurrence during the winter months in Lahore, primarily due to temperature inversions and the high concentration of aerosols in the atmosphere.

The fog appears as a smooth and uniform layer, distinct from the more textured appearance of clouds and dust. This uniformity is a characteristic feature of fog, indicating its formation close to the ground. The presence of aerosols, such as pollutants from industrial activities and vehicular emissions, contributes to the formation and persistence of fog in the region. These particles provide surfaces for moisture to condense, leading to the dense fog observed in the imagery. The dense fog significantly reduces visibility in the region, impacting transportation and daily activities. This reduction in visibility is a critical factor to consider when formulating mitigation strategies.

The meteorological conditions prevailing during the observation period include low temperatures and high relative humidity, creating an environment conducive to fog formation. The interaction between these meteorological factors and the high

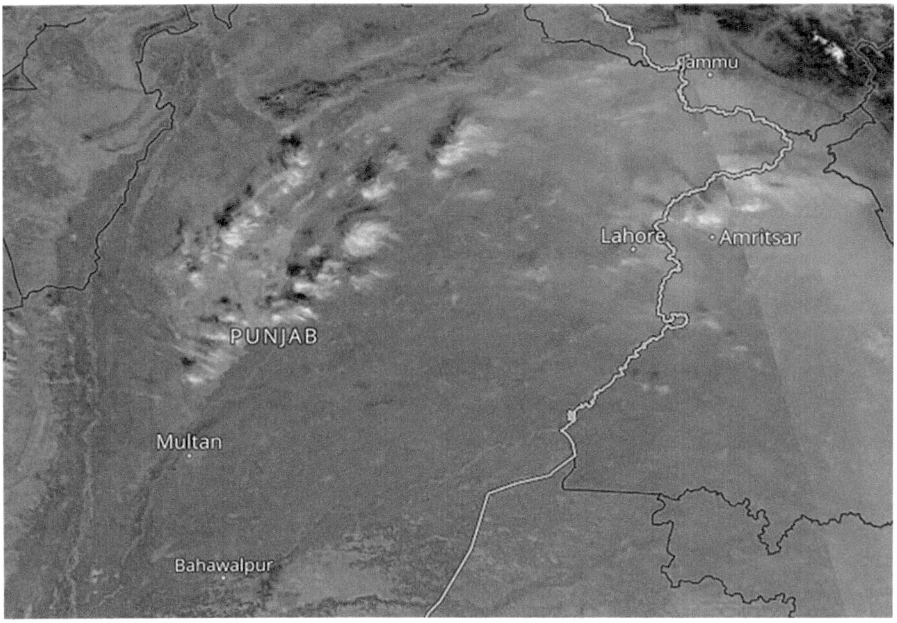


**Fig. 12.7** VIIRS images from Suomi NPP on 21 December 2023

concentration of aerosols led to the persistent fog observed in Lahore during this period.

Understanding the meteorological conditions and the role of aerosols in fog formation is essential for developing effective policies and mitigation strategies. Measures such as regulating industrial emissions, promoting cleaner transportation options, and enhancing public awareness can help mitigate the impact of fog and improve air quality in Lahore.

## 12.5 Socioeconomic Impacts of Fog in Lahore

Fog in Lahore has profound socioeconomic consequences at the regional level, affecting healthcare, economic productivity, transportation, and agriculture. The most immediate impact is on public health. Prolonged exposure to fog-related pollution, which is often laden with harmful particulate matter and industrial pollutants, exacerbates respiratory and cardiovascular diseases (Majeed et al. 2023). Hospitals and clinics in Lahore, particularly during the winter fog season, face overwhelming patient loads, increasing both the strain on healthcare workers and medical costs for individuals and families. These healthcare expenses are a significant financial burden for many households, especially those from lower-income backgrounds, who may lack access to proper medical insurance (Mushtaq et al. 2024).

Economically, fog and its by-product, smog, diminish workforce productivity as employees either fall ill or face transportation challenges (Riaz and Hamid 2018). This not only reduces individual productivity but also has a cascading effect on businesses, causing delays in operations, lower output, and ultimately hindering the local economy. Transportation networks are severely disrupted due to poor visibility, leading to frequent road accidents, flight cancellations, and delays in the logistics sector (Educator 2024). These interruptions increase operational costs for businesses reliant on timely deliveries, compounding economic losses. Additionally, agriculture, a significant sector in the Punjab region, suffers as persistent fog damages crops, particularly during the crucial growing season. Prolonged exposure to pollutants also depletes soil quality, further diminishing agricultural yields and affecting food supply chains.

On the national scale, the socioeconomic impacts of fog in Lahore ripple across various sectors, amplifying challenges that go beyond regional boundaries. One of the most concerning effects is the significant public health burden (Mushtaq et al. 2024). With millions of people suffering from respiratory diseases like asthma and bronchitis, the national healthcare system is stretched thin, requiring increased government spending on healthcare infrastructure and medical supplies. This diverts funds from other critical areas such as education and infrastructure development, straining national resources. The demand for specialized healthcare and treatments escalates, further exacerbating health disparities, especially among the economically disadvantaged population.

Economic losses due to fog extend beyond the local economy and have national ramifications. Reduced productivity in major urban hubs like Lahore leads to a dip in the country's overall economic output (Jabeen et al. 2021a, b), directly affecting national GDP growth. The cost of transportation disruptions and supply chain inefficiencies also hamper trade and commerce at the national level. Furthermore, persistent fog contributes to long-term environmental degradation, deteriorating the quality of air, water, and natural habitats across the country. This accelerates the depletion of biodiversity and poses significant risks to natural resources, which are crucial for sectors like agriculture and tourism. Finally, the pervasive nature of fog-related issues often ignites social unrest. As people experience the adverse effects of fog on their health and daily lives, there is a growing sense of frustration over government inaction or inadequate measures to control air pollution. This social discontent can lead to protests, political instability, and demands for more stringent environmental regulations, placing pressure on national governance.

## 12.6 Strategic Measures to Address the Impact of Fog in Lahore

Below is the proposed policy responses intended to alleviate the socioeconomic and environmental impacts of fog:

- In light of the reported rise in aerosol optical depth (AOD), especially during heightened industrial activity, policymakers must prioritize more stringent regulation and oversight of industrial emissions. This includes the implementation of emission caps, the mandatory use of scrubbers and filters, and the promotion of cleaner technologies through incentives. Tajammul (2023) indicates that implementing stronger limits on industrial emissions has demonstrably decreased aerosol levels in comparable situations.
- The automotive emissions in Lahore considerably contribute to elevated aerosol levels. Enforcement of more stringent vehicular pollution regulations, promotion of electric vehicle (EV) adoption, and establishment of low-emission zones can alleviate the impact. World Bank (2024a, b) demonstrates that these approaches can significantly diminish particulate matter in metropolitan environments.
- Shifting to cleaner fuels (such as CNG, electric, or hybrid vehicles) can diminish particulate matter and AOD values, especially during peak winter fog conditions. Ittefaq and Kamboh (2023) advocate for the adoption of cleaner fuels to enhance air quality.
- The incineration of agricultural remnants in Punjab elevates AOD levels, especially in the winter season. Policies ought to encourage alternate techniques for crop residue management, such as composting or bioenergy generation, accompanied by stringent sanctions for incineration. The research conducted by Kaskaoutis et al. (2014) about biomass combustion and its impact on air quality underscores the necessity for such interventions.
- The implementation of proactive air quality management strategies before winter, emphasizing the mitigation of agricultural fires, may diminish the aerosol load. This would also entail predicting probable fog occurrences and notifying the public in advance. WHO (2005) encourages proactive steps to address seasonal air quality concerns.
- Enhancing urban green spaces can facilitate the absorption of particulate matter and diminish aerosol concentrations. Trees and vegetation function as natural filters, sequestering pollutants and enhancing air quality, potentially mitigating the intensity of fog situations. Nowak et al. (2008) conducted research indicating that urban forestry enhances air quality in urban environments.
- Establishing green buffer zones surrounding industrial regions might diminish the dissemination of pollutants into urban areas, hence lessening particulate matter that contributes to fog formation.
- Enhancing public awareness about the detrimental impacts of open-air biomass combustion, automobile emissions, and industrial pollutants can stimulate community initiatives to mitigate pollution during peak fog periods. Programs must incorporate efforts to mitigate household and individual emissions.
- Establish real-time air quality monitoring systems alongside public alert mechanisms to notify the populace of elevated pollution levels. These systems may also offer recommendations for minimizing exposure and mitigating actions that contribute to fog formation. These activities are corroborated by the verdicts of UNEP (2021).

- The investigation of fog dissipation technologies, including water spraying systems on highways, fog lamps, and fog warning systems, may alleviate the detrimental impacts of fog on transportation. Improving city design to enhance airflow (e.g., broader roadways, optimal building spacing) can decrease pollution concentrations, hence mitigating the prevalence of dense fog conditions. Insights on urban planning from World Bank (2024a) underscore the significance of infrastructure in regulating air quality.
- Techniques for weather modification, such as cloud seeding to stimulate precipitation, may diminish atmospheric aerosols. Despite being costly and resource-demanding, they may be utilized under extreme conditions where sight is significantly impaired. National Research Council (2003) examines the possibilities and obstacles associated with these technologies.
- The production of fog in Lahore is affected by atmospheric stability, humidity, and aerosol optical depth (AOD) levels. Establishing enhanced meteorological monitoring systems and models to predict fog events based on real-time AOD, temperature, and humidity can improve preparedness and policy response quality.
- Since transboundary pollution from neighboring regions like India (particularly biomass burning) significantly contributes to Lahore's AOD levels, regional cooperation and coordinated efforts are essential. Joint air quality management programs can be developed to reduce shared sources of aerosol pollution. Insights from Prabhakar et al. (2018) highlight the benefits of regional cooperation. Formulating agreements with neighboring countries to address seasonal pollutants through better communication, shared data, and mutual air quality targets would mitigate aerosol influxes that exacerbate fog in Lahore.
- Reducing reliance on fossil fuels (such as coal and diesel) by promoting renewable energy sources (solar, wind, and hydropower) can lower AOD levels and improve air quality. This transition would have a long-term impact on reducing aerosol concentrations during fog-prone months. Burrett et al. (2009) provide extensive data supporting the benefits of renewable energy. Policies focusing on urban cooling strategies—such as reflective roofing, green roofs, and increasing vegetation—can reduce temperature fluctuations, helping to mitigate the atmospheric conditions that contribute to dense fog formation.
- Fog elevates particulate matter concentration, presenting significant health hazards. Emergency health response strategies include temporary air filtration systems, complimentary mask distribution, and mobile health units in high-risk locations that can mitigate negative impacts on public health. The recommendations from Bernstein et al. (2004) emphasize the significance of these interventions. Establish thorough transportation management strategies to reduce accidents and delays during dense fog conditions. This includes enhancing fog warning systems at highways, airports, and train stations, as well as modifying schedules when required.

Through the implementation of these comprehensive policy measures, Lahore may successfully mitigate the detrimental impacts of fog, enhance air quality, and safeguard public health.

## 12.7 Conclusion and Recommendations

The research conducted on fog in Lahore from 2010 to 2024, utilizing AERONET values of aerosol optical depth (AOD) and Angstrom exponent (AE), reveals several key findings. During the winter seasons, there is a notable increase in both AOD (ranging from 0.4 to 0.8) and AE (ranging from 1.0 to 1.4), reflecting the higher concentration of fine particulate matter. These fine particles primarily originate from biomass burning, industrial emissions, and vehicular pollution, which are prevalent during this period.

Meteorological parameters play a significant role in the formation and persistence of fog in Lahore. The study finds that low temperatures (ranging from 6 °C to 9 °C), high relative humidity (often exceeding 70%), and low wind speeds during winter create favorable conditions for fog formation. These conditions lead to elevated static stability and the presence of a boundary-layer inversion, which traps pollutants near the surface, resulting in diminished visibility and elevated pollution concentrations.

The correlation between AOD and meteorological parameters indicates that higher AOD values are associated with increased relative humidity and lower temperatures. This relationship underscores the importance of addressing both pollution sources and meteorological conditions in developing effective strategies to mitigate the adverse impacts of fog.

Overall, the study concludes that implementing stricter emission controls, promoting cleaner fuels, enhancing public transportation, and increasing urban green spaces are essential to reducing aerosol concentrations and improving air quality in Lahore. Additionally, public awareness campaigns, advanced meteorological models for fog prediction, and regional collaboration on transboundary pollution are crucial for mitigating the health and socioeconomic impacts of fog.

By addressing these factors, Lahore can take significant steps toward improving air quality, reducing the incidence of fog, and protecting public health during the winter seasons. The findings provide a comprehensive understanding of the interplay between aerosols, meteorological conditions, and fog formation, offering valuable insights for policymakers and researchers working to combat this environmental challenge.

## References

Abdel-Aty M, Ekram AA, Huang H, Choi K. A study on crashes related to visibility obstruction due to fog and smoke. Accid Anal Prev. 2011;43(5):1730–7. https://doi.org/10.1016/J.AAP.2011.04.003.

Abu-Hamdeh NH. Measurement of the thermal conductivity of sandy loam and clay loam soils using single and dual probes. Journal of Agricultural and Engineering Research. 2001;80(2):209–16. https://doi.org/10.1006/JAER.2001.0730.

Ali G, Bao Y, Ullah W, Ullah S, Guan Q, Liu X, Li L. Spatiotemporal trends of aerosols over urban regions in Pakistan and their possible links to meteorological parameters. Atmosphere. 2020; https://www.mdpi.com/2073-4433/11/3/306

Ali O. How does fog harvesting provide sustainable water solutions? 2024. https://www.azocleantech.com/article.aspx?ArticleID=1782.

Aslam M. Fog hazards in Punjab. Pakistan J Meteorol. 2012;8

Ba S, Wang WJ, Sun H, Bao SG, Zhang H, He HS. The cooling and warming effects of potential forest transition on local land surface temperature in Northeast China. Ecol Indic. 2024;159:111645. https://doi.org/10.1016/J.ECOLIND.2024.111645.

Bernstein JA, Alexis N, Barnes C, Bernstein IL, Bernstein JA, Nel A, Peden D, Diaz-Sanchez D, Tarlo SM, Williams PB. Health effects of air pollution. J Allergy Clin Immunol. 2004;114(5):1116–23. https://doi.org/10.1016/J.JACI.2004.08.030.

Bharali C, Barth M, Kumar R, Ghude SD, Sinha V, Sinha B. Role of atmospheric aerosols in severe winter fog over the Indo-Gangetic Plain of India: a case study. Atmos Chem Phys. 2024;24(11):6635–62. https://doi.org/10.5194/ACP-24-6635-2024.

Bulbul G, Shahid I, Chishtie F, Shahid M, Hundal R. PM10 sampling and AOD trends during 2016 winter fog season in the Islamabad region. 2018; https://repository.kaust.edu.sa/bitstream/10754/627274/1/Articlefile1.pdf

Burrett R, Clini C, Dixon R, Eckhart M, El-Ashry M, Hamilton K, Hoskyns SJ, Koch H-J, Larrain S, Sustentable C, Junfeng L, Galàn EM, Mubiru P, Nassiep K, Pachauri R, Palz W, Pitka-Kangas L, Radka M, Rae P, et al. Renewable energy policy network for the 21st century REN21 Steering Committee. Update; 2009.

Bushahab A, Suwaidi AA, Ghedira H. Fog forecasting, detection and monitoring in the UAE using seviri-msg data. IEEE international geoscience and remote sensing symposium, 2009. 2009. https://ieeexplore.ieee.org/abstract/document/5418137/

Collier CG. Fog at Manchester. Weather. 1970;25(1):25–9. https://doi.org/10.1002/J.1477-8696.1970.TB03225.X.

Dejmal K, Repal V. Implementation of methods for the radiation fog prediction; 2010. http://www.unob.cz.

Dione C, Haeffelin M, Burnet F, Lac C. Role of thermodynamic and turbulence processes on the fog life cycle during SOFOG3D experiment. AcpCopernicusOrg. n.d.; Retrieved October 22, 2024, from https://acp.copernicus.org/articles/23/15711/2023/

Domen JK, Stringfellow WT, Camarillo MK, Gulati S. Fog water as an alternative and sustainable water resource. Clean Techn Environ Policy. 2014;16(2):235–49. https://doi.org/10.1007/S10098-013-0645-Z.

Du P, Zeng Z, Zhang J, Liu L, Yang J, Qu C, Jiang L, Liu S. Fog season risk assessment for maritime transportation systems exploiting Himawari-8 data: a case study in Bohai Sea, China. Remote Sensing. 2021;13(17):3530. https://doi.org/10.3390/RS13173530.

Dutta D, Chaudhuri S. Nowcasting visibility during wintertime fog over the airport of a metropolis of India: decision tree algorithm and artificial neural network approach. Nat Hazards. 2015;75(2):1349–68. https://doi.org/10.1007/S11069-014-1388-9/METRICS.

Educator S. Fog in Lahore, mysterious and drastic issue since 4 years- SU Educator; 2024. https://subhanedu.com/fog-in-lahore-mysterious-and-issue-since-4-years/.

Gautam S, Elizabeth J, Gautam AS, Singh K, Abhilash P. Impact assessment of aerosol optical depth on rainfall in Indian rural areas. Aerosol Sci Eng. 2022;6(2):186–96. https://doi.org/10.1007/S41810-022-00134-9/FIGURES/4.

Gebhart M, Leitgeb E, Muhammad SS, Flecker B, Chlestil C, Naboulsi MA, de Fornel F, Sizun H. Measurement of light attenuation in dense fog conditions for FSO applications. Proc SPIE Atmos Opt Model Meas Simul. 2005;5891:175–86. https://doi.org/10.1117/12.614830.

Gultepe I, Tardif R, Michaelides SC, Cermak J, Bott A, Bendix J, Müller MD, Müller M, Pagowski M, Hansen B, Ellrod G, Jacobs W, Toth G, Cober SG. Fog research: a review of past achievements and future perspectives. Pure Appl Geophy. 2007;164(6–7):1121–59. https://doi.org/10.1007/s00024-007-0211-x. Springer

Guo B, Wang Y, Zhang X, Che H, Zhong J. Temporal and spatial variations of haze and fog and the characteristics of PM2. 5 during heavy pollution episodes in China from 2013 to 2018. Atmos Pollut Res. 2020; https://www.sciencedirect.com/science/article/pii/S1309104220302105

Hameed S, Ishaq Mirza M, Ghauri BM, Siddiqui ZR, Javed R, Khan AR, Rattigan OV, Qureshi S, Husain L. On the widespread winter fog in northeastern Pakistan and India. Geophys Res Lett. 2000;27(13):1891–4. https://doi.org/10.1029/1999GL011020.

Husain L, Dutkiewicz V, Khan A, Ghauri B. Characterization of carbonaceous aerosols in urban air. Atmos Environ. 2007; https://www.sciencedirect.com/science/article/pii/S1352231007003810

Igawa M, Kamijo K, Nanzai B, Matsumoto K. Chemical composition of polluted mist droplets. Atmos Environ. 2017;171:230–6. https://doi.org/10.1016/J.ATMOSENV.2017.10.029.

Isil S, Collett J, Lynch J, Weiss-Penzias P, Rogers CM. Cloud and fog deposition: monitoring in high elevation and coastal ecosystems. The past, present, and future. Atmos Environ (Oxford, England: 1994). 2022;274:1. https://doi.org/10.1016/J.ATMOSENV.2022.118997.

Ittefaq M, Kamboh S. Communities of care for air pollution policies in Pakistan: an integrated approach to public engagement and policy implementation. Environ Commun. 2023;18(1–2):49–55. https://doi.org/10.1080/17524032.2023.2295784.

Jabeen F, Ali Z, Maharjan A, Assessing Health A. Assessing health impacts of winter smog in Lahore for exposed occupational groups. Atmosphere. 2021a; https://doi.org/10.3390/atmos12111532.

Jabeen F, Ali Z, Maharjan A, Assessing Health A. Assessing health impacts of winter smog in Lahore for exposed occupational groups. Atmosphere. 2021b; https://doi.org/10.3390/atmos12111532.

Kabanov D, Ritter C, Sakerin S. Interannual and seasonal variations in the aerosol optical depth of the atmosphere in two regions of Spitsbergen (2002–2018). Atmos Measure Techniq. 2020; https://amt.copernicus.org/articles/13/5303/2020/amt-13-5303-2020.html

Kaskaoutis DG, Kumar S, Sharma D, Singh RP, Kharol SK, Sharma M, Singh AK, Singh S, Singh A, Singh D. Effects of crop residue burning on aerosol properties, plume characteristics, and long-range transport over northern India. J Geophys Res Atmos. 2014;119(9):5424–44. https://doi.org/10.1002/2013JD021357.

Klemm O, Lin NH. What causes observed fog trends: air quality or climate change? Aerosol Air Qual Res. 2016;16(5):1131–42. https://doi.org/10.4209/AAQR.2015.05.0353.

LaDochy S. The disappearance of dense fog in Los Angeles: another urban impact? Phys Geogr. 2005;26(3):177–91. https://doi.org/10.2747/0272-3646.26.3.177.

Lakra K, Avishek K. A review on factors influencing fog formation, classification, forecasting, detection and impacts. Rendiconti Lincei Scienze Fisiche e Naturali. 2022;33(2):319–53. https://doi.org/10.1007/s12210-022-01060-1.

Landrigan PJ, Fuller R, Fisher S, Suk WA, Sly P, Chiles TC, Bose-O'reilly S. Pollution and children's health. Sci Total Environ. 2018; https://doi.org/10.1016/j.scitotenv.2018.09.375.

Li Y, Eugster W, Riedl A, Westerhuis S, Buchmann N, Scholder-Aemisegger F. Identifying key stages of radiation fog evolution using water vapor isotopes. Agric For Meteorol. 2023;334:109430. https://doi.org/10.1016/J.AGRFORMET.2023.109430.

Liu C, Chung C, Yin Y. The absorption Ångström exponent of black carbon: from numerical aspects. Atmos Chem Phys. 2018;18:6259–73. https://doi.org/10.5194/acp-18-6259-2018.

Liu DY, Niu SJ, Yang J, Zhao LJ, Lü JJ, Lu CS. Summary of a 4-year fog field study in northern Nanjing, part 1: fog boundary layer. Pure Appl Geophys. 2012;169(5–6):809–19. https://doi.org/10.1007/S00024-011-0343-X/METRICS.

Liu D, Yang J, Niu S, Li Z. On the evolution and structure of a radiation fog event in Nanjing. Adv Atmos Sci. 2011;28(1):223–37. https://doi.org/10.1007/S00376-010-0017-0/METRICS.

Majeed R, Anjum MS, Imad-ud-din M, Malik S, Anwar MN, Anwar B, Khokhar MF. Solving the mysteries of Lahore smog: the fifth season in the country. Front Sustain Cities. 2023;5 https://doi.org/10.3389/FRSC.2023.1314426/FULL.

Mkoma SL, Maenhaut W, Chi X, Wang W, Raes N. Characterisation of PM10 atmospheric aerosols for the wet season 2005 at two sites in East Africa. Atmos Environ. 2009;43(3):631–9. https://doi.org/10.1016/J.ATMOSENV.2008.10.008.

Mushtaq A, Mahmood S, Jr I. Analyzing the impact of smog on human health in District Lahore, Pakistan; 2024. https://www.researchgate.net/publication/383611314_Analyzing_the_Impact_of_Smog_on_Human_Health_in_District_Lahore_Pakistan

Muslehuddin M, Mir H, Faisal N. Recent occurrence of fog over Pakistan (1997 to 2000). J Meteorol. 2004;1(2) https://www.researchgate.net/profile/Nadeem-Faisal/publication/267374532_Recent_occurrence_of_fog_over_Pakistan_1997_to_2000/links/551bd0a30cf20d5fbde21360/Recent-occurrence-of-fog-over-Pakistan-1997-to-2000.pdf

National Research Council. Critical issues in weather modification research. Crit Issue Weather Modif Res. 2003; https://doi.org/10.17226/10829.

Niu F, Li Z, Li C, Lee K-H, Wang M, Niu F, Li Z, Li C, Lee K-H, Wang M. Increase of wintertime fog in China: potential impacts of weakening of the eastern Asian monsoon circulation and increasing aerosol loading. J Geophys Res Atmos. 2010;115(D7):0–20. https://doi.org/10.1029/2009JD013484.

Nowak DJ, Crane DE, Stevens JC, Hoehn RE, Walton JT, Bond J. A ground-based method of Assessing urban Forest structure and ecosystem services. Arboriculture & Urban Forestry (AUF). 2008;34(6):347–58. https://doi.org/10.48044/JAUF.2008.048.

Pagowski M, Gultepe I, King P. Analysis and modeling of an extremely dense fog event in southern Ontario. J Appl Meteorol Climatol. 2004;43(1):3–16. https://doi.org/10.1175/1520-0450(2004)043.

PMD. FOG forecast; 2024. https://nwfc.pmd.gov.pk/fog/home.html

Prabhakar SVRK, Shivakoti BR, Scheyvens H. Transboundary impacts of climate change in Asia: making a case for regional adaptation planning and cooperation; 2018.

Rattigan OV, Mirza MI, Ghauri BM, Khan AR, Swami K, Yang K, Husain L. Aerosol sulfate and trace elements in urban fog. Energy Fuels. 2002;16(3):640–6. https://doi.org/10.1021/EF010199G.

Riaz R, Hamid K. Existing smog in Lahore, Pakistan: an alarming public health concern. Cureus. 2018;10(1):e2111. https://doi.org/10.7759/CUREUS.2111.

Roach WT, Brown R, Caughey SJ, Garland JA, Readings CJ. The physics of radiation fog: I—a field study. Q J R Meteorol Soc. 1976;102(432):313–33. https://doi.org/10.1002/QJ.49710243204.

Rosenberg NJ, Blad BL, Verma SB. Microclimate: the biological environment, 2nd edn; 1983. https://books.google.com/books/about/Microclimate.html?id=c6MI6VvTqTsC

Russell P, Bergstrom R, Shinozuka Y. Absorption angstrom exponent in AERONET and related data as an indicator of aerosol composition. Atmos Chem Phys. 2010; https://acp.copernicus.org/articles/10/1155/2010/

Sachweh M, Koepke P. Radiation fog and urban climate. Geophys Res Lett. 1995;22(9):1073–6. https://doi.org/10.1029/95GL00907.

Saikh SR, Das SK. Fog-induced alteration in airborne microbial community: a study over central Indo-Gangetic Plain in India. Appl Environ Microbiol. 2023;89(1):e01367–22. https://doi.org/10.1128/AEM.01367-22.

Saikh SR, Mushtaque MA, Pramanick A, Prasad JK, Roy D, Saha S, Das SK. Fog caused distinct diversity of airborne bacterial communities enriched with pathogens over central Indo-Gangetic plain in India. Heliyon. 2024;10(4):e26370. https://doi.org/10.1016/J.HELIYON.2024.E26370.

Saraswati. Contribution of fog in changing air quality: extremities and risks to environment and society. Singapore: Springer; 2022. p. 87–111. https://doi.org/10.1007/978-981-16-7727-4_5.

Sawaisarje GK, Khare P, Shirke CY, Deepakumar S, Narkhede NM. Study of winter fog over Indian subcontinent: climatological perspectives. Mausam. 2014;65(1):19–28. https://doi.org/10.54302/MAUSAM.V65I1.858.

Saxena M, Sharma A, Sen A, Saxena P, Saraswati MTK, Sharma SK, Sharma C. Water soluble inorganic species of PM10 and PM2.5 at an urban site of Delhi, India: seasonal variability and sources. Atmos Res. 2017;184:112–25. https://doi.org/10.1016/J.ATMOSRES.2016.10.005.

Singh A. Urban air pollution and its effect on fog occurrence: a view from Northern India|Request PDF; 2014. https://www.researchgate.net/publication/281241566_Urban_air_pollution_and_its_effect_on_fog_occurrence_a_view_from_Northern_India.

Smith DJT, Harrison RM, Luhana L, Pio CA, Castro LM, Tariq MN, Hayat S, Quraishi T. Concentrations of particulate airborne polycyclic aromatic hydrocarbons and metals collected in Lahore, Pakistan. Atmos Environ. 1996; https://www.sciencedirect.com/science/article/pii/1352231096001070

Syed FS, Körnich H, Tjernström M. On the fog variability over South Asia. Clim Dyn. 2012;39(12):2993–3005. https://doi.org/10.1007/S00382-012-1414-0/METRICS.

B. SFS, A. Y. Variation in fog intensity/duration and El Nino. Pakistan J Meteorol. 2004;1:1.

Tajammul KI. Critical review and comparative analysis of the Government of Punjab's 'policy on controlling smog, 2017' with counterpart strategies in London, Beijing and Los Angeles. J Develop Policy Res Pract (JoDPRP). 2023;7(1):181–202. https://doi.org/10.59926/JODPRP.VOL07/09.

Tiwari S, Payra S, Mohan M, Verma S. Visibility degradation during foggy period due to anthropogenic urban aerosol at Delhi, India. Atmos Pollut Res. 2011; https://www.sciencedirect.com/science/article/pii/S1309104215305237

Toon OB, Tabazadeh A, Browell EV, Jordan J. Analysis of lidar observations of Arctic polar stratospheric clouds during January 1989. J Geophys Res Atmos. 2000;105(D16):20589–615. https://doi.org/10.1029/2000JD900144.

Trenberth KE, Fasullo JT, Kiehl J. Earth's global energy budget. Bull Am Meteorol Soc. 2009;90(3):311–23. https://doi.org/10.1175/2008BAMS2634.1.

Trenberth K, Smith L. The mass of the atmosphere: a constraint on global analyses. J Clim. 2005; https://journals.ametsoc.org/view/journals/clim/18/6/jcli-3299.1.xml?tab_body=abstract-display

UNEP. Actions on air quality: a global summary of policies and programmes to reduce air pollution|UNEP - UN Environment Programme; 2021. https://www.unep.org/resources/report/actions-air-quality-global-summary-policies-and-programmes-reduce-air-pollution

Wan N, Lin X Sr, R. P., & Zeng, X. Global total precipitable water variations and trends over the period 1958–2021. Hydrol Earth Syst Sci. 2024; https://hess.copernicus.org/articles/28/2123/2024/hess-28-2123-2024.html

Wang Y, Niu S, Lu C, Lv J, Zhang J, Zhang H, Zhang S, Shao N, Sun W, Jin Y, Song Q. Observational study of the physical and chemical characteristics of the winter radiation fog in the tropical rainforest in Xishuangbanna, China. Sci China Earth Sci. 2021;64(11):1982–95. https://doi.org/10.1007/S11430-020-9766-4/METRICS.

Weathers KC, Ponette-González AG, Dawson TE. Medium, vector, and connector: fog and the maintenance of ecosystems. Ecosystems. 2020;23(1):217–29. https://doi.org/10.1007/S10021-019-00388-4/FIGURES/3.

WHO. Air quality guidelines global update. 2005:2005. https://www.who.int/publications/i/item/WHO-SDE-PHE-OEH-06.02

WHO. Air pollution; 2024. https://www.who.int/health-topics/air-pollution#tab=tab_1

WMO. Aviation|Hazards|Low Visibility and Low Cloud|World Meteorological Organization; 2020. https://community.wmo.int/en/activity-areas/aviation/hazards/visibility

World Bank. Cities and climate change; 2024a. https://www.worldbank.org/en/programs/cities-and-climate-change-platform/overview.

World Bank. Urban air quality management strategy in Asia: guidebook 171; 2024b. https://documents.worldbank.org/en/publication/documents-reports/documentdetail/881391468740671903/Urban-air-quality-management-strategy-in-Asia-guidebook

Yan S, Zhu B, Huang Y, Zhu J, Kang H. To what extents do urbanization and air pollution affect fog? Atmos Chem Phys. 2020; https://acp.copernicus.org/articles/20/5559/2020/acp-20-5559-2020.html

Yao W, Zheng Z, Zhao J, Wang X, Wang Y, Li X, Fu J. The factor analysis of fog and haze under the coupling of multiple factors—taking four Chinese cities as an example. Energy Policy. 2020;137:111138. https://doi.org/10.1016/J.ENPOL.2019.111138.

Yasmeen Z, Rasul G, Zahid M. Impact of aerosols on winter fog of Pakistan. Pakistan J Meteorol. 2012;8. https://www.researchgate.net/publication/270589514

Zhang X, Jiang H, Mao M. Absorption angstrom exponent of dust aerosols over the Tarim Basin. Pure Appl Geophys. 2021;178(11):4549–60. https://doi.org/10.1007/S00024-021-02874-0.

Zhang Z, Li Y, Li L, Zhang C, Sun G. Characteristics of advection fog at Qingdao Liuting International Airport. Atmosphere. 2023;14(8):1310. https://doi.org/10.3390/ATMOS14081310.

# Glossary

**Advection Fog** Advection fog forms when warm, moist air moves horizontally over a cooler surface, causing the air to cool to its dew point, leading to condensation and fog formation

**Aerosol** Consists of very finely subdivided liquid or solid particles dispersed in and surrounded by a gas.

**Aerosol Optical Depth (AOD)** Aerosol Optical Depth (AOD) is the measure of aerosols (e.g., urban haze, smoke particles, desert dust, sea salt) distributed within a column of air from the instrument (Earth's surface) to the top of the atmosphere.

**Anthropogenic Emissions** The emissions which have their origin from man-made activities like vehicular exhaust, industrial activities, and domestic activities.

**Atmosphere** A gaseous envelope surrounding the Earth. Atmosphere also contains clouds and aerosol.

**Atmospheric Phenomena** As commonly used in weather observing practice, an observable occurrence of particular physical (as opposed to dynamic or synoptic) significance within the atmosphere.

**Atmospheric Processes** Atmospheric processes play the lead role in determining such fundamental properties of climate as the disposition of incoming solar radiation, temperatures at the Earth's surface, the spatial distribution of water in the terrestrial biosphere, and the distribution of nutrients in the euphotic zone of the ocean.

**Bioaerosols** Airborne particles produced from microbial, viral, fungal, and actinomycete, as well as microbes from organic materials, humidifiers, vaporizers, heating, ventilating, and air conditioning systems (HVAC) lead to allergies, infections, and poisoning.

**Biodiversity** Number and relative abundances of different genes (genetic diversity), species, and ecosystems (communities) in a particular area.



P. Saxena et al. (eds.), *Fog: Formation, Chemistry, Dynamics, Impact and Dissipation*, https://doi.org/10.1007/978-981-96-4422-3

**Bioindicators** Bioindicators are living organisms such as plants, planktons, animals, and microbes, which are utilized to screen the health of the natural ecosystem in the environment. They are used for assessing environmental health and biogeographic changes taking place in the environment.

**Biomass** Total mass of living organisms in a given area or volume; recently dead plant material often included as dead biomass.

**Biosphere** The part of Earth system comprising all ecosystems and living organisms in the atmosphere.

**Blue-Green Infrastructure (BGI)** BGI is an interconnected network of natural and designed landscape components, including water bodies and green and open spaces.

**Carbon Sequestration** The process of storing carbon dioxide. This can happen naturally, as growing trees and plants turn $CO_2$ in to biomass (wood, leaves, and so on). It can also refer to the capture and storage of $CO_2$ produced by industry.

**Chronic Obstructive Pulmonary Disease (COPD)** Chronic obstructive pulmonary disease, commonly referred to as COPD, is a group of progressive lung diseases. The most common of these diseases are emphysema and chronic bronchitis. Many people with COPD have both of these conditions.

**Cloud** A visible mass of particles of condensed vapor (such as water or ice) suspended in the atmosphere of the Earth.

**Coarse Particulate Matters** They are also known as $PM_{10-2.5}$; particles with diameters generally larger than 2.5 μm and smaller than, or equal to, 10 μm in diameter. Note that the term **large coarse particles** in this course refers to particles greater than 10 μm in diameter.

**Criteria Air Pollutants** The criteria air pollutants include particle pollution, ground-level ozone, carbon monoxide, sulfur dioxide, nitrogen dioxide, and lead. These pollutants can harm your health and the environment and cause property damage. Of the six pollutants, particle pollution and ground-level ozone are the most widespread health threats. EPA calls these pollutants "criteria" air pollutants, because it regulates them by developing human health-based and/or environmentally based criteria (science-based guidelines) for setting permissible levels.

**Dew** Moisture condensed upon the surfaces of cool bodies especially at night.

**Disaster Risk Management** Disaster risk management is the application of disaster risk reduction policies and strategies to prevent new disaster risk, reduce existing disaster risk, and manage residual risk, contributing to the strengthening of resilience and reduction of disaster losses.

**Dispersion Modeling** Dispersion modeling uses mathematical formulations to characterize the atmospheric processes that disperse a pollutant emitted by a source.

**Dry Deposition** Primary and secondary particulates are removed by the gravitational force from the atmosphere over time without precipitation, this process is termed as dry deposition.

**Environmental Hazard** Environmental hazards are generally measured as exposure to pollution such as the amount of dust or sulfur dioxide in the air or the presence/amount of soil or groundwater contamination on the property.

**Epidemiology** Epidemiology is the branch of medical science that investigates all the factors that determine the presence or absence of diseases and disorders. Epidemiological research helps us to understand how many people have a disease or disorder, if those numbers are changing, and how the disorder affects our society and our economy.

**Evaporation Fog** Evaporation fog occurs when there is a significant temperature contrast between the water surface and the overlying air

**Fine Particulate Matter** Also known as $PM_{2.5}$; particles generally 2.5 μm in diameter or smaller. This group of particles also encompasses **ultrafine** and **nanoparticles** which are generally classified as having diameters less than 0.1 μm.

**Fog** Fog is a complex meteorological phenomenon defined as a suspension of minute, typically microscopic, water droplets in the atmosphere near the Earth's surface which reduces visibility to less than 1 km.

**Ice Fog** Ice fog, often termed pogonip, generally forms at low temperatures and consists solely of ice crystals. Thus, it most often occurs in high latitudes or high elevations, being most common in Arctic and subarctic regions during winter and can be seen in populated areas where moisture sources such as exhaust from vehicles or industry are present.

**Indo-Gangetic Plain (IGP)** Also called **North Indian Plain**, extensive north-central section of the Indian subcontinent, stretching westward from (and including) the combined delta of the Brahmaputra valley and the Ganges (Ganga) River to the Indus River valley.

**Indoor Air Pollution (IAP)** Indoor air pollution is dust, dirt, or gases in the air inside a building such as your home or workplace that harms us if we breathe it in.

**Ion Chromatography** Ion chromatography, a form of liquid chromatography, measures concentrations of ionic species by separating them based on their interaction with a resin. Ionic species separate differently depending on species type and size.

**Mitigation** An anthropogenic intervention for lessening, reducing, and diminishing the climatic stress and harm caused by adverse effects of climate change. It includes strategies to reduce air pollutants sources and emissions.

**Montreal Protocol** Montreal Protocol is regarded as one of the most successful international environmental treaties which achieved its objective of prevention of ozone depletion by banning the production and use of ozone depleting substances that include CFCs, HCFCs, and halons.

**National Clean Air Program (NCAP)** It was launched by the Ministry of Environment, Forests and Climate Change, Government of India in 2019 as an urgent response to curtail and control growing air pollution levels in the country.

**Particulate Matter (PM)** PM stands for particulate matter, and it is a mixture of solid particles and liquid droplets found in the air.

**Planetary Boundary Layer (PBL)** PBL is the lowermost portion of the atmosphere, adjacent to the Earth's surface, where maximum interaction between the Earth surface and the atmosphere takes place, leading to maximum exchange of physical properties like momentum, heat, and moisture.

**Precipitation Fog** Precipitation fog, also known as frontal fog, occurs when precipitation falls into a layer of cooler air near the surface.

**Radiation Fog** Radiation fog forms typically during the end of the night or early morning hours, when the air near the ground cools through radiative heat loss.

**Radiative Forcing** Radiative forcing is the change in the net, downward minus upward, radiative flux (expressed in $Wm^{-2}$) at the tropopause or top of atmosphere due to a change in an external driver of climate change such as a change in the concentration of carbon dioxide or the output of the Sun.

**Rain** Water falling in drops condensed from vapor in the atmosphere.

**Secondary Organic Aerosol (SOA)** Secondary organic aerosol (SOA), formed in the atmosphere by oxidation of organic gases, represents a major fraction of global submicron-sized atmospheric organic aerosol.

**Sink** Any process, activity, or mechanism which involves a pollutant gas or aerosol from the atmosphere.

**Source** Any process, activity, or mechanism which releases a pollutant gas, an aerosol, or a precursor of a pollutant gas or aerosol into the atmosphere.

**Thermal Inversion** Thermal inversion is a natural phenomenon that involves a change in the normal tendency of the air to cool down with altitude.

**Tracheobronchial Region** The tracheobronchial tree is composed of the trachea, the bronchi, and the bronchioles that transport air from the environment to the lungs for gas exchange.

**Ultrafine Particulate Matter** Ultrafine particles (UFPs) are aerosols with an aerodynamic diameter of 0.1 μm (100 nm) or less. There is a growing concern in the public health community about the contribution of UFPs to human health.

**Upslope Fog** Upslope fog forms when moist air is forced to ascend along the slope of a mountain or elevated terrain. This process resembles cloud formation but occurs near the ground, adhering to the slopes of the terrain.

**Wet Deposition** Wet deposition is the process whereby atmospheric gases mix with suspended water in the atmosphere and are then washed out through rain, snow, or fog.